Environmental and Social Justice in the City

Environmental and Social Justice in the City: Historical Perspectives

Edited by

Geneviève Massard-Guilbaud and Richard Rodger

The White Horse Press

The White Horse Press, 10 High Street, Knapwell, Cambridge, CB23 4NR, UK

Set in 10 point Adobe Garamond Pro
Printed by Lightning Source

British Library Cataloguing in Publication Data
A catalogue record for this book is available from the British Library

ISBN 978-1-874267-61-4 (HB)

Acknowledgements

The production of any book is dependent on many people, and on considerable goodwill. We would particularly like to thank our editor, Sarah Johnson, for her responsive and supportive approach to our project. It meant a lot to both of us as editors to have Sarah's good sense and flexible 'can do' attitude. We are also grateful to Vincent Lemire for the image on the jacket cover. Insofar as any single image can, his atmospheric photograph captures perfectly an environment saturated with inequalities and injustices such as form the subject matter of this book. Bringing together materials from a variety of contributors from different countries, intellectual disciplines, languages and publishing conventions is never straightforward, but our team has been attentive to our requests for further details and in answering our queries and that has made the project achievable in a short time. From the date of our first meeting to final production has taken a little more than two years. We thank you all.

Geneviève Massard-Guilbaud and Richard Rodger
Paris and Edinburgh
Summer 2010

Note: Unless otherwise stated, in the ensuing chapters all translations from languages other than English are authors' own.

Contents

Illustrations and Tables

Contributors

WANDA BALCERS is a postgraduate student of the *Université Libre de Bruxelles* (ULB) and has been a Research Fellow of the *Fonds National de la Recherche scientifique* (FNRS) in Belgium since 2008. She is currently working on her PhD thesis, dedicated to socio-economic and cultural aspects of the history of beer production and consumption in Brussels (1860–1914). (W. Balcers, 'Résistance et Mondialisation. Le cas des bières bruxelloises', in *Les Cahiers de la Fonderie: Boire et Manger. L'alimentation à Bruxelles*, t. 40, 2009, 8–13.)

CHRISTOPH BERNHARDT has studied history and German studies at the Free University, Berlin. He worked as an Assistant Professor in the Institute of Historical Sciences at the TU, Berlin (1994–1998), where he was also awarded his doctorate (Dr. phil.) with a book on the history of housing and the building industry in Berlin before the First World War. He is now research director at the Leibniz-Institut for Regional Development and Structural Planning in Erkner near Berlin. In 2007 he completed his Habilitation on the environmental history of the Upper Rhine (1800–2000). He teaches urban and environmental history at the Technical Universities of Berlin and Darmstadt, is co-editor of *Informationen zur modernen Stadtgeschichte* ['Information on Modern Urban History'] and has published extensively on the social and environmental history of European cities. More information may be found at www.irs-net.de

CHRISTOPHER BOONE is a professor in the School of Sustainability at Arizona State University. His research examines the social and environmental drivers and consequences of urbanisation, especially within an ecologically-informed environmental justice framework. He is an active member of the US Long Term Ecological Research network and the Urbanization and Global Environmental Change project of the International Human Dimensions Programme on Global Environmental Change. He is currently co-editing a volume for Springer Press entitled *Linking Urban Ecology, Environmental Justice, and Global Environmental Change: A Framework for Urban Sustainability.*

GEOFFREY L. BUCKLEY is an associate professor in the Department of Geography at Ohio University. He received his BA from Connecticut College, his MA from the University of Oregon, and his PhD from the University of Maryland. His research interests include resource conservation and sustainability; management of public lands and urban green spaces; and environmental justice. His articles have appeared in the *Annals of the Association of American Geographers, Geographical Review, Historical Geography*, and the *Encyclopedia of Energy*. His book, *America's*

Conservation Impulse: A Century of Saving Trees in the Old Line State was recently published by the Center for American Places.

CRAIG E. COLTEN is the Carl O. Sauer Professor in the Department of Geography and Anthropology at Louisiana State University. His interests span urban hazards and environmental geography. His latest book is *Perilous Place, Powerful Storms: Hurricane Protection in Coastal Louisiana* (2009) and he is currently working on a project that will examine water in the American south-east.

JOANNA DEAN is Associate Professor of History at Carleton University in Ottawa, Canada, where she teaches environmental history and gender history. She has published several articles on the history of the urban forest, including 'Seeing trees, thinking forests: urban forestry at the University of Toronto in the 1960s', in *Method and Meaning in Canadian Environmental History*, edited by Alan MacEachern and William Turkel (2009); and is currently working on a monograph on the history of street trees in the Ottawa and Gatineau area.

CHLOÉ DELIGNE is a Research Associate of the *Fonds National de la Recherche Scientifique* (FNRS) in Belgium, working in the Department of History, Arts and Archaeology of the *Université Libre de Bruxelles* (ULB). As a historian, geographer and environmental scientist, she has dedicated several works to the history of water management in Belgium. (C. Deligne, *Bruxelles et sa rivière. Genèse d'un territoire urbain (12e–18e siècle)*, Studies in European Urban History, 1, Turnhout, 2003 and more recently C. Deligne, *Brussels and Its Rivers, 1770–1880: Reshaping an Urban Landscape*, University of Pittsburgh Press 'History of the Urban Environment' Series, in press, 2010).

STÉPHANE FRIOUX teaches at the École Normale Supérieure de Lyon. His doctoral dissertation (Université de Lyon, 2009), *Les réseaux de la modernité. Diffusion de l'innovation et amélioration de l'environnement dans la France urbaine (fin XIXe siècle années 1950)*, examines the municipal policies of environmental sanitation in provincial French cities in the first half of the 20th century, focusing on the diffusion of innovative sanitation devices. In 2003, he received the first ESEH Publication Prize. He has published 'Villes et entreprises. Acteurs de l'assainissement urbain en France (fin XIXe siècle–milieu XXe siècle)', in *Histoire Urbaine*, 18, 2007, 125–140.

JANET GREENLEES is a Lecturer in Health History and member of the Centre for Social History of Health and Healthcare at Glasgow Caledonian University. She has published a number of works on the textile industries of the United States and Great Britain, including *Female Labour Power: Women Workers' Influence on Business Practices in the British and American Cotton Industries, c. 1780–1860* (Ashgate, 2007) and several journal articles, including items in *Urban History*, *Social History* and *Historical Studies in Industrial Relations*, as well as edited collections. Her research interests currently focus on nineteenth and twentieth century British and American history, particularly employers' healthcare provision and maternal and child health and welfare.

Jonas Hallström, PhD, Associate Professor, is a historian of technology and education, and is currently working at the Department of Behavioural Sciences and Learning, Linköping University, Sweden. His research includes studies at the intersection between the history of technology, environmental history and urban history, such as his dissertation, *Constructing a Pipe-Bound City*, Linköping, 2002. Hallström also does research within the history of education and technology education, including 'Technical Knowledge in a Technical Society: Elementary School Technology Education in Sweden, 1919–1928', *History of Education* 38:4, 2009, 455–474.

Geneviève Massard-Guilbaud is Professor at the École des Hautes Études en Sciences Sociales, Paris, where she holds a chair in Environmental History. Her last book, *Histoire de la pollution industrielle, France 1789–1914,* was published in 2010 (Paris: EHESS). She has just edited, with Stephen Mosley, a book entitled *Common Ground. Integrating the Social and Environmental in History* (Newcastle: CSP, 2011). Since 2007 she has been the president of the European Society for Environmental History. In France, she initiated the RUCHE (Scholarly Network of Researchers in Environmental History). At the beginning of 2010 she was appointed by the Humanities and Social Sciences Institute and the Ecology and Environment Institute of the CNRS to lead their common interdisciplinary research network in Environmental History.

Richard Oram is Professor of Medieval and Environmental History and Director of the Centre for Environmental History at the University of Stirling, UK. He is the author of 'Abondance inépuisable? Crise de l'approvisionnement en combustible et réactions en Écosse du Nord environ 1500 et environ 1800' in J-M. Derex and F. Grégoire, eds. *Histoire économique et sociale de la tourbe et des tourbières: Actes du deuxième colloque international du Groupe d'histoire des zones humides* (Cordemais, 2009) and, with Paul Adderley, of 'Lordship and Environmental Change in Central Highland Scotland c.1300–c.1450', *Journal of the North Atlantic*, 1, 2008.

Richard Rodger is Professor of Economic and Social History at Edinburgh University. He has published widely on the economic, business and urban history of Britain since 1800, was editor of *Urban History* from 1987–2007, and joint editor for a series of forty books under the title of 'Historical Urban Studies', published by Ashgate. His book *The Transformation of Edinburgh: Land, Property and Trust in the Nineteenth Century* was awarded the Frank Watson Prize for works on Scottish history. Ongoing research includes a project on the development of public and environmental health in Victorian Scotland. In recognition of his contributions to the study of economic and social history, Rodger was elected to the Academy of Social Sciences in 2004.

Tim Soens is lecturer in Medieval and Environmental History at the University of Antwerp, Belgium. He studied Medieval History at the University of Ghent, Belgium, obtaining his PhD in 2006 with a thesis on water management and the interaction of man and nature in the late medieval and early modern Flemish coastal plain. He has presented papers at numerous international conferences and his publications include *De spade in de dijk? Waterbeheer en rurale samenleving in de Vlaamse kustvlakte (1280–1580)*, Gent, Academia Press, 2009. ['Giving in to the water? Water management and rural society in the Flemish Coastal Plain 1280–1580'] and 'Explaining Deficiencies of Water Management in the Late Medieval Flemish Coastal Plain (13th–16th centuries)', *Jaarboek voor Ecologische Geschiedenis*, 2005–06, 35–62.

Marcus Stippak holds an MA and PhD in history from Darmstadt University of Technology, Germany. His fields of research are the history of technology, urban history, environmental history, and contemporary church history. Currently, he is working as a freelance historian at the Central Archive of the Protestant Church in Hesse and Nassau. His publications include *Beharrliche Provisorien, Städtische Wasserversorgung und Abwasserentsorgung in Darmstadt und Dessau 1869–1989*, Münster *et al.* 2010 ('Persistent Provisional Arrangements. Urban Water Supply and Sewage Systems in Darmstadt and Dessau, 1869–1989', originally a PhD thesis at Darmstadt University of Technology, Germany, 2008).

Chapter One

Reconsidering Justice in Past Cities: When Environmental and Social Dimensions Meet

Geneviève Massard-Guilbaud and Richard Rodger

> We, cities and towns, are aware that the poor are worst affected by environmental problems (such as noise and air pollution from traffic, lack of amenities, unhealthy housing, lack of open space) and are least able to solve them. Inequitable distribution of wealth both causes unsustainable behaviour and makes it harder to change. (Ålborg Charter, 1994, 1.7)

In 1994, European city officials convening in Ålborg acknowledged that environmental problems do not affect the poor and the wealthy equally.[1] The idiosyncrasy of the Ålborg charter was the fact that it was signed by cities, not states. As far as states were concerned the United Nations Environmental Program, created by the Stockholm conference, recognised as early as 1972 that environmental policies should not curtail the ability of populations to benefit from the best possible living conditions. Indeed, it has become increasingly obvious that the issues of social justice, human rights and environmental quality of life are inextricably linked. Since then, many texts have tried to define what constitutes a just environmental policy[2] and to identify the negative, sometimes disastrous, social consequences of well-meaning policies. Despite these good intentions, it has been shown that the earliest environmental policies implemented have led to a worsening of inequalities.[3]

1. First European Conference on Sustainable Cities and Towns, Ålborg, Denmark, 24–27 May 1994.
2. A list of these texts is available in C.L. Fortwangler, 'The Winding Road. Incorporating Social Justice and Human Rights into Protected Area Policies', in S.R. Brechin, P.R. Wilshusen, C.L. Fortwangler, and P.C. West, eds. *Contested Nature. Promoting International Biodiversity with Social Justice in the Twenty-first Century*, Albany, 2003, 25–40.
3. For France, see J-M. Offner and C. Pourchez, 'La ville durable, perspectives européennes et françaises'. *Problèmes économiques et sociaux*, La Documentation française, 933, 2007. See also W. Diebolt, A. Helias, D. Bidou, and G. Crepey, *Les inégalités écologiques en milieu urbain, rapport de l'inspection générale de l'environnement*, 2005, http://www.ecologie.gouv.fr/Les-inegalites-ecologiques-en.html Consulted 16 Nov 2009. For a wider view, see M. Dowie, *Conservation Refugees. A Hundred-Year Conflict between Global Conservation and Native People*, Cambridge, MA, 2009; M. Humphrey, *Preservation versus People? Nature, Humanity and Political Philosphy*,

Four years after the Ålborg charter, the Århus convention, adopted in 1998 and completed in 2003, clearly asserted the right to environmental justice.[4] In February 1994, an Executive Order from President Clinton addressed issues of environmental justice in minority and low-income populations.

In spite of these various national and international instruments, sixteen years after the Ålborg charter, in 2010, the situation remains pretty much unchanged. As Scottish First Minister Jack McConnell recognised in 2002:

> Too often the environment is dismissed as a concern of those who are not confronted with bread and butter issues. But the reality is the people who have the most urgent environmental concerns in Scotland are those who cope daily with the consequences of a poor quality of life, and live in a rotten environment, close to industrial pollution, plagued by vehicle emissions, streets filled by litter and walls covered with graffiti. This is true for Scotland and also true elsewhere in the world. These are circumstances which would not be acceptable to better off communities in our society, and those who have to endure such environments in which to bring up a family, or grow old themselves are being denied environmental justice.[5]

Despite official recognition in many countries, neither local studies nor much data concerning inequalities have been forthcoming in Europe. While a number of studies are now available in the USA, and to a lesser extent in Britain, they remain rare elsewhere, particularly in France. Where they do exist, all studies agree on one conclusion: not only do environmental inequalities exist, but they may well be stronger than social inequalities.[6] In the USA, three out of every five African Americans live close to facilities emitting toxins.[7] In the south east district of Los Angeles, mainly inhabited by Latinos, the ratio of income levels to those of the agglomeration overall was 1:2, and the unemployment ratio varied between 1–2:1; however the overcrowding ratio was 3:1, toxin emissions were 25:1, and the ratio

Oxford, 2002; Brechin *et al. Contested Nature*; M. Gadgil and R. Guha, *Ecology and Equity. The Use and Abuse of Nature in Contemporary India*, London, 1995.

4. Convention on Access to Information, Public Participation in Decision-Making and Access to Justice in Environmental Matters, adopted 25 June 1998, came into force on 30 October 2001 and completed by directives 2003/4 and 2003/35 which contain provisions on access to justice.

5. Scottish First Minister's speech at the ERM conference at Dynamic Earth Edinburgh, 18 February 2002, http://www.scotland.gov.uk/News/News-Extras/57 Accessed 28 October 2009.

6. J. Theys, 'Quand inégalités sociales et inégalités écologiques se cumulent', *Développement durable, villes et territoires,* Note du centre de prospective et de veille scientifique, 13, Paris, quoted in J. Theys, 'Pourquoi les préoccupations sociales et environnementales s'ignorent-elles mutuellement. Un essai d'interprétation à partir du thème des inégalités écologiques', in P. Cornut, T. Bauer and E. Zaccaï, eds. *Environnement et inégalités sociales,* Bruxelles, 2007, 26, fn 6.

7. J. Byrne, L. Glover, and C. Martinez, eds. *Environmental Justice. Discourses in International Political Economy, Energy and Environmental Policy,* vol. 8, New Brunswick/London, 2002, quoted by C. Emelianoff, 'Connaître ou reconnaître les inégalités environnementales?', ESO, *Travaux et Documents,* no.25, December 2006, 39.

of the SELA district to the metropolitan area in terms of proximity to incinerators was 30:1.[8] In France, according to the ONZUS,[9] half of the so-called '*zones urbaines sensibles*' [sensitive urban areas] are exposed to loud noise problems, and the cities where they are situated are twice as likely to be sited in the vicinity of a factory with dangerous materials than the rest of the country. In England, using both pollution data from the official Environment Agency and the Government's Index of Multiple Deprivation, a Friends of the Earth Report dated 2001 showed that 66 per cent of carcinogen emissions are in the most deprived ten per cent of wards, and 82 per cent in the most deprived twenty per cent of wards.[10]

Environmental justice issues can be dealt with at varied scales. These range from the most local (neighbourhood, street) to the global (twenty per cent of the world population consumes eighty per cent of resources), taking in, for example, unfairness of exchanges (southern hemisphere countries bear the brunt of the consequences of global warming induced by northern hemisphere industry) and biological piracy (eighty per cent of biological resources are concentrated between the tropics, but they mainly benefit northern corporations and populations).

A report commissioned by the European Community and edited by the Indian Pavan Sukhdev demonstrates clearly how the relationship between poverty and environmental degradation works, and how the poorest on the planet bear the greatest consequences of this degradation.[11] Even before the 2010 earthquake, ninety per cent of Haitian children suffered intestinal parasites due to pollution of drinking water, in turn due to floods resulting from erosion that has followed from the 97 per cent decline in forest cover. Just to survive, the disenfranchised have to degrade ever more of what remains of their environment. Though many publications point an accusing finger at the poor as being responsible for world environment degradation, in reality this degradation results overwhelmingly from the way industrialised countries have themselves developed in the course of two centuries. Considered globally, the ecological footprint of the disenfranchised is far more limited.

Furthermore, the poor are not always passive when confronted with the degradation of their environments. In South Africa, Brazil and India, an 'environmentalism of the poor', as Catalan historian and economist Joan Martinez-Alier has termed it, was born in defence of the land, of traditional ways of using local

8. Quoted in Theys, 'Quand inégalités sociales', 27.

9. Observatoire National des Zones Urbaines Sensibles [Sensitive Urban Areas National Observatory], http://www.onpes.gouv.fr/Observatoire-National-des-Zones.html Accessed April 2010.

10. See www.foe.co.uk/resource/.../pollution_and_poverty.pdf Accessed 30 October 2009.

11. P. Sukhdev, dir. *L'économie des écosystèmes et la biodiversité. Rapport d'étape,* Communauté Européenne, 2008 http://ec.europa.eu/environment/nature/biodiversity/economics/index_en.htm Accessed April 2010. See also É. Laurent, 'Écologie et Inégalités', *Revue de l'OFCE,* 109, April 2009, 3.

resources, and in opposition to multinational corporations. Environmental action and concern has not been the preserve of the wealthy northern hemisphere.[12] Though the chapters in this book focus on case studies and the local scale, and particularly on the past urban experience of northern industrialised countries, the global background should also be borne in mind.

While scholarship on environmental inequalities issues is progressively growing everywhere, with studies in sociology, social geography, public policy and law, historians have not yet devoted much attention to the topic, at least not in Europe. Our main purpose, in editing this series of essays, is to take stock of existing historical research on environmental inequalities, and to gather together scarce studies in order to give them exposure to a wider readership.

Another of our purposes is to help create bridges between perspectives that should never have been separated: the social and the environmental dimensions of inequalities. Indeed, in the last third of the twentieth century, with the development of an international environmentalist consciousness, far from being reunited, environmental and social inequalities have been artificially separated.[13] In a recent article, Jacques Theys listed factors that contributed to this split: the mainstream conception of the environment is bio-centric; analyses produced by social sciences have wrongly characterised environmentalism as a middle class problem; and green parties constructed their identities and political platforms around the issues of consumerism and technology, and not social issues.[14] To these explanations might be added the abandonment of social issues to traditional labour or socialist parties that have cared little for the environment, and failed to recognise it as a priority for all sections of society. Most of these factors can be located in the birth of ecology as a scholarly discipline which, according to some, has been founded on an 'exclusion of man'.[15] The result is that the social and environmental have been wrongly separated. Happily, today, there is a growing recognition that a reunification is timely.

12. J. Martinez-Alier, *The Environmentalism of the Poor. A Study of Ecological Conflicts and Valuation,* Cheltenham, 2002. For a short synthesis in French, see J. Martinez-Alier, 'Conflits écologiques et languages de valorisation', in 'Des Inégalités écologiques parmi les hommes', *Écologie et Politique. Sciences, Cultures et Sociétés,* 35, 2007, 91–107. See also Gadgil and Guha, *Ecology and Equity*; R. Guha and J. Martinez-Alier, *Varieties of Environmentalism. Essays North and South,* London, 1997; D. McDonald, ed. *Environmental Justice in South Africa,* Cape Town, 2002. The themed issue of *Geoforum,* 37, 5, 2006, 'Geographies of Environmental Justice'; F. Giraut, S. Guyot, and M. Houssay-Holzschuch, 'La nature, les territoires et le politique en Afrique du Sud', *Annales, Sciences Sociales,* 2005:4, 697–717; D. Blanchon, S. Moreau, and Y. Veyret, 'Comprendre et construire la justice environnementale', *Annales de Géographie,* 665–66, 2009, 35–60.

13. Actually, they were separated earlier in the twentieth century with the birth of the discipline of 'ecology', and the divergence became more marked at the end of the century.

14. Theys, 'Pourquoi les préoccupations'.

15. See R. Larrère, 'L'écologie ou le geste de l'exclusion de l'homme', in A. Roger and F. Guéry, eds. *Maîtres et protecteurs de la nature,* Seyssel, Champvallon, 1991, 172–96.

This, then, is why the contributors to this book have been invited to explore not only the historical dimension of environmental inequalities and the way they were built, but also, as far as possible, their interactions with social inequalities. Did these two kinds of inequalities always overlap, and to what extent? Did the presence of one enhance the other? These are some of the questions explored here.

In preparing this book we paid attention to specific terms from the very start. Should we use the word *inequalities* (most often used in the plural), or the word *injustice* (usually employed in the singular)? Initially the difference was not obvious to us. *Inequalities* seemed detached from the philosophical and moral connotations that surround the word *justice*, rarely used by historians. But we soon discovered that this semantic issue goes far beyond disciplinary habits; it reflects a perception of the environment that differs from one country, or culture, to the other. We shall return to this.

The book gathers authors from various countries and two continents. For reasons we will explore, scholarship on environmental inequalities has reached very different levels and taken different paths on either side of the Atlantic. The gap that has resulted even complicates this introduction since readers start from very different knowledge bases. We have endeavoured to write for as wide a readership as possible, and to provide the elements needed by readers from different backgrounds. To French historians, for instance, it is necessary to explain what the *movement for environmental justice* has involved, since there is a strong probability that they will never have heard of it. For American readers, living in a country where thousands of publications have been devoted to the topic, no such explanation is necessary, though they will probably be unaware of the fact that, with the exception of Britain, in European countries the concept of *environmental justice* is hardly used, and that in its place, if at all, the terms *environmental* or *ecological inequalities* are employed. Only once this difference is realised may readers on both side of the Atlantic explore the significance of the terminology more fully, and what its usage reveals and obscures. We hope this book will help them.

Approaches to Environmental Justice: A Pioneer, US-based, Urban and Bottom-up Movement

The issues we are interested in here were first raised in the United States at the end of the 1970s and were immediately considered as issues of *justice*. The discovery of tons of toxic waste illegally dumped by Hooker Chemical in Love Canal in New York in 1976 resulted in the complete evacuation of the site, and is often considered to be the first episode in the story.[16] It was a minister of the United Church of Christ and former head of the National Association for the Advancement of Colored People

16. A. Dobson, *Justice and the Environment, Conceptions of Environmental Sustainability*, Oxford, 2003, 17 (first published New York, 1998).

(NAACP), Benjamin Chavis Jr. who racialised the problem and coined the terms *environmental justice* and *environmental racism,* while taking part in the protest against a toxic waste dump in Warren County, North Carolina. Firmly connected to the Civil Rights Movement, the emerging movement for environmental justice was a grassroots one, based on a radical eco-populism in which women were key leaders; it was not an outcome of the mainstream environmental movement.[17] Its principal argument was that African Americans and other minorities shouldered a disproportionate share of the so-called LULUs (Locally Undesirable Land Uses) and were consequently more affected than others by pollution and environmental health problems. In 1987, the United Church of Christ published the first national enquiry showing that minorities were the victims of what they categorised as environmental racism.[18]

In January and March 1990, the Gulf Coast Tenant Leadership Development Project, then the Southwest Organising Project, sent letters to the so-called 'Group of Ten', the nickname for the major environmentalist organisations like the Sierra Club, the National Audubon Society, and others.[19] These letters gained the attention of the press and the public. Their authors accused the Group of 'ignorance, ambivalence and complicity with the environmental exploitation of communities of color',[20] and of being unable to take working people's and coloured people's environmental needs into account.

In October 1991, the First National People of Color Environmental Leadership Summit was held in Washington, D.C. At this summit, the delegates adopted the so-called 'Principles of Environmental Justice', which have since become a point of reference. These Principles give a definition of environmental issues that encompasses social, political and ethical elements.[21] Members of some mainstream environmental organisations attended the congress and pleaded for a new partnership.

The enquiry of the United Church of Christ, the accusations lodged against the mainstream organisations and the Washington summit signalled the emergence of a national awareness and stimulated debate on the place that race and class should have in environmentalism. In opposing its traditional eco-centrist values, the movement for environmental justice challenged the conventional environmental

17. M. Melosi, 'Environmental Justice, Political Agenda Setting and the Myths of History', *Journal of Policy History,* 12:1, 2000, 43 *et seq.*

18. Commission for Racial Justice–United Church of Christ, *Toxic Wastes and Race in the United States: A National Report on the Racial and Socio-economic Characteristics of Communities with Hazardous Waste Sites*. New York, 1987.

19. A list can be found in R. Sandler and P.C. Pezzulo, eds. *Environmental Justice and Environmentalism. The Social Justice Challenge to the Environmental Movement,* Cambridge, Ma. 2007, 19, fn.4.

20. Sandler and Pezzulo, eds. *Environmental Justice and Environmentalism,* 4.

21. Available on http://www.ejnet.org/ej/principles.html Accessed April 2010.

organisations and accused them of paying more attention to wilderness and animals than to human beings and the health hazards with which they were confronted. The emerging debate opened the door to a stream of research, including the pioneering, though not uncontested works, of Robert Bullard, a sociologist from the University of California, Riverside.[22]

From that point in the early 1990s, scholars began to investigate themes like race, class, and gender in contemporary environmental problems and in the history of conservation.[23] One of the intriguing issues discussed by scholars was the pertinence of the racial variable to environmental injustice, though it was difficult to distinguish this from the income variable. In a paper full of insights, the legal scholar Vicki Been insisted on the importance of longitudinal studies so as to reveal not only the social composition of an area when a hazardous waste facility is built but also, more significantly, what happens afterwards. Once the plant has been constructed, the wealthy flee and this virtually guarantees a change of status for the area. This exposes the mechanism by which inequalities develop and contradicts, at least partly, the argument that the siting of the LULUs close to the poorest is premeditated or racist. In one of many articles on the topic, Been concluded:

> Multivariate analysis, however, suggests that race is a better predictor than income. In total, the analysis also reveals that environmental injustice is not a simplistic PIBBY, 'put it in Blacks' backyards'[24]. It suggests, instead, a much more ambiguous entanglement of class, race, educational attainment, occupational patterns, relationships between the metropolitan areas and non-rural or non-metropolitan cities, and possibly market dynamics.[25]

By the beginning of the 1990s, the movement for environmental justice had become very influential. The pressure led President Clinton to sign Executive Order 12898, 'Federal Actions to Address Environmental Justice in Minority Populations and Low Income Populations', and to set up the National Environmental Justice Advisory Council (NEJAC), the Environmental Protection Agency, and the

---

22. R.D. Bullard, *Dumping in Dixie: Race, Class and Environmental Quality*, Boulder, 1990; R.D. Bullard, ed. *Confronting Environmental Racism: Voices from the Grassroots*, Boston, 1993. See also R.D. Bullard, 'Environmental Justice in the 21st Century', available on the Environmental Justice Center website, http://www.ejrc.cau.edu/ejinthe21century.htm Consulted 28 October 2009.

23. See, for example, C. Merchant, 'Shades of Darkness: Race and Environmental History', *Environmental History*, 8:3, 2003.

24. Robert D. Bullard, *Dumping in Dixie: Race, Class and Environmental Quality*, 5 (1990) [note by V. Been].

25. V. Been, 'Analyzing Evidence of Environmental Justice', *Journal of Land Use and Environmental Law*, 11:1, 1995, 1–36. See also V. Been, 'What's Fairness Got to Do With It? Environmental Justice and the Siting of Locally Undesirable Land Uses', *Cornell Law Review*, 78, 1992–93, 1001–85. See also her critic, C. Crawford, 'Analyzing Evidence of Environmental Justice: A Suggestion for Professor Been', *Journal of Land Use and Environmental Law*, 12:1, 1996, 103–20.

Inter-agency Working Group on Environmental Justice.[26] Unsurprisingly, President Bush's administration tried to redefine environmental justice in a way that did not refer to environmental inequities.

With time, and in spite of the wide-ranging scholarly literature that has developed around it, the environmental justice movement 'has come to be identified with the social movement that protests toxic waste siting, excessive pesticide use, contamination of air and water on Indian reservations, among Latino farm-workers and in poor white and African American communities'.[27] Individually these issues were symptomatic of a much wider multi-racial movement with specific concerns like lead contamination, pesticides, workplace safety, transportation, and more recently sprawl and smart growth, sustainability and climate justice.[28] A special journal has even been created, entitled *Environmental Justice.*

It is not possible, nor is it the aim here, to draw conclusions on the impact of such an expansive movement within the United States, but it is worth noting that some scholars still seek to return to a strictly naturalist environmentalism, three decades after this environmental justice movement arose. Kevin DeLuca writes: 'I propose a wilderness environmentalism because I think that it is crucial that the environmental movement be grounded in wilderness, not humanism.' He continues, 'Putting human first dilutes the focus and efforts of environmental groups. Further, many human issues involve abstractions, such as social justice and human rights, they are Sisyphean tasks with no clear way to even define victory.'[29] DeLuca states that the environmental justice movement is 'a good thing', but a regrettable one since it has diverted environmentalism from its initial and valuable goal, protecting wilderness. At least these views show how far the environmental justice movement questioned and even disturbed the previously limited perspectives of an early environmentalism focused exclusively on the wilderness and the natural world.

How Did the Concept Cross the Atlantic?

Born in the cities of a country where race has played and still plays a significant role in matters of inequality, how would the concept of environmental justice cross the Atlantic? Unsurprisingly, it was in Britain that it first landed, as promoted by the environmentalist organisation The Friends of the Earth, which, in 1999, published a Report entitled 'The Geographic Relation Between Household Income and Pol-

26. Executive Order 12898, 11 February 1994.

27. D. Jamieson, 'Justice, the Heart of Environmentalism', in Sandler and Pezzulo, eds. *Environmental Justice*, 88.

28. J. Agyeman and B. Evans, '"Just Sustainability": The Emerging Discourse of Environmental Justice in Britain', *The Geographical Journal,* 170:2, 2004, 155–164.

29. K. DeLuca, 'A Wilderness Environmental Manifesto' in Sandler and Pezzulo, eds. *Environmental Justice*, 40, 45.

luting Factory'.[30] Using postcode information this Report was based on two sets of data: household income distribution and the location of industrial facilities registered under the Integrated Pollution Control framework between 1992 and 1996. It showed that ninety per cent of the most polluting factories in London were located in communities of below average income and that a similar, although less extreme pattern, was true throughout England and Wales. 'This is a clear cut issue of environmental justice, in which poorer people are subjected to greater risks and impacts of pollution, and have less control over their environment while the benefits of industrial activity largely accrue elsewhere', wrote the authors of the report.[31] Even before the Friends of the Earth report was published, their sister branch in Scotland, led by Kevin Dunion, launched a campaign in 1997 with the slogan 'no less than our right to a decent environment; no more than our fair share of Earth's resources'.

At the same time, a so-called Black Environmental Network (BEN) was created, with the aim 'to promote equality of opportunity with respect to ethnic communities in the preservation, protection and development of the environment'.[32] Since then, the BEN has developed many projects and produced an impressive number of reports. Also at the end of the 1990s, the Staffordshire Institute for Environment, Sustainability and Regeneration began to publish research on environmental justice.[33] Many more reports followed, coming from either Non-Governmental Organisations (NGOs) or from official agencies.[34] One of them, entitled 'The Relationship between Socio-economic Indicators and Air Pollution in England and Wales: Implications for Environmental Justice', published in an environmental science journal, referred in its introduction to the American movement

---

30. M. Duncan, O. Cottray, M. Taylor, S. Pipes and S. Bullock, The Geographic Relation between Household Income and Polluting Factories, A Report for Friends of the Earth, 1999. Report available on http://www.foe.co.uk/resource/reports/income_pollution.html Accessed 30 October 2009.

31. Duncan *et al.* The Geographic Relation, 1.

32. See http://www.ben-network.org.uk/ Accessed October 2009.

33. See http://www.staffs.ac.uk/schools/sciences/geography/links/IESR/themes_envJust.shtml Accessed April 2010.

34. For example see B. Boardman, S. Bullock and D. McLaren, *Equity and the Environment: Guidelines for Green and Socially Just Government,* Catalyst pamphlet 5, 1999; J. Agyeman, 'Environmental Justice: From the Margins to the Mainstream?' Town and Country Planning Association, 'Tomorrow' Series, 2000; K. Dunion, *Troublemakers. The Struggle for Environmental Justice in Scotland,* Edinburgh, 2003; K. Lucas, G. Walker, M. Eames, H. Fay and M. Poustie, Environment and Social Justice: Rapid Research and Evidence Review, Final report, SDRN, DEFRA, 2004. See also UK Environment Agency Reports: Addressing Environmental Inequalities: Flood Risk. Science Report: SC020061/SR1, 2006; Addressing Environmental Inequalities: Waste Management. SC020061/SR3, 2006; Addressing Environmental Inequalities: Water Quality. Science Report SC020061/SR2, 2006. More references can be found in J. Agyeman and B. Evans, '"Just Sustainability"', 158–59.

and the scholarly studies that followed.[35] It investigated the distribution of three common air pollutants – sulphur dioxide, nitrogen dioxide and fine particulates (PM_{10}) – in England and Wales with respect to social class, and stated that it was the first study of its kind to be performed in the United Kingdom. It provided precise correlations at different scales – regions, districts, rural and urban – and made clear recommendations, underlining that ill-designed policies could increase the already existing inequalities: 'National legislation introduced to reduce air pollution levels may give rise to environmental injustice, with geographical and social differences in the costs and benefits arising to the population due to such legislation.'[36]

The variety of the agencies producing these reports, including the Economic and Social Research Council, the Rowntree Foundation, the Sustainable Development Commission – set up to advise the British government – and the Centre for Law and the Environment at London University, confirm a pattern for the British movement based on top-down, research-led, policy-driven agencies that contrasted with the way the American movement developed. The two forms now co-exist: environmental justice is both an activist claim and a policy concept, associated loosely with sustainability.[37] It is also noticeable that, while the term *environmental justice* is known and used in Britain, and references to the American justice movement can be found there, British official reports prefer the term *inequalities*.[38] Would injustice be a harder problem to deal with than inequalities? To a degree, then, Britain provides an intermediate position between the USA, where the term environmental justice is widely used, and the rest of Europe, where the term is rarely used, with the exception of European Union circles where some agencies employ it.

Indeed, if the concept of environmental justice crossed the Atlantic, it found it harder to reach the Continent. In France, it is still almost unknown to the general public. In the academic milieu, it is used by a few geographers, sociologists and economists, but is unknown to historians, including environmental historians. Actually, the notion that terminology has crossed the Atlantic and then the Channel is somewhat misleading; it assumes that all the thinking about what is called environmental injustice originated in the USA, which is not the case. As early as the 1970s, reports can be found in France on something called *ecological inequalities*.[39] However, these reports remained confidential. Environmental, rather than ecological,

35. H. McLeod, I.H Langford, A.P. Jones, J. R. Stedman, R.J Day, I. Lorenzoni, and I.J. Bateman, 'The Relationship between Socio-economic Indicators and Air Pollution in England and Wales: Implications for Environmental Justice', *Regional Environmental Change*, 1:2, 2000, 78–85.

36. McLeod, *et al.* 'The Relationship between Socio-economic Indicators', 78–85.

37. Though authors disagree on this point we do not pursue it here. Readers interested in this matter should compare Agyeman's and Dobson's theories. For references, see above.

38. See note 28.

39. D. d'Ollone and J. Theys, *Note préparatoire au 7e plan, ministère de l'environnement et du cadre de vie*, 1978, quoted by Theys in 'Pourquoi les préoccupations', 26, fn. 5.

inequalities appeared in a document produced by the French Committee for the preparation of the Johannesburg Conference on sustainable development in 2002. This committee defined environmental inequalities as between individuals or groups (including social classes but also generations, women, the elderly and children), and as regards a wide range of problems, including exposure to natural and technological risks, degradation of quality of life, and the deprivation of access to some common goods, facilities and resources.[40] It stressed that environmental inequalities could be positive, since some geographical territories have specific non-transferable advantages, and that inequalities are also a problem of power. Some people simply had louder voices in the contest for goods and services.

In France, as in Britain, the initiative came from the top.[41] The Ministry of l'Équipement[42] showed an interest in these inequalities, and launched a call for research on coastal and urban environmental inequalities in 2003. A report, *Les enjeux sociaux et environnementaux du développement urbain: la question des inégalités écologiques* ['Social and environmental issues in urban development: the question of ecological inequalities'] was published in 2004.[43] In it, the authors, sociologists Lydie Laigle and Valérie Oehler, identified four dimensions to what had become again *ecological* inequalities: territorial inequalities; inequalities in access to urbanity and a pleasant lifestyle; inequalities concerning nuisances and risks; and inequalities in the ability to engage with public authorities. As in the American literature, the boundaries between social and environmental drawbacks were unclear. Using official data on the region Nord-Pas-de-Calais, the report mapped the relationship between low income and soil pollution.[44] Predictably, no racial variable was considered here. Emphasis was put on the territorial dimension of inequalities and on

40. Comité français pour la préparation du sommet mondial sur le développement durable, *Inégalités écologiques, inégalités sociales*, Rapport du Groupe de Travail 9, 2002.

41. It is possible that activists have raised the issue of environmental justice – though, if they have, their campaigns have been limited since little has reached the mainstream media, and the terms remain unknown to the public. Unlike the Friends of the Earth website, on that of their French counterpart, Les Amis de la Terre, no campaign, article or report on environmental justice (or inequalities) is available.

42. The term does not translate well but, the closest is probably 'planning' or 'land settlement', and the functions are mostly concerned with the building of public infrastructural facilities (roads, tunnels, dams) and their strategic distribution across the country.

43. L. Laigle and V. Oehler, *Les enjeux sociaux et environnementaux du développement urbain: la question des inégalités écologiques*, Rapport final, Centre Scientifique et Technique du Bâtiment, Paris, 2004. Since the first version of this report is difficult to find today, the version used here is that published in 2006 by the IFEN, Institut Français de l'Environnement [French Institute for the Environment], '*Les inégalités environnementales*', *Les Synthèses*, IFEN, 2006. The IFEN was disestablished in 2008.

44. On social and environmental health inequalities in the Region Nord-Pas-de-Calais, see also P. Aïach, M. Marseille and Y. Theis, coor. *Pourquoi ce lourd tribut payé au cancer? Le cas exemplaire du Nord-Pas-de-Calais*. Rennes, 2004.

changing perceptions of how quality of life was shaped by income levels. Indeed, studies have shown how wealthy people sometimes willingly accept a relatively high level of flood risk, for example, in order to enjoy the amenity of proximity to a river. In other cases, people who have insufficient resources to move away can deny the negative feelings they sense from danger, as has been shown with the case of the reprocessing plant for nuclear waste in La Hague, Normandy.[45] This does not mean, however, that under-privileged populations are insensitive to environmental degradation. As Jacques Theys has shown, this myth has been built and propagated by liberal economic theory, just as, at a global level, it has been claimed that the poor were responsible for world environmental degradation.[46]

The definition of ecological inequalities by Laigle and Oehler was criticised by several French authors, which was predictable as, in France as in Britain, practically every author has formulated his or her own individual definition of these inequalities – or inequities. In the conclusion of their report, McLeod and his co-authors wrote:

> Lavelle (1994)[47] suggests that environmental equity implies an equal sharing of risk burdens, but not necessarily a reduction in the total burden of pollution. Cutter (1995)[48] argues that environmental justice implies much more, including remedial action, to correct an injustice imposed upon a specific sub-group of society, while Perlin *et al.* (1995)[49] broaden this concept into 'the goal of achieving adequate protection from the harmful effects of environmental agents for everyone, regardless of age, culture, ethnicity, gender, race or socio-economic status'.[50]

The physicist, environmental activist and editor of the journal *Écologie et Politique* [Ecology and Politics], Jean-Paul Deléage, provides what can only be called an obscure definition of ecological inequalities, in a themed journal issue on 'Ecological inequalities amongst men'. According to him, *ecological* inequalities refer to the

45. On these aspects see F. Duchêne and C. Morel-Journel, *De la culture du risque. Paroles riveraines à propos de deux cours d'eau péri-urbains.* La Tour d'Aigues, 2004; F. Zonabend, *La Presqu'île au nucléaire,* Paris, 1989; G. Walker, J. Fairburn, and G. Mitchell, *Environmental Quality and Social Deprivation, Phase II. National Analysis of Flood Hazard, IPC Industries and Air Quality, R&D Project.* Report available at http://www.staffs.ac.uk/schools/sciences/geography/links/IESR/themes_envJust.shtml Accessed April 2010.

46. J. Theys, *Les inégalités écologiques, dimension oubliée de l'action publique: entre raisons politiques et explications épistémologiques,* Paris, 2005, 3. See also, Theys, 'Pourquoi les préoccupations', 23–36.

47. M. Lavelle, 'Environmental Justice', in World Resources Institute, ed. *The 1994 Information Environmental Almanac,* Boston, 1994, 183–92.

48. S. Cutter, 'Race, Class and Justice', *Progress in Human Geography,* 19:1, 1995, 111–22.

49. S. Perlin, R. Setzer, J. Creason, and K. Sexton, 'Distribution of Industrial Air Emissions by Income and Race in the United States: An Approach Using the Toxic Release Inventory', *Environmental Science Technology,* 29:1, 1995, 69.

50. McLeod *et al.* '*The Relationship between Socio-economic Indicators*', conclusion.

unequal confrontation between individuals and social groups on the one hand and supposedly natural elements (air, water, biodiversity) on the other. *Environmental* inequalities widen the confrontation between the concrete and imaginary 'supra-nature' [*la surnature*] 'deeply worked by the technological and human hold on world ecosystem'.[51] In the same issue, geographer Cyria Emelianoff proposes a somewhat clearer definition, with *ecological* inequalities having a wider meaning beyond that of *environmental* inequalities. Ecological inequalities would embrace individuals both as victims and also as sources of environmental inequalities, so that when the balance sheet is drawn up it shows pollution received against pollution emitted. Under these circumstances it becomes more obvious that those who suffer less from their environment are also those who pollute more.[52] There is also a temporal difference: while *environmental* inequalities tend to refer to current visible inequalities, *ecological* inequalities include long-run intergenerational inequalities.

Another point of view emerged from a group of academic ecologists who protested against the use of the term *ecological* inequalities by social scientists. These ecologists consider the term to refer to the invasion of indigenous species by exotic ones. The term *ecological* should, in their view, only be used for biodiversity, heritage, species and spaces of particular interest, and protected areas.[53] It is a scientific view consistent with that of Jacques Theys[54] who explains that *ecological* inequalities do not actually refer to *physical* situations but to *social* inequalities *in relation to the environment.*[55]

In 2007, a jointly authored article discussing 'The multiple facets of ecological inequalities' insisted, rightly, that in France the terminology has not yet been agreed.[56] But if uncertainty remains as regards the use of the words *ecological* or *environmental*, there is a consensus of sorts on the use of the word *inequalities* in

51. 'Des Inégalités écologiques parmi les hommes', *Écologie et Politique. Sciences, Cultures et Sociétés*, 35, 2007, introduction, 13.

52. This argument was advanced as early as 1998 in A. Lipietz, 'Économie politique des écotaxes' Rapport au Conseil d'analyse économique du premier ministre, *La Documentation Française*, 1998. See also S. Pye, I. Skinner, N. Meyer-Ohlendorf, A. Leipprand, K. Lucas and R. Salmons, 'Addressing the Social Dimensions of Environmental Policy. A Study of the Linkages between Environmental and Social Sustainability in Europe', European Commission Directorate-General 'Employment, Social Affairs and Equal Opportunities', 2008.

53. G. Bellan, D. Bellan-Santini and J-C. Dauvin, 'À propos de quelques utilisations des termes "Inégalités écologiques": simples impropriétés de langage ou accaparement abusif?' *Développement durable et territories*, Dossier 9: Inégalités écologiques, inégalités sociales, online 3 May 2007, consulted 30 October 2009, see http://developpementdurable.revues.org/index3426.html

54. As a political science graduate with a doctorate in Mathematics and Economics, Jacques Theys has held high office in the Ministry of Ecology. His articles in scholarly journals are very thought provoking.

55. See footnote 39.

56. L. Charles, C. Emelianoff, C. Ghorra-Gobin, I. Roussel, F.-X. Roussel and H. Scarwell, 'Les multiples facettes des inégalités écologiques', *Développement durable et territoires* Dossier 9:

preference to *injustice*. Although it is impossible to extend this discussion to the terminology used in other European countries, it is worth noting that, as far as we can determine, the evolution in the use of the vocabulary has been the same in Belgium,[57] Spain and Germany where, as in France, the concept of environmental justice is not used.

Injustice, Inequalities: What Are We Talking About?

Although the same kind of problems existed on both sides of the Atlantic – people living in inadequate or dangerous conditions, exposed to pollutants of various kinds or deprived of pleasant amenities and life styles, and usually unable to move away from them – there were, then, three divergences between the USA and continental Europe. The first one, already mentioned, is that what was, in the USA, a grassroots social movement has become, in Europe, a top-down policy concept and an analytical tool for scholars. The other two divergences concern the vocabulary: on the continent, *injustice* has most often become *inequality*; and, though it is not totally settled yet, there is a tendency to replace the adjective *environmental* by *ecological*. These shifts are meaningful; they have not occurred by chance.

Before exploring the reasons why they have happened, it is worth noting that those who talk and write about *injustice* in fact make a claim for *justice*, a right that can hardly be contested, while those who talk about *inequalities* are not demanding *equality*, which would probably be difficult to justify either from a legal or from a philosophical point of view, and no more from an ecological point of view, as *ecological equality* does not make any sense – there have always been, and will always be, places more pleasant or convenient to live in than others, even if values change over time. Actually, those who talk about inequalities rarely specify their objective precisely. They sometimes call for policies aimed at a 'curbing of inequalities', or likely to promote *equity*. But what does *equity* mean as far as environment is concerned? This question would deserve an article of its own. We offer just a few remarks, in order to underline the complexity of the problem.

The liberal theory of justice, as expressed by the moral and political philosopher John Rawls, deems that an inequality becomes an injustice when it does not benefit everyone, and specifically the poorest.[58] When inequalities do not benefit the poor as much as the wealthy (the situation Rawls calls 'the maximin'), injustice must be remedied, and *unjust* inequalities must be turned into *just* inequalities. This

Inégalités écologiques, inégalités sociales, online 28 August 2007, consulted 30 October 2009. See http://developpementdurable.revues.org/index3892.html

57. In francophone Belgium at least. Flemish speaking scholarship may also have undergone the same evolution.

58. J. Rawls, *A Theory of Justice*, Oxford, 1971. See also the later *Justice as Fairness: A Restatement*, Cambridge, Ma. 2001, which modifies the earlier work.

is what Rawls calls 'justice as fairness'.[59] Rawls' theory has been both extended and challenged by several authors. Neo-marxist David Harvey also regards inequalities as a question of distribution, but deems that they cannot be solved in the context of a capitalist economy.[60] According to him, equity will come with a shift to a new system of production that will allow a fairer distribution of resources. Feminist philosopher Iris Marion Young considers that classical theories of justice cannot address adequately the problems of equity because they do not take domination and oppression into account and pretend to manage injustice according to white, European, male norms of reason and respectability.[61] Amartya Sen, the Nobel prize-winner in economics, has also been influential in this debate in developing his concept of 'capabilities'.[62] Though environment is not central to Sen's work, this concept of capabilities is a convenient one for our topic. Environmental injustices are not premeditated most of the time, if ever, but some people have a greater capability than others to protect themselves from, or flee, unexpected nuisances.

While most of these theoreticians of justice do not focus particularly on *environmental* justice, they have influenced those who do. As early as 1996, Hampson, Laberge and Reppy wrote:

> It has become clear that concepts of social justice rooted in traditional liberal theory are not adequate for addressing the kinds of competing moral claims that have arisen over environmental issues. The interests of disenfranchised majorities – like the world's poor who live at or below the subsistence level or those of unborn generations – are problematic from the point of view of liberal theories, which start with the assumption that individual preferences are equal and should simply be taken at face value. This stance is too passive when it comes to the environment, not only because some parties are not represented at the bargaining table, but because the preferences of those who are represented are often at the root of the problem.[63]

Grounded in a concern for individual rights, liberal theories do not recognise claims based on the value of group identity, while the environmental justice movement is rooted in communities and collectivities.

Rawls' theory of justice as applied to the environment is also challenged by deep ecologists who claim 'rights (or justice) for nature'. These radical ecologists also accuse environmental justice of using an old, outdated frame of thinking, that of the distribution of goods (Rawlsian justice is a distributive justice) which instrumentalises nature, reducing it to a set of resources to be used for human

59. 'Fairness' has no direct equivalent in French, and is translated in the French version of Rawls' books by the word *équité* – equity.

60. D. Harvey, *Justice, Nature, and the Geography of Difference*, Oxford, 1996.

61. I. M. Young, *Justice and the Politics of Difference*, Princeton, 1990.

62. For a recent synthesis of Sen's ideas, see A. Sen, *The Idea of Justice*, Harvard, 2009.

63. F.O. Hampson, P. Laberge and J. Reppy, 'Introduction: Framing the Debate', in F.O. Hampson and J. Reppy, eds. *Earthly Good. Environmental Change and Social Justice*, Ithaca, 1996, 5.

consumption and distributed fairly for this purpose. To move away from this distributive justice, Will Kymlicka pleads for a perspective 'that emphasizes stewardship rather than possession of the environment'.[64] But some authors, such as French geographer Bernard Bret, still believe that Rawls' theory of justice is an appropriate tool to interpret inequalities, including socio-spatial dimensions.[65] Following Rawls' principles, these authors propose compensation for unjust environmental inequalities by means of appropriate distributive policies.[66] On the whole, however, the mainstream within the field of environmental ethics is more interested in nature and the fate of the planet than in the lot of humans, and so tends to minimise or ignore social and economic inequalities.[67]

In a recent study focusing on the concept of justice in the movement for environmental justice, David Schlosberg underlines that an evolution is taking place:

> Even within the movement, there is an understanding of the limitations of the equity approach. Between the early movement activities in the early 1980s and the beginning of reflection on the movement in the 1990s, both activists within, and scholars writing about, the environmental justice movement replaced the term equity with that of justice. Both practitioners and researchers began to understand that justice was broader than the singular question of equity.[68]

He adds, quoting Bryant: 'The former concept was too limiting for the job that needed to be done. By making connections between environmental and social issues, environmental justice provides an opportunity for building broad-based coalitions in order to make profound changes to enhance the quality of life of people.'[69]

This tentative panorama shows how difficult it is to identify and define what is at stake in the idea of environmental justice. However, it still does not explain why the concept of *environmental justice*, so widely employed in the USA, is hardly used

---

64. Kymlicka, 'Concepts of Community and Social Justice', 36. For a longer discussion on the various concepts of justice see W. Kymlicka, *Contemporary Political Philosophy: An Introduction*, Oxford, 1999, translated into French as *Les théories de la justice. Une introduction*, Paris, 2003.

65. B. Bret, 'Interpréter les inégalités socio-spatiales à la lumière de la Théorie de la justice de John Rawls', in *Annales de Géographie*, 2009, no. 665–66, 16–34; A. Dobson, *Justice and the Environment, Conceptions of Environmental Sustainability*, Oxford, 2003.

66. Blanchon *et al.* 'Comprendre'.

67. For a discussion on the relationship between environmental justice and environmental ethics, see B. Gagnon, N. Lewis and S. Ferrari, 'Environnement et pauvreté: regards croisés entre l'éthique et la justice environnementale', in *Écologie et Politique*, 79–90. For a first introduction in French to the theories of justice applied to environment, see Blanchon *et al.* 'Comprendre'.

68. D. Schlosberg, *Defining Environmental Justice. Theories, Movements and Nature*, Oxford, 2007, 58.

69. B. Bryant, *Environmental Justice: Issues, Policies and Solutions*, Covelo, 1995, 7.

in Europe.[70] According to Emelianoff, the shift from *injustice* to *inequalities* could result from an inability of the republican French people to contemplate, far less accept, differences. Also, it might arise from their tendency to hide discrimination by renaming it *inequality*, a term that may suggest there is something intrinsically natural in the situation, rather than an injustice directed towards specific groups.[71] It is true that, in French, *inequalities* sounds 'less serious' than *injustice*. It connotes something manageable with a well-designed policy. But it is worth noting that the term *environmental justice* has also become *environmental inequalities* in Belgium and Spain, countries that are not republican. This shows the limitation of the argument that French republicanism has prevented thinking about differences.

In 2007, Lydie Laigle and Mélanie Tual explored the conception of ecological inequalities in five European countries and the role they play in sustainable urban development policies.[72] More than half of their article is devoted to current policies, far removed from our concerns here, but it begins with an attempt to describe and explain the nature and origins of the difference between the prevailing conceptions in the USA and continental Europe. According to Laigle and Tual, the shift from injustice to inequalities results from historically different visions of the environment, law and justice on the one hand, and a different conception of the state and its role on the other. In the USA, the prevailing idea of the environmental justice movement is that any group subjected to environmental discrimination should be able to contest this situation. The federal government would ultimately address citizens' concerns by passing a new law. The same situation would apply in Britain, based on the idea that individuals are granted natural rights that are derived from a moral order 'which is embodied independently from the society'. In this context, the duty of the state and the basis of public policies are circumscribed to the protection of human rights, in response to pressure from citizens. This trend also finds its roots in an interactive vision of the relationship between societies and their environments.[73]

70. The French journal of social and political philosophy *Raison Publique* recently devoted a themed issue to environmental justice (*Raison Publique, Éthique, Politique et Societé,* 8, 2008). The three articles it contains deal with the concept of justice from a philosophical point of view but do not attempt a comparison between the Anglo-Saxon and the French conceptions.

71. Emelianoff, 'Connaître ou reconnaître les inégalités environnementales?'

72. L. Laigle and M. Tual, 'Conceptions des inégalités écologiques dans cinq pays européens: quelle place dans les politiques de développement urbain durable?' *Développement durable et territories,* Dossier 9: Inégalités écologiques, inégalités sociales, online 28 August 2007, consulted 30 October 2009. See http://developpementdurable.revues.org/index3892.html

73. Laigle and Tual, 'Conceptions', 8, translation into English is ours. On this last point, Laigle and Tual refer to philosopher Lionel Charles, who has offered an analysis of the origins of the differing Anglo-Saxon and Latin perceptions of nature of the environment. While in Latin countries the relationship of man to environment retains a degree of externality, British philosophers, from the seventeenth century, developed an opposite vision in which man is part of his environment and interacts with it. L. Charles, 'Du milieu à l'environnement', *L'Environnement, question sociale,*

Despite their efforts to solve the difficult question of the change that has apparently happened between Anglo-Saxon and continental European countries, it does not seem possible to conclude, as Laigle and Tual do, that 'anglophone countries' have been more advanced 'regarding their thinking on and protection of the environment and the right of individuals to a safe environment'.[74] Indeed, in Europe today, Britain can hardly be regarded as a leader in this field and, although criteria should be defined before medals are allocated, Nordic and German speaking countries would seem stronger competitors. It is even more difficult to see in this explanation the origin of the shift from *injustice* to *inequality*.

Laigle's and Tual's paper suggests yet another answer to the question. In France, since the Revolution, the idea of justice has rested on 'equality in law' of all citizens, while in Britain and other Anglo-Saxon countries it is based on 'equality of rights'. Thus, in France, public policies tend to aim at redistributive and remedial actions towards inequalities, while British and US policies aim at repairing situations regarded as unjust. All these attempts to explain the difference in terminology, as presented by geographers and philosophers, are valuable but they seem insufficient, in their current form, to really explain what we observe. Perhaps historians will sooner rather than later shed some additional light.

The second shift in the vocabulary is that from *environmental* to *ecological*, and is somewhat easier to explain. In France, it happened at the very time when the right-wing government that took office in May 2002 decided to rename the former Ministry of Aménagement et Environnement (roughly equivalent to 'Planning and Environment'), as the Ministère de l'écologie et du développement durable [Ministry of Ecology and Sustainable Development]. Several authors have emphasised that this replacement of the word *Environment* by the word *Ecology* reflects the power and influence of naturalism within the environmental movement. This perceives nature as independent from man, attacked by him and in need of protection from human preoccupations. The ideology that lies behind this terminological shift is unable, or unwilling, to think in terms of inter-relationship between man and his surroundings, and rejects any link between the social and the environmental.[75] A parallel can be drawn with what happened to the concept of sustainable development, which was initially inspired by the theory of eco-development, as elaborated

---

Dix ans de recherches pour le ministère de l'environnement, Paris, 2001, 21–28. See also L. Charles and B. Kalaora, 'Pensée, sensibilité et action dans la société française autour de la question de la nature', *Annales de Géographie*, 5:2008, 663, 3–25; and L. Charles and B. Kalaora, 'Prégnance et limites d'une approche esthétique de l'environnement', in M. Bédard, dir. *Le paysage, un projet politique*, Québec, 2009.

74. Laigle and Tual, 'Conceptions', 8.

75. L. Charles, 'Une appréhension interculturelle des inégalités environnementales', in 'Des inégalités écologiques parmi les hommes', *Écologie et Politique. Sciences, Cultures et Sociétés*, 35, 2007, 79–90.

by Ignacy Sachs,[76] and had a strong social component. But it was diluted, after the Rio conference in 1992, to a much more naturalist concept that no longer questions the very character of economic development.[77]

Writing about *ecological* inequalities also raises the question of whether they are independent from man. For some, ecological inequalities arise directly from the uneven distribution of natural amenities, not from the complex relationship between man and nature or the use man makes of this nature. Though apparently a simpler, clearer concept, it is one that shrinks the problem by reducing ecological inequalities to 'natural' problems and catastrophes, which they usually are not. According to Charles, it seems also to betray the French difficulty in thinking about the environment.[78]

Proponents of the term *ecological* inequalities often have a static vision, while those who favour the phrase *environmental* inequalities have more dynamic one, and are more interested in pointing to a process that, as Vicki Been has shown, is necessary to understand the *result* in the long term and the responsibilities involved in the process. The reference to territories, ('territorial ecological inequalities'), omnipresent in France, tends to obscure the social by aiming only at better territorial policies.[79]

As we mentioned at the beginning of this chapter, our first move in preparing this book was to use a title that included social and environmental *inequalities,* not *justice.* This seemed to us more consistent with the tradition of historical writing, which refers only occasionally to ethics. So contributors to the book were asked to think in terms of inequalities. But the more we thought of the issues, the more we understood we should, and wanted to, use the word *justice* in the title of the book. We do not want to situate our contribution alongside those that privilege natural environmental problems and marginalise man and social issues. On the contrary, our conception of the environment conforms to the definition given by geographer Pierre George as early as 1971: 'a system of relations between social, economic and spatial dynamics and a physico-chemical and biological field of forces'.[80] This definition integrates man and society, and matches the purpose that was ours, considering simultaneously social and the environmental dimensions that have been artificially separated for too long. Accordingly, the title of our book became as it is: *Environmental and Social Justice in the City: Historical Perspectives.*

76. I. Sachs, *L'écodéveloppement, stratégie pour le 21e siècle*, Paris, 1980.

77. On the history and evolution of the concept of sustainable development, see Y. Veyret, *Le développement durable*, Paris, 2007, 13–51.

78. Charles, 'Une appréhension interculturelle'. See also L. Charles, 'Entre milieu et environment', in M. Boyer, G. Herzlich, and B. Maresca, *L'environnement, question sociale*, Paris, 2001.

79. On this point see Theys, 'Pourquoi les préoccupations'.

80. P. George, *L'Environnement*, Paris, 1971.

Environmental Justice and History

The previous paragraphs have outlined the history of a concept that was born a quarter of a century ago. This is, in itself, history. By this reminder we aim to explore to what extent the concept of environmental justice is useful in understanding a more distant past. In *Who Pays The Price?* anthropologist Barbara Rose Johnston and other authors explored, in 1994, the relationship between human rights, environmental crisis, environmental policies, and what she called 'social justice environmentalism' using the methods and subjects associated with social anthropology.[81] The aim of our book is, similarly, to seize the concept of environmental justice and use it with the tools and methods of historians. This raises epistemological problems, as the concept of justice has obviously changed with time.

In a paper first published in the *Environmental History Review* and recently re-issued in a fascinating collection, *'To Love the Wind and the Rain': African Americans and Environmental History*, historian Martin Melosi summarised rightly the points raised by the movement for environmental justice which are particularly interesting for historical analysis: '(1) environmental equity, especially as it relates to race, class and gender; (2) environment as a cultural construct; (3) the clash between anthropocentrism and ecocentrism; (4) the importance of urban environmental problems, especially as they impact human life; (5) and the nature of the environmental movement itself.'[82] All these points do matter to the historian. The fact, clearly revealed by the environmental justice movement, that the environment is a social construction, can also be applied to past perceptions. How did people, at different moments in time, perceive what we call today our environment? When neither the word nor the concept existed, what did people enjoy or dislike among the elements that surrounded them, whether built or natural, in a given time and given circumstances? What did they regard as an amenity, and what a nuisance? What struck them, if at all, as unfair, and did they complain? As historians, which elements of past environments can we define today, with our view of the world, as unfair? What can historical studies bring to an understanding of the way environmental injustices were constructed, intentionally or otherwise? These are amongst the first set of questions we should try to answer.

Also, as urban historians, we can only be sensitive to the fact that the environmental justice movement has put the city at the front of the picture again. Where historians looked with economic, social and cultural history lenses, we would like them to look again, adding a new focus to include the environmental, and especially environmental justice, in their vistas. In other words, we would

81. B.R. Johnston, *Who Pays the Price? The Sociocultural Context of Environmental Crisis.* Washington DC, 1994.

82. M. Melosi, 'Environmental Justice, Ecoracism and Environmental History', in D. Glave and Mark Stoll, eds. *'To Love the Wind and Rain': African Americans and Environmental History*, Pittsburgh, 2006, 120–32, and an earlier publication in *Environmental History Review*, 19, 1995, 194–211.

like them to reconsider the past by looking at both the environmental inequalities people could be affected by, and the way these inequalities were constructed. A single example may help. Public hygiene has conventionally been conceived as a movement that almost always improved the living conditions of the poor, by progressively addressing the conditions of slum life.[83] Some negative side effects have also been revealed: the break-up of communities and greater distance from work, for example. But, using a wider angle, has the question been posed as to the responsibility public health incurred by increasing environmental inequalities through the specialisation of spaces, or the expense of over-specifying sanitary standards and so increasing costs and thus rents, or the disadvantages of additional water supply if water disposal facilities are deficient? There are many new environmental perspectives of this kind to be explored.

Despite the importance we have identified, and the bulk of publications devoted to the topic of environmental justice, history as a discipline has not played a major role in the field, even in the USA, with a few noticeable exceptions. Andrew Hurley is one of first to have grasped the topic from an historical point of view. In *Environmental Inequalities. Class, Race and Industrial Pollution in Gary*, he focused on the post-Second World War steel mill community of Gary, Indiana, showing the inability of capitalism to reconcile industrial priorities, social justice and quality of life, and how the poor and African Americans have borne more than their share of pollution.[84] Ten years later, in *Packing Them In*, Sylvia Hood Washington explored the historical and philosophical foundations of environmental racism and environmental justice, and focused, in the empirical part of her book, on the Chicago 'jungle' between 1865 and 1954, in a way that is sadly more descriptive than analytical.[85] More recently, the book edited by Dianne Glave and Mark Stoll, '*To Love the Wind and the Rain': African Americans and Environmental History*, although not strictly devoted to environmental justice, as the title indicates, has gathered a compelling collection of essays.[86] The American journal *Environmental History* devoted a special issue to the subject in which Maureen A. Flanagan's introduction announced: 'Environmental justice in the city: a theme for urban environmental history.' In spite of the three strong articles it contained and the editor's encouragement to write more, environmental justice issues have not yet developed as a major element in environmental history in America.

---

83. An exception is A. Mayne, *The Imagined Slum: Newspaper Representation in Three Cities 1870–1914*, Leicester, 1993.

84. A. Hurley's *Environmental Inequalities. Class, Race and Industrial Pollution in Gary, Indiana, 1945–1980*, Chapel Hill, 1995.

85. S.H. Washington, *Packing Them In: An Archaeology of Environmental Racism in Chicago, 1865–1954*, Lanham, 2005.

86. Glave and Stoll, eds. '*To Love the Wind and Rain*'.

In Europe, though contributions might have been published in languages we cannot read, the historical literature seems still more limited. True, the main European contribution is a significant one: Joan Martinez-Alier, whose works have already been quoted, has coined phrases and promoted concepts like 'the environmentalism of the poor' and the 'ecological debt', and written widely about them, sometimes with the Indian scholar Ramachandra Guha.[87] But Martinez-Alier's contribution to the field, however important it is, deals with the global scale.

As far as the local, and the urban in particular, is concerned, the historiography is sparse. A careful exploration of the tables of contents in journals over the last twenty years produces only a handful of first class contributions. Here and there have been scattered occasional papers dealing with environmental justice topics. Naomi Williams' study of infant mortality in nineteenth century Sheffield is a case in point.[88] Her excellent paper showed that the mortality rate varied not only according to the social status and the season, but also to a specific environmental factor, the proximity of the river. More recent papers have called, in one case, for a deeper engagement of environmental history with social theory[89] and, in another, as an example of common ground, for explorations in 'the interconnections between social inequality and environmental degradation'.[90] These are interests shared by the contributors to this book and they bring fresh perspectives to issues of environmental justice.

Exploring Past Inequalities

Each of the essays in this volume illuminates different facets of environmental inequalities, using a range of time periods, locations and forms to reveal how inequalities developed, were sustained or altered. Sometimes the environmental inequality arose inadvertently, as the unintended consequence of human actions on the urban environment; on other occasions the environmental inequality was imposed, knowingly and with a clear understanding that the wellbeing of others was impaired. Based on deep archival and primary research, the contributors to this book consider efforts to address and manage environmental inequalities in particular historical settings while showing an awareness of the underlying processes

87. Martinez-Alier, *The Environmentalism of the Poor*; Guha and Martinez-Alier, *Varieties of Environmentalism.*

88. N. Williams, 'Death in Its Season: Class, Environment and the Mortality of Infants in Nineteenth-century Sheffield', *Social History of Medicine*, 5:1, 1992, 71–94.

89. S. Sörlin and P. Warde, 'The Problem of the Problem of Environmental History: A Re-reading of the Field', *Environmental History*, 12:1, 2007, 107–30.

90. S. Mosley, 'Common Ground: Integrating Social and Environmental history', *Journal of Social History*, 39:3, 2006, 915–33. See also G. Massard-Guilbaud and S. Mosley, eds. *Common Ground. Integrating the Social and Environmental in History*, Newcastle, 2011.

that have more general relevance to the subject of environmental inequalities and injustice. The essays illuminate how natural resources and topographical features, patterns of landownership, markets and a class-based industrial capitalism, jurisdictional boundaries, municipal intervention, rural–urban interactions, scientific and technical knowledge, as well as voluntary organisations functioned, often in different combinations, to produce different environmental inequalities, according to local circumstances and historical settings.

Inequalities and Risks

Nowadays, in the context of concerns about climate change, insurers recognise that natural disasters do not affect all elements of the population equally. Environmental agencies develop 'at risk' maps, identifying where floods are most likely to occur and insurers follow these topographical characteristics to increase premiums and even to refuse cover.[91] Potential house buyers can also check the areas most susceptible to flooding and may well decide to purchase property; in so doing, they are adjusting their preferences to some implicit risk profile. This is not new. During the twentieth century, as Craig Colten shows, the Delta region of the Mississippi River was subject to several serious floods. Despite the risks, minority and low-income inhabitants chose to remain in high-risk areas and were disproportionately affected when the lèvees were breached. In the preparations for future floods, and responses to them, minorities and the poor were less well placed, either in terms of lower sea-wall defences, or of early warnings about an impending hurricane, or in relation to access to transport for evacuation. Employment opportunities and the social and cultural capital associated with their communities were sufficiently strong short-term factors for low-income groups to counter the potential damage and death associated with floods. This rationale was unattractive to wealthier inhabitants who applied political pressure to obtain higher levees in their residential districts and had sufficient wealth, educational background and alternative employment opportunities to retreat from areas of high risk.

The trade-off between damage to assets and loss of life compared to the transaction costs of relocating further from the most vulnerable areas produced a different risk profile for these middle class residents and their strategies heightened both the environmental and social inequalities in the Delta area. At first glance it might seem that this was also the case in the Low Countries and eastern England – in Romney Marsh, the Flemish Polders, and the Dutch peatlands – where susceptibility to flood damage was not just an occasional event, as in the Delta, but an omnipresent reality. Tim Soens' study, spanning several centuries from medieval times to 1800, shows that floods, though a common denominator in the

91. See http://www.environment-agency.gov.uk/homeandleisure/floods/default.aspx and http://www.sepa.org.uk/flooding.aspx Accessed 15 May 2010.

environmental history of these three areas, prompted divergent responses and highly specific social and economic inequalities as a result of the interactions between towns and their hinterlands. As in the Mississippi Delta, where slave labour was used to construct the levees, in the three North Sea areas dikes were built and maintained by cheap or indentured labour. As long as this heavy investment in sea defences was maintained then all three regions were able to maintain an environmental equilibrium with no groups particularly shouldering the adverse consequences of flooding. Indeed, the areas were transformed in the twelfth, thirteenth and fourteenth centuries into some of the most densely populated in northern Europe based on the harbours, fishing, salt-production and woollen industries that were features of the delicate balance between agricultural and urban production in the marshlands. In response to the buoyancy of the local economy, land consolidation in the hands of urban and mostly absentee gentry landowners in both the Romney and Flemish marshes led to the decline of smallholdings. A radical restructuring of landownership contributed to a reduction in the available labour supply, vital to the maintenance of flood defences, and as a result fundamentally changed the delicate ecological balance in these two areas. The environmental impact fell most severely on tenant farmers because it was they who had least scope for adjustment as a result of changed landholding patterns, but it adversely affected the towns too. By contrast, the continuation of peasant smallholders in Holland for at least two centuries longer was largely because of the presence of peat as an energy source, which required labour supplies on the land to cut, store and transport the peat. So land consolidation made less economic sense and, while the labour supply lasted, so did the prosperity of the rural and urban settlements.

What emerges then from the study of wetlands management is that environmental risk was initially spread amongst the various elements of society because of the interdependencies involved. A more equitable social distribution of risk existed in the medieval period, largely because of the institutional underpinning of collective responsibility that was involved in maintenance of the sea walls. Thereafter, with land consolidation, the interests of owners and workers were no longer in alignment and the environmental inequalities fell both on the smallholders, since their interests were directly prejudiced by floods, and on the urban craftsmen whose livelihoods depended on marsh products. Environmental management depended most on mutual interests, and when this mutuality broke down and the financial strategies of the absentee landowners assumed greater significance, environmental inequalities became increasingly skewed, with smallholders most affected. In the Dutch peatlands ecological equilibrium was the most durable under these circumstances.

Energy and Fuel Poverty

Peat was an energy source of considerable importance in many parts of early modern Europe. Fuel supplies for industrial processes that required smelting or boiling are usually presented as a sequence from peat or wood to coal, with coal-fired steam engines, crucially, also replacing water-powered textile mills. The pace of this technological shift depended much on the regional availability, cost and quality of coal, and the details of the timing of the transition have filled many pages in accounts of the industrial revolution in Europe. However, Richard Oram focuses attention on the impact this transition in energy supplies had on the domestic user and notes that, in many cases, it was far from positive. The association of coal with modernity, it is claimed, meant that fashionable households preferred this fuel to peat; it was seen as progressive, efficient, and thus consistent with 'Enlightenment' and 'Improvement' values. Peat, conversely, was a primitive source, associated with earlier modes of production and consumption. The shift in domestic demand for coal in Scotland was achieved during the late eighteenth and early nineteenth centuries when the middle classes enjoyed rising disposable incomes so that in real terms coal represented a falling proportion of their budgets. At the same time, domestic demand was further boosted as towns and cities expanded. Here the environmental inequality resulted from a collapse in the market for traditional fuels. This had a particularly serious impact on the poorer segments of society and on towns distant from sources of coal supplies, since transport costs for coal were punitive. More serious, perhaps, was the removal of traditional sources of supply, firstly by landowners who had less interest in the management of peat mosses and woods as they sought to modernise their agricultural lands and secondly by the burghs, where communal rights to collect fallen wood were rescinded and sold to the highest bidders. It was a double penalty on the poor: less wood for collection and payment for what was formerly free. Increasingly, then, the rising numbers of urban poor were deprived of their source of fuel. It was they who were the casualties in a process of modernisation and they exhibited their resistance to the withdrawal of traditional rights by stealing and scavenging for wood wherever possible. The current British definition of fuel poverty, the expenditure of more than ten per cent of household income to maintain a living room at 21°C, remains a contemporary political issue of some importance both to pensioners and the poor and Oram's chapter is a timely reminder of how environmental impacts vary according to place.

Fuel poverty was an environmental inequality closely tied to social class and the focus on class pervades several contributions in the book. It is clear that in the early stages of issuing written notices about flood warnings, the poor and poorly-educated of New Orleans were less easily able to prepare for and respond to the emergency. Melosi's research also shows that, in Houston in the 1930s, incinerators were mostly located in black or Hispanic areas of the city and this can be generalised to the location of public facilities such as water treatment plants, sewage disposal, pumping stations and later refuse and recycling centres and electricity generating stations in areas dominated by low income groups.[92] There are, however, convincing explanations as to why this was so. It would have been perverse, for example, for Joseph Bazalgette's main drainage scheme for London not to follow the downhill gradient towards an outlet near existing East End working class housing. It is also worth remembering that all manner of industrial activities, some of them associated with slaughtering and the disposal of animal waste, were often centrally located, close to markets, so the intrusion of a tramway depot or sewage pumping station in the nineteenth century was logical. Whitehand explains this issue of the location of public investment projects by relating the land requirement of public facilities to the cost of acquisition and rental charges.[93] He concludes that it made sense for major land-hungry activities such as hospitals and public utilities – gas works, electricity generating stations and water treatment plants – to be located on low cost marginal land, some of which was on the urban fringe and in low-lying areas less attractive to house builders. However, as the nineteenth century advanced and cities expanded, even these low-lying marginal lands also became relatively attractive because of their central position and proximity to employment opportunities, and so gasworks and waste destructors, among other facilities, were co-located with workers' housing.

Land Use, Path Dependency and the Housing Market

The hypothesis that waste disposal facilities were located in working class areas is explored by Stéphane Frioux as part of a study of class relations in a large number of French cities. He discovers quite the reverse before 1940. In the early twentieth century, incinerators were often located centrally and Frioux explains this convincingly, arguing that the faith placed by hygienists and engineers in the sanitary efficiency of incinerators disarmed public anxiety; after all, everything was burned to a cinder. Expert knowledge, then, was used to reassure the public and far from such facilities being located in the poorest, politically weakest districts, it was the middle classes that willingly accepted the arguments of efficiency and economy. Incineration saved using labour intensive garbage dumps and the offensive practice of carting refuse through middle class districts to the outskirts. International examples of efficient incinerator plants sited beside the Prince of Monaco's palace, an up-market hotel in London, and a large flour mill in Edinburgh are just a few of the instances quoted to show that urban development was not impeded by the

92. M.V. Melosi, 'Sanitary Services and Decision Making in Houston, 1876–1945', *Journal of Urban History*, 20:3, 1994, 365–406.

93. J.W.R. Whitehand, *The Changing Face of Cities: A Study of Development Cycles and Urban Form*, Oxford 1987.

location of such facilities.[94] Such was the confidence in incinerator technology that in Villeurbanne, an independent socialist city adjacent to Lyon, a new incinerator was built in the very heart of the civic centre and provided heating for over 200 social housing apartments within 300 metres of the plant. French town planning legislation in 1919 and 1924 encouraged zoning to develop and incinerators were in many cases located in these emerging industrial zones. However, it seems that this was more because of the earlier existence of insalubrious industrial activities, to which incinerators were logically added, rather than as a result of complaints about the environmental impact of incinerators themselves.[95] Frioux concludes, therefore, that the evidence does not convincingly show environmental discrimination.

Rising land prices and urban rents were direct consequences of rapid urban growth in many cities. Initially the result was greater density of building in the town or city centre, followed by a centrifugal effect that rarely, if ever, corresponded exactly with the Chicago model of urban development with its concentric inner rings of working class housing encircled by middle class suburbs. Land use varied according to topographical features and local circumstances. Often the result, as in the Swedish towns of Norrköping and Linköping, was to propel the working class beyond the periphery of the city and thus beyond the jurisdiction of the municipality. This jurisdictional issue is critical to understanding environmental inequalities within many towns and cities in Europe and North America.[96] In a number of countries, towns and cities could legally only administer the geographical areas for which they had statutory responsibility. To do otherwise transgressed their jurisdictional reach and rendered city officials liable for legal actions for maladministration and the misuse of funds. Indeed, the city was composed of many different mosaics for each of several jurisdictions – local political wards, national constituencies, port and harbour authorities and ecclesiastical parishes were increasingly complicated by several overlays of police, fire, water supply and sanitary districts, to name just a few municipal jurisdictions. Indeed, in parts of Edinburgh these working class districts were simply called 'the colonies' and, until embraced within the boundaries of the city, had, in some cases, a semi-detached juridical status. Östra Eneby, a suburb of Norrköping, was designated a municipality and so came under national building codes while remaining a rural district administratively, judicially, and ecclesiastically. The history of annexations of such outlying settlements and minor townships is a topic for further exploration and the issue of the legitimacy of the

94. J. Clark, '"The Incineration of Refuse is Beautiful": Torquay and the Introduction of Municipal Refuse Destructors', *Urban History*, 34:2, 2007, 255–77 argues (262) that the fact that these sensitive locations were mentioned at all is suspicious.

95. As in Houston, there seemed to be few complaints except from the middle classes.

96. The concept of 'city limits' is not the same for all countries. In France or Spain, for example, when leaving one city the traveller enters another commune with exactly equivalent powers.

municipal authority acting to redress environmental injustice is central to a number of the chapters here.

One element that is crucial, therefore, in the investigation of environmental injustice is the extent to which the scope for independent municipal action was often severely circumscribed; to cross the jurisdictional boundary line was to invite legal challenges and create administrative precedents, and so to condemn town councils for a lack of intervention to mitigate inequalities is inappropriate if they were not empowered to do so. Accordingly, until municipal authorities obtained statutory powers, or confirmation of an extension to their existing powers, they tended to be conservative in relation to appeals from suburban districts to extend municipal services. Of course, some municipal services were initially operated as private concerns and these were not necessarily bounded by the city's jurisdictional territory. So tramway lines sometimes ran beyond the city limits to a terminus in an outlying settlement, creating amenity mainly for middle class passengers and yielding greater revenues for the operating company.[97]

This raises another intriguing aspect of environmental inequalities, the effect of 'legacies' – that is to say, how heavy capital investment in public facilities affected amenities for city residents who, in terms of class, ethnicity and the social composition of a district, changed considerably over the course of a few decades. This 'fixed and immobile capital'[98] in buildings and land use was both a blight and a benefit to city residents. City incinerators were fixed in position for many decades; the consequences of locating waste management in a central part of an exclusive resort such as Biarritz, therefore, had to be endured. Where municipal systems of water supply and disposal were concerned, networks of pipes for municipal services resulted in rigidities that were difficult to alter.[99] A century later, Victorian pipes still carry water in many British cities and what was a boon for those first supplied is now a penalty for those who have to live with encrusted cast-iron water pipes and the contaminants they convey. Conversely, in Baltimore, as in other American cities, deliberate planting policies in a socially exclusive part of the city provided a legacy of green space for generations of black residents who were the beneficiaries of 'white-flight' to the outer suburbs. A process of 'filtering-down' was, therefore, at work in the housing market, meaning that Blacks inherited an environment that

97. A.D. Ochojna, 'The Influence of Local and National Politics on the Development of Urban Passenger Transport in Britain 1850–1900', *Journal of Transport History*, 4, 1978, XX.

98. See D. Harvey, 'The Geography of Capitalist Accumulation: A Reconstruction of the Marxian Theory', in R. Peet, ed. *Radical Geography*, Chicago, 1977, 263–92 for an extended discussion of this term.

99. The grid-like pattern of many town extensions in Europe has as much to do with the cost saving associated with building and laying down pipes and public utilities in straight lines as it does with town planning.

had earlier enjoyed substantial capital investment on a scale unimaginable had it been undertaken when black households eventually moved in.

The housing market tended to reinforce social inequalities through the price mechanism.[100] The poor could afford only limited space and that affected both morbidity and mortality rates. However, an environmental dimension was sometimes superimposed on the social inequality, as was the case in the Swedish cities of Linköping, Norrköping, Malmö and Stockholm. This resulted from population pressure that outpaced the capacity of towns and cities to absorb it in central districts, with the consequence that speculative housing developed in peripheral areas where building regulations were non-existent; in a seller's market, and beyond the jurisdiction of the city, house builders could refuse to provide water or drainage in these often already marshy marginal areas.[101] The ambiguous status of the peripheral areas allowed different technical projections of future demand for water to develop and this gave town councils an excuse to prevaricate over the issue of supply to such districts. It was in these crowded suburban areas that epidemic disease was most pronounced, which eventually persuaded councillors that water supply should be provided alongside an improved system of drainage. Not for the first time, fear of epidemic disease prompted intervention. For the mercantile and professional classes there was little point in accumulating wealth during a lifetime if the early death of their offspring rendered legacies futile.[102] In denying water supplies and drainage for some years, a fortress mentality developed whereby Swedish town officials sought to keep the suburban working class, their soiled water and epidemic disease out of the city. Only when it suited them and when there was little choice did the official policy of incorporation of the suburbs within the municipal system proceed, with public utilities provided on an equal footing in 1886. This was exactly the case in Göteborg, too, except that there was a real need as early as 1872 to include the working class suburbs, Masthugget and Majorna, into the planned municipal water system because, unlike the situation in Linköping and Norrköping, they were considered necessary for the future commercial and industrial growth of the city as they both lay near the harbour, which was essential for the city's commercial life.

Water Management and Justice

A crucial transition from traditional patterns of uneven and fragmented water supply and management to one of technological advancement and systematic modern

100. See, for example, R. Rodger, 'Slums and Suburbs: The Persistence of Residential Apartheid', in P.J. Waller, ed. *The Oxford History of the British Landscape*, Oxford, 2000, 233–68.

101. See the important exception of France, noted above.

102. R.J. Morris, *Men, Women, and Property in England, 1780–1870: A Social and Economic History of Family Strategies amongst the Leeds Middle Classes*, Cambridge, 2005.

management forms the basis for one section of this book. The improved availability and disposal of water and sewage might seem to represent a breakthrough in environmental justice by contributing significantly to the quality and length of life in late nineteenth century cities.[103] However, this was no simple linear progression. The examples drawn from water provision in Swedish cities illustrate the problems, legal and financial, that confronted cities as they encountered an entirely new scale of demand for water. No less a matter for anxiety was the disposal of soiled water, particularly in the many low-lying districts of western European cities. Levels of expenditure and the local taxation necessary to raise funding for major capital projects, and issues of equity concerning those paying for and benefiting from them did not normally coincide. As a result there were political pressures that it was difficult for city managers to overlook. What is also evident from studies on water supply and disposal is that the transition was often slow, hesitant and inconsistent. Dessau councillors seemed unable to deal with their local circumstances and were pilloried in the national press for their incompetence. In Darmstadt, for example, estimates of the capacity needed always seemed to lag reality and the policy of spreading sewage on nearby fields continued into the 1960s, when the acreage treated in this manner was thirty times greater than a century earlier.

It might reasonably be expected that in the imperial capital, Berlin, modernity would be on show after 1870. This was indeed the case with a degree of town planning combined with several strands of centralised public health initiatives that included drinking wells, an underground piped sewage system with compulsory connections from all properties and the amalgamation of private waterworks under municipal management, also with obligatory connections to the public supply. Traditional practices of using cesspools were banned. A political consensus endorsed the initiatives and their egalitarian nature resulted in falling mortality rates, particularly those associated with the water-borne diseases, typhoid and cholera. The public health policies contributed to a reduction in environmental inequalities and set technical standards not just for German cities but more widely, as visiting municipal delegations to the capital confirmed.[104]

Scale, as so often in historical analysis, was crucial. Many large and medium sized German cities did not pursue the assault on environmental inequalities with the same vigour as did Berlin. The practices of traditional water and sewage disposal before 1870 – discharges into the river system in Dessau and spreading sewage on agricultural fields around Darmstadt – were not dissimilar to many other German

103. S. Szreter, *Fertility, Class and Gender in Britain 1860–1940*, Cambridge, 1996.

104. P-Y. Saunier and S. Ewen, eds. *Another Global City: Historical Explorations into the Transnational Municipal Moment 1850-2000* Basingstoke, 2008; M. Hietala, *Services and Urbanization at the Turn of the Century: The Diffusion of Innovation*, Helsinki, 1987.

and indeed west European cities.[105] Traditional methods of disposal based on cesspools and carting came increasingly under pressure and a new vista of environmental equality appeared possible. From 1880 Darmstadt opened its new waterworks, fully operational in 1894; Dessau replaced its sewage system in 1872, its water supply in 1874 and again in 1886, and upgraded its sewage works in 1888. The language of hyperbole then took over. The cities were both considered by officials to be beautiful, healthy and convenient places to live and the impression was given through statistics about WCs and piped water that earlier environmental inequalities had been largely eliminated, an interpretation challenged by Marcus Stippak. The frequency of WCs and other amenities in Darmstadt was overstated and, more seriously, the area devoted to sewage fields increased dramatically; in Dessau, sewage continued to be released into the rivers. These traditional methods of disposal continued into the inter-war period and if both cities were slow to adopt biological and chemical methods of treatment so, too, were the 266 German cities that continued into the 1960s to transport sewage across cities and into nearby fields.

The evidence is that, in a very large number of German cities, municipal water and sewage management was less than forward-looking. It is not surprising, then, that Berlin represented an island of centralised modernity in a sea of conservatism, as represented by the 35 smaller towns and cities of Brandenburg and their methods of traditional waste treatment. In small towns such as Zehdenick, north of Berlin, farmers insisted on the use of urban waste as fertilisers and their political leverage meant that compulsory sewer connections proved impossible to establish, far less enforce. If the 'sanitary city' was something of a mirage, a number of villages in the Greater Berlin area still sought to develop connections to the public water supply, despite financial austerity after the First World War. Rates of water usage escalated; river levels fell considerably, to a point considered to be environmentally damaging. Water shortages prompted a search for alternative supplies, often far removed from Berlin itself. Perversely, a new type of environmental inequality developed: that of a water-consuming and waste-water-exporting metropolis in contrast to its suburban and rural periphery which was exploited for its water resources and sewage farms. The ecological footprint of Berlin was very extensive and produced environmental inequalities on a regional level. This necessitated comprehensive planning for water resources from 1929 to ensure that the interests of municipalities and rural areas were coordinated at technological, legal and institutional levels.

This ecological footprint, as we would term it today, was rarely a matter of concern for most urban authorities. In Berlin, a self-obsessed municipality seemed almost oblivious to the wider impact of its actions on the neighbouring districts in the 1920s, and industrial waste discharged by Dessau factories into the Elbe and Mulde rivers had a serious effect on the ecology of those water systems. Indeed,

105. J. Sheail, 'Town Wastes, Agricultural Sustainability and Victorian Sewage', *Urban History*, 23:2, 1996, 189–210.

poisoning from lead pipes affected the inhabitants of Dessau directly and the industrial oil and grease spread on surrounding farms around Darmstadt altered the water table and made absorption of the liquid sewage more problematic. This may be have been a blessing in disguise, since roundworm flourished on the sewage farms around the city, affecting the fruit and vegetables consumed by Darmstadt inhabitants. Further downstream from Darmstadt, communes pressed for relief from the waste disposal methods of the city since they suffered most from the effluents that repeatedly killed fish stocks in the river system.

Voluntary Societies: Class, Race and Justice

German examples of pollution and ecological damage were, of course, already well known to inhabitants of British and French towns and cities. Wanda Balcers and Chloé Deligne draw attention through their research to the impact of chemical and metallurgical industries on river systems and water quality in Belgium. They show that, out of individual protests against industrial polluters at a local level, a more organised protest movement emerged from anglers, who were incensed at the deterioration of water quality as result of the industrial effluents poured into the river systems. Significantly, though, because of the interconnected nature of rivers and streams, the ecological impact of pollution was shown to be far from local and so a wider national protest movement emerged after 1870, devoted to reversing the impact on fish stocks and the damage to the recreational interests of local people. Much of the impetus for the protest movements originated in the suburbs and this is a reminder, as in Hallström's study of Norrköping and Linköping, that these were working class districts, directly affected by the emerging factory system in continental Europe and the increasing scale of the units of production. Balcers and Deligne also provide a useful reminder that the Belgian industrial system was both an early and spatially confined one, with major nodes at Ghent, Liege, Namur, Charleroi, Verviers, and Brussels. Over the border, in France, Lille, Roubaix and Tourcoing combined with their neighbouring Belgian cities to form a major early European industrial region of the type also found in the West Midlands of England and, later, in the Ruhr valley in Germany. This intense geographical concentration of heavy industry produced environmental problems and injustices beyond those of individual cities, as downstream and down-wind inhabitants, through no fault of their own, suffered the contamination caused by their near neighbours.

The Liege Anglers' Society, formed in 1888, was quickly followed by similar organisations in Ghent and Brussels, and soon afterwards by the foundation in 1890 of the Société Centrale pour la Protection de la Pêche Fluviale, which marked the emergence of a national federation of anglers. This might be seen as a narrow pressure group but angling straddled many sections of society and middle class support, including that of lawyers, scientists, academics and hygienists, widened its membership base and provided specialist knowledge and networks. Support was further strengthened by links with emerging working class labour organisations and, in 1903, at a location favoured by Belgian socialists, a nationwide society, the Anglers' Society of Belgium, was formed. The aim of this organisation was to persuade political representatives to put water pollution on the legislative agenda of parliament, a resolution also supported by the Belgian Labour Party. In a further development, in 1906, that reflected a broadening of the political agenda, the Anglers' Society of Belgium was re-badged as the League against Water Pollution with several regional branches and support enlarged by prominent Catholic, socialist and liberal figures. This signalled the move from a focus on issues closely associated with fishing to ones more broadly based on water quality and pollution, and this in turn was further extended in 1913 to include air pollution and ground contamination.

The politicisation of environmental inequalities as perceived by different elements of society lies at the heart of the Belgian study. At each of local, regional and national levels, alliances and mediation were essential in achieving political objectives in the environmental arena. The pace of development from the foundation of the Liege Anglers' Society in 1888 to the national league against all forms of pollution in 1913 was remarkable, not only for its speed but for its ability to build a consensus contesting environmental pollution and the associated inequalities.

The Liege Anglers' Society and its successors are a reminder that remediation of environmental inequalities is not confined to the actions of private individuals or public intervention. There is, indeed, a third way, through organisations and associations, that environmental inequalities can be deconstructed. This theme is one that Buckley and Boone also engage with in their study of environmental inequalities in Baltimore neighbourhoods. In some ways, their study revolves around the defence of property, creating environmental inequalities by excluding individuals and groups perceived as hostile to private residential investment and its associated amenity. This resonates with concepts of civil policing in an earlier age, whereby efforts to define, exclude and punish certain categories of anti-social and environmentally damaging behaviour were undertaken on the grounds of the defence of the common good.[106]

In Baltimore, as in many American cities, environmental inequalities were constructed on the basis that 'sites [were] to be protected from outside interests' and because members of neighbourhood organisations were compelled to act when they believed 'spatial boundaries are threatened' and 'order is jeopardized'.[107] Fear of the 'other', the perception of the alien, was the issue around which local organisation was mobilised. 'White privilege' provides a framework for exploring uneven

106. See for example, D.G. Barrie, *Police in the Age of Improvement: Police Development and the Civic Tradition in Scotland 1775–1865*, Cullompton, 2008.

107. M. England, 'When "Good Neighbors" Go Bad: Territorial Geographies of Neighborhood Associations', *Environment and Planning A*, 40, 2008, 2880.

power relations, issues of inclusion and exclusion, and the many and varied ways in which these relations express themselves in the landscape.[108] If Geoff Buckley and Chris Boone employ an explicitly American reference frame in their chapter, it is worth remembering that restrictive covenants, gated communities, ghettoes, and currently the 'Peace Lines' in Belfast have provided an assortment of practices in the European historical experience designed to exclude and thus to perpetuate inequalities of various kinds.

Many Neighborhood Improvement Associations in Baltimore were concerned with infrastructural improvements – in street lighting, electricity, sewerage, tree-planting, traffic lights and in generally attempting to improve the material condition of an area. More specifically, the South Baltimore Improvement Association identified the severe shortage of sports and recreational facilities in the neighbourhood and sought to address this inequality in the provision of amenities from 1934. However, others conceived 'environment' differently. Mount Royal Improvement Association, formed in 1930, was based in an exclusive suburb that perceived beautification of the neighbourhood as consistent with white occupancy. In its 1908 Constitution, the Peabody Heights Improvement Association sought to promote parks, plant trees, prevent smoke pollution and commercial developments, and oppose tenancies going to Blacks. The power and influence wielded by the influential members of the Peabody Heights Improvement Association suggest that, coupled with aggressive enforcement of segregation, it contributed to an inequitable distribution of amenities, with a disproportionate share of resources and services flowing into wealthy white districts like Peabody Heights at the expense of other neighbourhoods.

In some respects, the actions of the Peabody Heights and Mount Royal Improvement Associations conform to a more general model of the creation of environmental inequalities, based on the relationship between private and public interests and the extent to which it is acceptable in different cultures, political contexts and time periods to privilege the former over the latter. This tension is inherent in all cities and was mediated in certain cases in Baltimore by the intervention of Improvement Associations; the actions of the prevailing local political culture were moderated by groups of individuals setting up their organisations to influence the resource allocation and social capital of the city according to different priorities. What is reasonable and acceptable in terms of amenities, as with social practices, varies over time, with knowledge and in comparison to other places.[109] Ultimately, it was institutional interventions that fundamentally changed the position of environmental inequalities in Baltimore. First, an extension of the city

108. L. Pulido, 'Rethinking Environmental Racism: White Privilege and Urban Development in Southern California', *Annals of the Association of American Geographers*, 90:1, 2000, 12–40.

109. N. Elias, 'Introduction' in N. Elias and J. L Scotson, eds. *The Established and the Outsiders: A Sociological Enquiry into Community Problems*, London, 1994, 15–52.

limits in 1918 created new residential possibilities and the resulting 'white flight' combined with block busting practices – whereby white properties were sold to Blacks – meant that the suburbs began to lose their exclusivity. This process was accelerated when the federal Home Owners Loan Corporation undertook a risk-assessment of districts based on the occupation of residents, average annual income, predominant nationality, percentage of 'negro families' to total number of families, percentage of families on relief and 'threat of infiltration of foreign born, negro or lower grade population'. As a result, the perception amongst federal agency staff and city officials was that some districts were in decline and therefore did not merit further investment of federal housing finance or city expenditure on parks and other civic amenities. Ironically, then, 'white flight' permitted the resident black population to spread into former exclusive suburbs and thereby gain access to amenities from which they previously had been excluded. So Baltimore's current African American population still enjoys relatively high access to amenities such as parks and playgrounds, by accident rather than by design; African Americans are the beneficiaries of an inherited landscape, one that was left behind when Whites quit the city in droves during the 1950s and 1960s.

Industrial Technologies and Environmental Justice

Factories interposed changed relationships between employers and employees. E. P. Thompson captured this brilliantly when he explored the changing relationship of workers to time, and in particular the transition from 'time passed' in traditional handicraft and rural activities to 'time spent' in the 'new universe' of factory-disciplined time.[110] Though neither total nor immediate, this transformation reconfigured areas of labour exploitation and resistance, and this is the theme that Janet Greenlees explores in her analysis of technological choices in the New England cotton textile industry between 1880 and 1930. Her central theme is that relationships between the business and workplace environment varied from place to place and between firms in the same sector, and that this made some places healthier to work than others. Through a forensic analysis of the different production structures of three leading Massachusetts cotton towns – Lowell, Fall River and Holyoke – Greenlees shows that technologies between and within the towns were uneven, despite the rigidities that governed mass production.[111]

The key technologies affecting the weaving environment were, firstly, the ventilation and humidifying equipment that controlled the hot and humid environment in the weaving rooms. The introduction of these machines raised concerns

110. E.P. Thompson, 'Time, Work Discipline and Industrial Capitalism' in E.P. Thompson, ed. *Customs in Common,* Harmondsworth, 1994, 352–403.

111. See also P. Scranton, *Endless Novelty: Specialty Production and American Industrialization, 1865–1925,* Princeton, 1997.

about the relationship between poor ventilation, sanitation, dust inhalation and disease contagion. Secondly, concerns were raised about dust inhalation and the spread of disease, since weavers shared shuttles that required them to suck the thread through the eye of the shuttle – a process they repeated hundreds of times a day.[112] In another example of institutional intervention to mitigate environmental inequalities, the State legislature sought to control these two technologies in an effort to improve both the working environment and the health of citizens.

The range of alternative outcomes in the textile industry was formidable. There were three towns; several actors – employers, the town council, the community, workers, physicians and the State – all of whom had different priorities; external factors such as markets and advertising; business connections with local machine-makers; individual company priorities; several alternative shuttle, ventilation and humidifying technologies; and workers' responses which included refusals to work on hot days and a switch to other local employers. In short, the range of possible interactions between workers' bodies, machines and the work environment was immense. To summarise these is problematic but what emerges is that the local community influence on technological choice was strongest in Lowell, though this extended only to improving the quality of the atmosphere and, hence, the collective wellbeing of workers, not the health of individuals. Local government pressure on industry was greatest in Holyoke, through its encouragement to employers to improve ventilation and air quality in their factories. The greatest workplace environmental inequalities were found in Fall River, where the dominance of mill-owners in local politics drove reformers to seek State assistance to improve the mill environment. While successful, the resulting legislation was limited in both strength and enforcement, allowing employers' individual preferences to prevail.

Greenlees' contribution, together with those of Frioux on incinerators, Stippak on mechanical and biological sewage treatment, Oram on thermal efficiency, and, severally, of Hallström, Soens and Colten on the role of water engineering are powerful reminders of how important it is to build the scientific and technological dimensions into explanations of environmental inequalities, whatever period is under study.[113] Because piped and diked networks of water and energy are involved, then systems of delivery and disposal are rarely in equilibrium if cities are expanding. Consequently there are gainers and losers in what is not necessarily a zero sum urban setting. Not surprisingly, then, sectional interests sought by various means to assert their preferences and so laid the basis of injustice and inequality.

---

112. J. Greenlees, '"Stop Kissing and Steaming!" Tuberculosis and the Occupational Health Movement, 1870–1918', *Urban History* 32:2, 2005, 225–28, 233–39.

113. J. Mokyr, *The Gifts of Athena: Historical Origins of the Knowledge Economy*, Princeton, 2002; P.M. Jones, *Industrial Enlightenment: Science, Technology and Culture in Birmingham and the West Midlands 1760–1820*, Manchester, 2010.

Trees, Urban Ecology and Environmental Injustice

The City Beautiful movement was a strand in the activities of some neighbourhood improvement schemes in Baltimore. Blacks inherited an aesthetic environment greened by a planting policy initiated decades before 'white flight' vacated the area. City trees provide many advantages: they reduce storm water run-off and summer heat, and moderate temperature fluctuations; they improve air quality and reduce exposure to ozone, sulphur dioxide, nitrogen dioxide and carbon monoxide. Joanna Dean shows that many of these benefits are enjoyed across an entire city but others are localised, demonstrating another dimension of environmental inequalities. Using an innovative method to quantify the tree canopy cover in four Ottawa neighbourhoods, based on aerial photography from the 1920s to 2005, Dean and her co-researchers identify complex inequalities in the distribution of the urban forest. An intriguing aspect of this is how certain species were chosen for their rapid growth and what the longer term implications of that have been, including, for example, radical felling in the inter-war years to manage the density, with the result that specific areas were denuded of their canopy. Broadly, the photographs show a darker, thus denser, tree canopy for wealthier districts like Sandy Hill, inhabited by civil servants, professionals and politicians; the neighbouring working class district of Lowerton had a visibly thinner tree canopy.

Though this correlation between wealth and tree cover might seem obvious, it was not an exact relationship. Indeed, the construction of an inter-provincial highway through Lowertown cleared trees while the effective political resistance to the highway extension by middle class New Edinburgh residents preserved tree cover in their area. Middle class residents have more money and more time to spend on landscaping their larger suburban plots, which they are likely to enjoy for longer since they move house less often. These conclusions correspond with others that conclude socio-economic factors to explain about fifty per cent of the variation in tree cover between districts. While some credence is given to the ethnic composition of neighbourhoods and their cultural preferences for trees, the Ottawa study puts more weight on ecological explanations for variations in tree cover. Thus species type and water table levels offer additional explanations and Dean provides insights into how the blend of maples, elms, and oaks gives a different character and life-cycle path to the treed nature of plots and neighbourhoods. But the ecological also extends to the human dimension, in terms of the life and death of districts. In this regard, inner city decline and regeneration schemes have had a dramatic effect on tree cover in older neighbourhoods and, as Buckley and Boone identified in Baltimore with 'redlining' or stigmatising older districts, so in Ottawa household data on the age of properties and the extent of green space condemned many Lowertown properties for demolition. Tree cover was diminished drastically as a result. Dean's conclusion, that when considering the history of environmental justice it is important to consider it as reinforced by other multiple environmental inequalities, is timely.

Power and Authority

By the standards of natural justice, the racial covenants, segregation ordinances, block busting, neighbourhood ranking system and diversion of park development and street plantings to white neighbourhoods in Baltimore were unjust. In Berlin, the Social Democratic Party advocated improved living conditions in the 1920s in the form of well-equipped and sanitary flats whose high rentals resulted in a voluntary emigration of Berliners to the margins of the city and 100,000 'wild settlers' squatting in makeshift accommodation. In Baltimore the environmental injustice was a deliberate act; in Berlin it was an indirect result of policy.

Decisions by the ruling elite at neighbourhood and city level exemplify the use, and, in terms of natural and environmental justice, the abuse of power. Whether New England mill-owners or North Sea landowners, the management of resources has conveyed to the ruling elite a level of decision-making and influence over the lives of those within their sphere of operations. In New Edinburgh (Ottawa) or Old Edinburgh (Scotland) middle class property owners defined the character of neighbourhoods when they planted trees or bought coal and so shaped fuel poverty in nearby small towns. These, and many other examples in this book, illustrate how environmental injustices were produced, managed, and sustained. Power and authority over land, water and air, and the resources each contained, were at the heart of the creation of environmental inequalities and since the market system was rarely, if ever, in equilibrium, there were always winners and losers.

The perpetuation of power through aristocratic and corporate dynasties, municipal oligarchies, political parties and voluntary associations meant that the winners tended always to be the same, though the losers changed according to technological changes and the ability of the environmentally excluded to negotiate concessions. Belgian anglers marshalled an alliance of interests to oppose, first, industrial pollution and, later, poor water quality; New Orleans minorities' exposure to hurricanes was not eradicated but political pressure meant their access to shelters was improved; and Dessau councillors learned to negotiate concessions with East German officials to obtain better water treatment facilities in the 1950s. In the complex mix of agents in the towns of Fall River, Holyoke and Lowell, workers' direct action regarding adverse working conditions saw them down tools on hot days or switch allegiance to another employer and the community or medical experts press for changes in the production structure and the implementation of State laws on working conditions. In different historical contexts and geographies, mediation of the scale and extent of environmental inequalities developed, though it seems its effectiveness was rarely more than partial and depended much on local conditions and the relative power of the actors involved.

Does this mean there was some long run convergence in environmental inequalities in ways that mirrored the path of incomes and political representation? There were some rigidities that complicated convergence. For instance, the topography of certain towns and cities rendered them more likely to experience natural disasters: the Gulf and seasonal hurricanes were fixtures for New Orleans residents; the North Sea communities were at or below sea level; and the river crossings that defined the low-lying geographical location of so many early European towns inclined them to dump sewage and waste either on nearby farms or in the river from which they derived a living as a transport and commercial hub. The dependence of farmers on sewage for fertiliser encouraged them to enforce contracts and, in the 1890s, to constrain Darmstadt council's ability to introduce water treatment plants.

Another rigidity was municipal investment. It is worth noting that from about 1870 the German urban system as a whole experienced an even faster rate of expansion than that of England in the years 1800 to 1830. In a number of other major European cities, too, if not in their entire urban systems, population growth was dramatic – Budapest was just such a case but there were many others. So, as was seen in Norrköping and Linköping, though the process of annexation brought outlying districts and suburbs within the orbit of administrative responsibility of larger municipal units and so, to a degree, redressed the environmental inequality for the citizens annexed, it brought the system as a whole nearer to breaking point. Indeed, the finite capacity of municipal water supply and treatment was considered a reason not to extend the public services in several chapters. To upgrade or replace municipal infrastructure as new technologies emerged also placed an enormous strain on local taxation, and thus on the political system which was so sensitive to tax matters. Economies of scale were attractive but the escalating up-front capital costs were formidable as cities doubled and tripled their areas and as the piping and pumping systems, power generating and engineering demands increased exponentially. Therefore, in some cities, old technologies produced a degree of inertia or 'path-dependency' and that constrained the municipalities' ability to reduce or remove some of the environmental inequalities.[114] So a dualism persisted: old technologies and methods co-existed with new ones. This was also the case in the private sector, as in the textile industry where the parallel existence of the ultra-modern Northrop looms with a mix of older machines was conspicuous. This dualism existed in many industries, with important implications for the workforce of the type Greenlees identifies.[115]

---

114. See for example, P.A. David, 'Clio and the Economics of QWERTY', *American Economic Review*, 75, 1985, 332–37.

115. C.F. Sabel and J. Zeitlin, eds. *World of Possibilities: Flexibility and Mass Production in Western Industrialization*, Cambridge, 1997.

Most conspicuously, disasters such as floods, fires and epidemic disease concentrated the minds of officials and property owners, and produced a political momentum that was supportive of improved dikes, town plans, and water supply and disposal. In such cases benefits accrued city-wide and, coincidentally, were disproportionately conveyed to the disadvantaged. Evidence is presented in several of the chapters of the political momentum injected in Norrköping, New Orleans, Baltimore, and German cities by natural disasters and public health panics.

In a 'Review of Progress on Environmental Justice', in 2005, a Scottish government enquiry identified many different forms of environmental inequalities.[116] These included flooding; waste; health, air quality; green space; housing standards; vacant land, quarries and mines; transport and traffic; regeneration; and environmental crime. The enquiry was concerned with both local street-level environmental problems and with larger-scale industrial pollution. The terms of reference for the enquiry recognised that the built environment is as significant as the issues of biodiversity and green places that normally capture the media headlines. Two interrelated and equally important strands of environmental justice were noted: firstly, fairness, or inequalities, in the distribution of the factors affecting environmental conditions and, secondly, the information and opportunities allowing people to participate in decisions about their environment. These parallel contemporary concerns are also displayed to differing degrees in the historical studies.

116. Review of Progress on Environmental Justice, Paper 2005/372005 http://www.scotland.gov.uk/Publications/2005/12/12161827/18283 Accessed 15 May 2010.

Part I

Constructing the Injustice

Chapter Two

'To Promote the Material and Moral Welfare of the Community': Neighbourhood Improvement Associations in Baltimore, Maryland, c.1900–1945

G.L. Buckley and C.G. Boone

Nearly 25 years have passed since the United Church of Christ's Commission for Racial Justice released its pioneering study, *Toxic Wastes and Race in the United States.*[1] In the intervening years, environmental justice, as both a field of inquiry for academics and a vehicle for activism, has evolved and matured. Indeed, social scientists conducting environmental justice research over the past decade have shifted the trajectory of the field in at least two important ways.

First, they have expanded the scope of the environmental justice project by introducing an 'amenities' component to the discourse. Once preoccupied almost exclusively with the inequitable distribution of environmental disamenities – for example, the siting of landfills, power plants and polluting industries near low-income minority neighbourhoods – geographers, sociologists, and others have increasingly turned their attention to the inequitable distribution of environmental amenities, such as parks, playgrounds, golf courses, and street trees.[2] According to Boone:

> By far, the vast majority of environmental studies have focused on the distribution of environmental disamenities. However, it is equally valid to examine the distribution of environmental amenities in relation to social groups and households. The fairness in distribution of environmental amenities such as parks and open space is

1. United Church of Christ–Commission for Racial Justice. *Toxic Wastes and Race in the United States: A National Report on the Racial and Socio-economic Characteristics of Communities with Hazardous Waste Sites*, New York, 1987.
2. R. Jensen, J. Gatrell, J. Boulton, and B. Harper, 'Using Remote Sensing and Geographic Information Systems to Study Urban Quality of Life and Urban Forest Amenities', *Ecology and Society*, 9, 2004, http://www.ecologyandsociety.org/vol9/iss5/art5/ Accessed 25 May 2009. H.A. Perkins, N. Heynen, and J. Wilson, 'Inequitable Access to Urban Reforestation: The Impact of Urban Political Economy on Housing Tenure and Urban Forests', *Cities*, 21, 2004, 291–99; M. Pastor, R. Morello-Frosch and J.L. Sadd, 'The Air Is Always Cleaner on the Other Side: Race, Space, and Ambient Air Toxics Exposures in California', *Journal of Urban Affairs*, 27, 2005, 127–48; J. Wells, G.L. Buckley, and C.G. Boone, 'Separate but Equal? Desegregating Baltimore's Golf Courses', *The Geographical Review*, 98, 2008, 151–70.

a legitimate and important environmental justice issue.[3]

Second, scholars in recent years have stressed the value of bringing a historical perspective to equity studies – one which links 'process' with 'pattern'. While advanced spatial analyses have contributed significantly to our understanding of patterns of inequity, they often fail to explain how those patterns were created or how they have changed over time. As Wolch *et al.* explain, present-day environmental inequities are often 'rooted in past histories of racial oppression and discriminatory service delivery patterns'.[4] Repeated calls to uncover the processes that produce the patterns observed, including the 'subtle and complex ways that these processes actually work', highlight the value of adopting a historical approach to environmental justice research.[5]

In this chapter, attention is focused on the activities of a particular set of groups that exercised considerable influence in Baltimore, Maryland during the first half of the twentieth century. Neighbourhood improvement associations, as they were commonly known, occupied a prominent position in the decision-making hierarchy prior to passage of strict zoning laws. They were especially effective when it came to attracting amenities such as parks and street trees and discouraging or eliminating disamenities such as unwanted land uses. Many of these associations, of which there were approximately seventy scattered across the city by about 1910, were also known for their restrictive covenants and exclusionary housing practices (Fig. 1). Using newspapers, government documents, neighbourhood association meeting minutes, and promotional literature, this research shows that the key issues

3. C.G. Boone, 'Environmental Justice as Process and New Avenues for Research', *Environmental Justice*, 1, 2008, 151.

4. J. Wolch, J.P. Wilson, and J. Fehrenbach, 'Parks and Park Funding in Los Angeles: An Equity-mapping Analysis', *Urban Geography*, 26, 2004, 4.

5. A. Hurley, 'Fiasco at Wagner Electric: Environmental Justice and Urban Geography in St. Louis', *Environmental History*, 2, 1997, 460–81; A. Szasz and M. Meuser, 'Environmental Inequalities: Literature Review and Proposals for New Directions in Research and Theory', *Current Sociology*, 45, 1997, 109; A. Weinberg, 'The Environmental Justice Debate: A Commentary on Methodological Issues and Practical Concerns', *Sociological Forum*, 13, 1998, 25–32; D.N. Pellow, 'Environmental Inequality Formation', *American Behavioral Scientist*, 43, 2000, 581–601; B. Bolin *et al.* 'The Ecology of Technological Risk in a Sunbelt City', *Environment and Planning A*, 34, 2002, 317–39; B. Bolin, S. Grineski and T. Collins, 'The Geography of Despair: Environmental Racism and the Making of South Phoenix, Arizona, USA', *Human Ecology Review*, 12, 2005, 156–68; C.G. Boone, 'An Assessment and Explanation of Environmental Inequity in Baltimore', *Urban Geography*, 23, 2002, 581–95; C.G. Boone, 'Zoning and Environmental Inequity in the Industrial East Side', in W. Deverell and G. Hise, eds. *Land of Sunshine: An Environmental History of Metropolitan Los Angeles*, Pittsburgh, 2005, 167–78; C.E. Colten, 'Basin Street Blues: Drainage and Environmental Equity in New Orleans 1880–1930', *Journal of Historical Geography*, 28, 2002, 581–95; C.E. Colten, 'Cities and Water Pollution: An Historical and Geographical Perspective', *Urban Geography*, 26, 2005, 435–58; R. Saha and P. Mohai, 'Historical Context and Hazardous Waste Facility Siting: Understanding Temporal Patterns in Michigan', *Social Problems*, 52, 2005, 618–48.

that resonated with prominent homeowners' groups in Baltimore were part and parcel of an overall spatial vision that sought, among other things, to attract urban amenities such as parks and street trees, discourage commercial development, and preserve the social and ethnic identity of particular neighbourhoods.

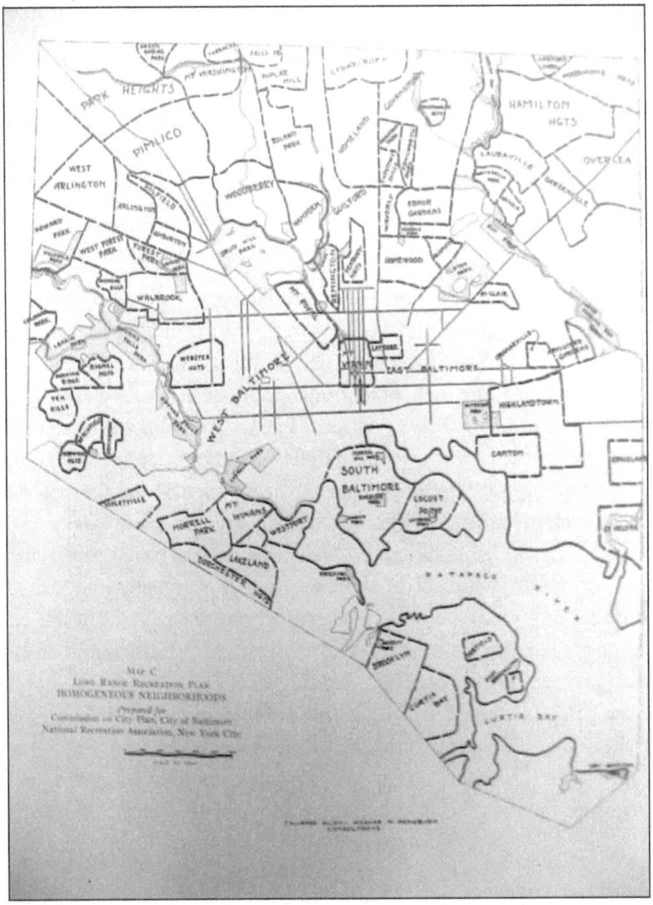

Figure 1. 'Homogeneous Neighborhoods'.

This map depicts Baltimore's neighbourhoods based on information contributed by the city's Neighbourhood Improvement associations.[6]

6. Source: Weaver W. Pangburn and F. Ellwood Allen, Long Range Recreation Plan, City of Baltimore, Maryland. Prepared for the Commission on City Plan by the National Recreation Association, 1943. Baltimore: Department of Public Recreation. Cartography by Michael Boruta, Ohio University Cartographic Center.

This chapter begins with a brief overview of the recent literature on neighbourhood associations, focusing in particular on the historical role these groups played in directing African American settlement. Next we describe Baltimore in the years immediately following the calamitous fire of 1904. It was at this time that African Americans from the South first began to migrate to Baltimore in large numbers. In the third section we track the rise of neighbourhood improvement associations. While many groups were assembled expressly for the purpose of lobbying government officials, others were convened primarily to enforce restrictive covenants and police boundaries. Finally, we examine some of the demographic shifts that took place in Baltimore during the 1930s when the effects of suburbanisation and 'white flight' were beginning to take hold.

Neighbourhood Associations

For most people, the idea that homeowners should band together and take definitive steps to 'secure' the geographic space they occupy is axiomatic. For decades, urban and suburban dwellers have done just that, forming neighbourhood associations and building walls around their neighbourhoods, both literally and figuratively.[7] The actions of neighbourhood associations, in particular, have attracted increasing attention in recent years. Since the mid-1990s, numerous scholars have contributed to our understanding of these groups: why they formed, how they operated, and how their activities continue to influence and shape the urban landscape today.[8]

As Hayden, Jackson, Stilgoe, Davis, and others have convincingly demonstrated, the 'suburban ideal' in America is a highly gendered, middle class, and homogeneously white landscape.[9] According to Purcell, 'Issues of class, race, gender, nature, sexuality, citizenship, regionalism, and so on are all embedded in the political project of homeowners.' Arguing that any explanation that privileges one of these categories over another is unduly limiting when it comes to understanding the motivations and actions of homeowners' groups, Purcell advocates a spatial approach: 'Homeowners ... do not think about their agenda in terms of social categories. Instead, they think about their agenda in terms of a spatial vision.

7. M. Davis, City of Quartz: Excavating the Future in Los Angeles, New York, 1992; S. Low, *Behind the Gates: Life, Security, and the Pursuit of Happiness in Fortress America*, New York, 2003.

8. E. McKenzie, *Privatopia: Homeowner Associations and the Rise of Residential Private Government*, New Haven, 1994.

9. D. Hayden, *Redesigning the American Dream: The Future of Housing, Work, and Family Life*, New York, 1984; K. Jackson, *Crabgrass Frontier: The Suburbanization of the United States*, New York, 1985; J. Stilgoe, *Borderland: Origins of the American Suburb*, New Haven, 1988; Davis, *City of Quartz*.

Their activism is designed to defend and proactively realise their spatial vision in the material space of their neighbourhoods.'[10]

Echoing Purcell, England maintains that neighbourhoods are often viewed by residents as 'sites to be protected from outside interests', and that the members of neighbourhood organisations are compelled to act when they believe 'spatial boundaries are threatened' and 'order is jeopardised'.[11] Community 'policing' is often carried out by any of a number of neighbourhood organisations, including block watch groups, homeowners' associations, and crime-prevention councils. 'In these neighbourhood associations', she writes, 'there exists a community of fear based on the threat of an "invasion by outsiders". Residents believe their neighbourhoods are under siege by outsiders, and they seek to 'take back' their community.'[12]

As the preceding paragraph suggests, neighbourhoods are often formed through the act of excluding. As Garb points out, neighbourhood formation in racially divided northern cities, in particular, 'needs to be understood through the social processes of enforcing exclusion' as opposed to the more common practice of conceptualising neighbourhoods as a collection of households bound together through 'internal networks of assistance and exchange'.[13] For many neighbourhood associations, including neighbours while excluding individuals 'not constructed as such' is necessary and acceptable to protect territory and maintain property values.[14]

Related to this, Purcell asserts that in their 'struggle for space' homeowners are confronted with 'a constant spatial mismatch between the geography [they] want and the geography they actually experience' – a condition that spurs them to exert political and economic power.[15] Although she does not focus specifically on the activities of neighbourhood associations, Pulido's pioneering work on 'white privilege' in Los Angeles has proven particularly useful as a framework for exploring uneven power relations, issues of inclusion and exclusion, and the many and varied ways in which these relations express themselves on the landscape.[16]

Historically, it has been middle class and affluent white homeowners who have spearheaded efforts to enforce exclusion.[17] In Chicago, white property owners,

---

10. M. Purcell, 'Neighborhood Activism among Homeowners as a Politics of Space', *Professional Geographer*, 53, 2001, 178.
11. M. England, 'When "Good Neighbors" Go Bad: Territorial Geographies of Neighborhood Associations', *Environment and Planning A*, 40, 2008, 2880.
12. England, 'When "Good Neighbors" Go Bad', 2881.
13. M. Garb, 'Drawing the "Color Line": Race and Real Estate in Early Twentieth-century Chicago', *Journal of Urban History*, 32, 2006, 774.
14. Garb, 'Drawing the "Color Line"'; England, 'When "Good Neighbors" Go Bad', 2882.
15. Purcell, 'Neighborhood Activism among Homeowners', 179.
16. L. Pulido, 'Rethinking Environmental Racism: White Privilege and Urban Development in Southern California', *Annals of the Association of American Geographers*, 90, 1, 2000, 12–40.
17. Garb, 'Drawing the "Color Line"', 774–75.

faced with growing numbers of black immigrants seeking housing in predominantly white neighbourhoods during the first decade of the twentieth century, attempted to stem the tide of 'invasion' through the actions of block clubs and neighbourhood associations.[18] In Kansas City, improvement associations mobilised to enter their districts into restrictive covenants, a tactic that was commonly employed in cities across the country.[19] In a further attempt to stave off the 'encroachment of negroes', officers of the Linwood Improvement Association asked the Kansas City Board of Commissioners for Parks and Boulevards in 1926 to condemn 'a sparsely populated spit of land occupied partly by black families' for the purpose of connecting two already existing parks, thus forming 'a solid boundary of green space along LIA's northern edge'.[20] In Baltimore, officers from several neighbourhood improvement associations supported passage of a municipal segregation ordinance designed to 'stop the negro invasion'.[21] Sometimes violence or the threat of violence was used to discourage African Americans from locating in white neighbourhoods.[22] In other cases, intimidation took on other forms, such as in Miami where one neighbourhood association, the Seaboard White Citizens Council, planted burning crosses in the lawns of African Americans who dared to move into predominantly white neighbourhoods.[23]

As African Americans gained a foothold in previously all-white neighbourhoods, speculators and real estate brokers engaged in the practice of 'block busting'. After convincing white property owners to sell quickly at low prices, they then turned around and resold the homes to black buyers at inflated prices. The end result was that in places like Chicago, brokers and speculators, who initiated the racial transition in the first place, 'used white fears and black aspirations to fill their bank accounts'.[24] Despite high prices, the practice of block busting offered certain advantages for the black community as Schirmer's research in Kansas City suggests:

> After all, the black family who moved into a 'busted' block on the eastside's fringe acquired a piece of the American Dream: security and pride of home ownership

18. Garb, 'Drawing the "Color Line"', 773, 779.
19. S.L. Schirmer, *A City Divided: The Racial Landscape of Kansas City 1900–1960*, Columbia, 2002; S. Hood Washington, 'Mrs Block Beautiful: African American Women and the Birth of the Urban Conservation Movement, Chicago, Illinois 1917–1954', *Environmental Justice*, 1, 2008, 13–23.
20. Schirmer, *A City Divided*, 112–3.
21. G. Power, 'Apartheid Baltimore Style: The Residential Segregation Ordinances of 1910–1913', *Maryland Law Review*, 1983, 289–328; C.H. Nightingale, 'The Transnational Contexts of Early Twentieth Century American Urban Segregation', *Journal of Social History*, 39, 2006, 673.
22. Garb, 'Drawing the "Color Line"'; Hood Washington, 'Mrs. Block Beautiful'.
23. R.A. Mohl, 'Race and Space in Miami', in A.R. Hirsch and R.A. Mohl, eds. *Urban Policy in Twentieth-Century America*, New Brunswick, 1993, 129.
24. Garb, 'Drawing the "Color Line"', 778.

> in a comfortable and healthy neighbourhood that still lay within easy reach of the commercial and social center at Eighteenth and Vine. If those neighborhoods quickly resegregated, so much the better, for there was a measure of safety in numbers. The lone black pioneer who penetrated a predominantly white neighbourhood, on the other hand, faced daunting risks.[25]

For African American residents unable to take advantage of such opportunities, the alternative was bleak. In Chicago, for example, restrictive covenants and the threat of white mob violence confined African Americans to the densely crowded 'deadly and decaying neighbourhoods of Chicago's Black Belt which became known for its notorious kitchenette apartments'.[26] Conditions were not much better in Baltimore. By 1910, the city's black population numbered approximately 80,000, most of whom were crowded into congested sections of town like Biddle Alley and Druid Hill Avenue.[27] (Fig. 2)

Baltimore in the Early Twentieth Century

On 7 February 1904, a catastrophic fire swept through Baltimore. Hampered by high winds and freezing temperatures, fire fighters were unable to control the blaze before it laid waste to 140 acres of the old downtown. Although there were no reported deaths and few injuries, the city's commercial district was reduced to ashes. Almost immediately, city officials, business owners, and local residents set about the task of rebuilding the city. In little more than a decade the city reclaimed its rank as the nation's seventh largest industrial centre, thanks to 'magnificent investments in port and rail facilities' and a booming wartime economy.[28] To accommodate the influx of new residents drawn to the city by a rejuvenated economy, as well as to capture a larger tax base, city officials approved a major annexation in 1918 that expanded the municipality's boundaries in all directions, especially to the north and east. By the close of the First World War, Baltimore had largely recovered from the fire and its citizens could look to the future with confidence.

The period from 1900 to 1918 in Baltimore's history should be remembered for more than just the fire, however. According to Olson, it was during this interval that the city 'managed to achieve an exceptional harmony of planning'.[29] Baltimore's grand plan to remake itself by installing a modern sewer system, improving its transportation infrastructure, and undertaking a variety of other major public

25. Schirmer, *A City Divided*, 101.
26. Hood Washington, 'Mrs Block Beautiful', 15.
27. Nightingale, 'The Transnational Contexts'.
28. S.H. Olson, *Baltimore: The Building of an American City*, Baltimore, 1997, 245, 291.
29. Olson, *Baltimore*, 245.

works projects, was delayed by the fire but not for long.[30] With the recovery well in hand, Baltimoreans soon returned to the unfinished business of reshaping their city. The desire to create a world-class system of parks serves as an important example. Following the lead of cities like Boston and New York, Baltimore, at the time of the great fire, was in the midst of rethinking its approach to park planning. To ensure that the rolling hills beyond the city's centre would not be adversely affected by suburban migration and to meet the recreational needs of a growing population, the Municipal Art Society of Baltimore, a civic group concerned primarily with city beautification, hired Olmsted Brothers of Brookline, Massachusetts to develop a comprehensive park plan.[31] Inspired by the rationalistic designs of the City Beautiful movement, the '1904 report', as it came to be known, served as a blueprint for park planning for decades to come.[32] Significantly, one of Frederick Law Olmsted, Jr's stated goals for Baltimore was to achieve 'a roughly equitable distribution' of parks and other outdoor recreation facilities 'for all its citizens'.[33] Government officials took yet another step toward 'greening' their city when they hired a professionally-trained forester in 1912 to care for and manage the city's trees. With a plan for park development in hand, a city forester on staff, and renewed interest in city beautification and planning, Baltimore endeavoured to show that it was more than just a centre of industry.

Not everyone benefited from the post-fire planning and prosperity. Like other large American cities at the time, Baltimore was sharply divided along racial lines.[34] Although segregation was not new to Baltimore, by the early years of the twentieth century it was manifesting itself in a new way. According to Olson, the pattern of 'street-and-alley segregation' that had dominated during the nineteenth century was giving way slowly 'to sizeable hemmed-in ghetto in East Baltimore, West Baltimore, and South Baltimore'.[35] Then, in 1911, Mayor J. Barry Mahool signed into law 'an ordinance for preserving peace, preventing conflict and ill feeling between the white and colored races in Baltimore city, and promoting the general welfare of the city by providing, so far as practicable, for the use of separate blocks by white and colored people for residences, churches and schools', making Baltimore

30. C.G. Boone, 'Obstacles to Infrastructure Provision: The Struggle to Build Comprehensive Sewer Works in Baltimore', *Historical Geography*, 31, 2002, 151–68.

31. The city later reimbursed the Municipal Art Society.

32. K. Zucker, 'Falls and Stream Valleys: Frederick Law Olmsted and the Parks of Baltimore', *Maryland Historical Magazine*, 90, 1995, 72–96; W.E. Orser, 'A Tale of Two Park Plans: The Olmsted Vision of Baltimore and Seattle 1903', *Maryland Historical Magazine*, 98, 2003, 467–83.

33. Olmsted Brothers, *Report Upon the Development of Public Grounds for Greater Baltimore*, Baltimore, 1904, 50; Olmsted Brothers, *Report and Recommendations on Park Extension for Baltimore*, Baltimore, 1926, 9.

34. H.A. McDougall, *Black Baltimore: A New Theory of Community*, Philadelphia, 1993; W.E. Orser, *Blockbusting in Baltimore: The Edmondson Village Story*, Lexington, 1994; Olson, *Baltimore*.

35. Olson, *Baltimore*, 372.

the first municipality in the United States to formally segregate its inhabitants.[36] Originally conceived by Milton Dashiell, a prominent lawyer and member of the McCulloh, Madison Avenue and Eutaw Place Improvement Association, ostensibly to protect real estate values, the West Ordinance, as it was called, stipulated that Blacks could not move into blocks that were half white, and vice versa. Exceptions were made for black servants living in white houses.[37] Soon the 'Baltimore idea' was copied in cities across the American South, including Atlanta, Birmingham, Louisville, Richmond and St. Louis. When the United States Supreme Court struck down Louisville, Kentucky's segregation ordinance in 1917, ruling that it violated the property rights law of the 14th amendment, the ordinances of other cities, including Baltimore's, were, in Power's words 'constitutionally eviscerated'.[38] In the end, however, little changed for Baltimore's African American residents for *de jure* segregation, as embodied by the ordinance, was quickly replaced by *de facto* segregation in its many and varied forms. It was during this period of rapid growth and change, and amidst this racial turmoil, that Baltimore's neighbourhood improvement associations rose to prominence.

The Emergence of Neighbourhood Improvement Associations in Baltimore

As Baltimore grew in size and population, neighbourhood improvement associations, working independently and in concert through a citywide congress, were formed to pressure city officials into addressing a wide range of problems and to provide residents with needed infrastructure and services. The Northeast Baltimore Improvement Association, for instance, fought for the extension of county electric and water lines.[39] The Arcadia Improvement Association lobbied for physical improvements such as modern sewers. They also pushed for the installation of traffic and streetlights, the planting of street trees, and improvements to Herring Run Park. Meanwhile, the Hamilton Improvement Association campaigned to introduce telephone service, gas lines, sewers and electric lights, and improve local roads and mail delivery.[40] On the other side of town, the Garrison Boulevard Association focused their attention on improved lighting, dangerous intersections,

36. Power, 'Apartheid Baltimore Style', 289.
37. *Baltimore News,* 10 December 1910; Power, 'Apartheid Baltimore Style'; Nightingale, 'The Transnational Contexts'.
38. Power, 'Apartheid Baltimore Style', 289.
39. The Northeast Baltimore Improvement Association of Baltimore City, *Constitution*, Baltimore, 1907. Copy on file at the Maryland Historical Society in Baltimore, Maryland.
40. E.L. Holcomb, *The City as Suburb: A History of Northeast Baltimore Since 1660*, Santa Fe, 2005, 140–1, 175–6.

the extension of streetcar lines and improvements to Hanlon Park.[41] According to the constitution of the Homestead League and Improvement Association, the purpose of this group, formed in 1910, was 'to promote the material and moral welfare of the community'.[42]

Other improvement associations had an entirely different set of objectives in mind. For Mount Washington, a suburban enclave located just beyond the northern city limits, the goal was to attract new residents. A promotional document published in 1900 shows that the Mount Washington Improvement Association hoped to lure urban dwellers to the suburbs by selling the idea that the 'country' was the best possible setting for raising children:

> The ideal of life for boys is the one that combines the intellectual stimulus of town life with the advantages of the country. To go to school and return after school hours to the shelter and the pleasures of a country home. A boy in a good neighbourhood, with social equals for his chosen friends, may, as it were, be turned loose in the country, yet when so turned loose, is far more under his parents' eyes, than in the streets of a great city. They know who he goes with, they know where he goes, they know, approximately, what he is doing, and for him the country is full of healthy interests and excitement ... Mothers, can you give yourselves to country life for the sake of your growing boys?[43]

Implicit in the preceding passage is the idea that the country – and not the city – is 'a place for boys'. The country, after all, is safe and predictable. In the country, one is less likely to encounter African Americans and foreign immigrants. Instead, one is surrounded by 'social equals'. According to the promotional literature, adults, too, stood to gain by relocating to the nearby suburbs, especially working men:

> The persistent city man who never goes beyond the brick walls and paved streets except for a few weeks of what he calls recreation, is entitled to pity very much upon the same ground as the caged animals of our menageries. Centuries of wrong living have evolved a people who stand confinement and bad air and foul odors, and increasing noise and the daily touch of vice and crime, wonderfully well, but Nature, sooner or later, takes revenge in one way or another. The city man's vital powers are drawn upon constantly ... He lives, consciously or unconsciously, in a constant state of nervous excitement, of depressed vitality and of fatigue. There is a constant wear of tissue and nerve ... Consider for a minute the effect of this upon your own health and happiness. To the body and mind, cramped by noxious work, nature is medicinal and tonic. Leave your store, your office, come away from the din of the street, the uproar and confusion of things, see the trees, the woods, the

---

41. Garrison Boulevard Association, *Meeting Minutes 1912–1957*. Copy on file at the Maryland Historical Society in Baltimore, Maryland.

42. Homestead League and Improvement Association, *Constitution*, 1907. Copy on file at the Maryland Historical Society in Baltimore, Maryland.

43. Mount Washington Improvement Association, *Picturesque Mount Washington*, Baltimore, 1900, 2–3. Copy on file at the Maryland Historical Society in Baltimore, Maryland.

> fields and let yourself come under the magical influence of sun and sky and clouds, and you are a man again.[44]

Although 'white flight' would not begin in earnest until after the Second World War, its roots can be found decades earlier in places like Mount Washington. Convinced that 'you can live as cheaply, get to your work as readily, and be nine parts more of a man than you are now' and that 'your children will be rosy, your wife happy, and life be worth living', tens of thousands of whites eventually joined the stampede to the suburbs.[45]

Nearly forty years later, the South Baltimore Improvement Association, which coincided with a large African American population, faced an entirely different problem. Made up of 'business, professional and civic minded individuals and organizations' that were 'pledged to the development of improved facilities for South Baltimore' the original aim of the members was to improve sports programmes for the neighbourhood's youth.[46] As time passed, and the organisation grew to 'maturity', its activities broadened to address a whole host of problems and deficiencies plaguing the neighbourhood. In a 1938 document published by the group, a great deal of space is devoted to describing the community's recreational facilities and the success of various youth teams. The authors noted, however, that much of the work had been accomplished 'under great handicaps'. Despite assistance from the Playground Athletic League and cooperation from local schools, recreational facilities in the area were woefully inadequate: 'Each year thousands of children flock to these places and it is to be regretted that many of them have great distances to walk before reaching an area that is safe for play.' 'This condition is true to even a greater extent with the colored children', the report continued. 'South Baltimore offers very meager facilities for the Negro youth.' With regard to the availability of indoor centres, the situation was even more dire: 'Here again, the Negro youth finds himself without adequate recreation space, since the small room in School No. 106 only serves a very small portion of the community in extremely limited activities.'[47]

Why did places like South Baltimore fare poorly when it came to attracting the amenities residents so clearly needed? Perhaps the fact that the South Baltimore Improvement Association was formed only in 1934, more than thirty years after the Mount Washington Improvement Association was established, played a role. Then again, perhaps there were other forces at work. Although one could argue that South

---

44. Mount Washington Improvement Association, *Picturesque Mount Washington*, 9–11.

45. Mount Washington Improvement Association, *Picturesque Mount Washington*, 16.

46. South Baltimore Improvement Association, *Souvenir Book of the Sesqui-Centennial Celebration of Federal Hill*, 1938, unpaginated. Copy on file at the Maryland Historical Society in Baltimore, Maryland.

47. South Baltimore Improvement Association, *Souvenir Book*, unpaginated.

Baltimore was in greater need of parks and playgrounds in 1938 than other, more prosperous, sections of town, this was not the sort of logic that was used when it came to deciding how resources should be allocated. Much as its members might struggle to make the 'reality' of their neighbourhood match their normative vision, circumstances beyond their control conspired against them, as we shall soon see.

To truly appreciate the power and influence of neighbourhood improvement associations in Baltimore at this time, it is imperative that we explore more thoroughly the activities of these organisations. Fortunately, the records of two prominent improvement associations are preserved in special collections at the Maryland Historical Society in Baltimore. Together, the meeting minutes of the Peabody Heights Improvement Association – 500 pages spanning the period 1909 to 1933 – and the promotional materials circulated by the Mount Royal Improvement Association, cast light on the mechanisms that produced an inequitable distribution of amenities in Baltimore in the years leading up to the Second World War.

Figure 2. Biddle Alley, c. 1911

During the first half of the twentieth century, Baltimore's African American population was confined to crowded sections of town just to the east, south, and west of downtown.[48]

48. Photographer unknown. Courtesy of the Maryland Historical Society.

The Peabody Heights Improvement Association

One group that exercised considerable influence in city affairs during the early years of the twentieth century was the Peabody Heights Improvement Association. According to Lewand:

> The major objective of the men who developed Peabody Heights was to profitably anticipate the growth of the City in that northern direction. They incorporated in 1870, shortly thereafter offering land for the erection of high quality residences. They hoped to attract some of Baltimore's wealthiest citizens to these 'suburbs' beyond the then northern City boundary at North Avenue.[49]

Absorbed by the city after the annexation of 1888, members of the newly formed Peabody Heights Improvement Association set out to protect what they had created, pledging in the constitution of the fledgling organisation: 'To promote the general welfare of that section of the city bounded by the centre of the following streets, viz: on the South by Twenty-fifth street; on the East by Guilford Avenue; on the North by the City Limits, and on the West by Maryland Avenue.'[50] The Association's constitution states that the purpose of the group was to 'secure a compliance with and to prevent a violation of any of the restrictions applicable to that portion of said section known as the Peabody Heights Tract'. These restrictions included 'requirements that buildings shall be set back 20' from the building line and that no lager beer saloons or places for sale of intoxicating drinks, slaughter houses, bone or glue factories of any kind *or nuisances of any description* shall be permitted.'[51]

The mission of the group revolved primarily around five major issues. First, the group was interested in promoting street tree-planting and enhancing the beauty of the neighbourhood. According to monthly meeting minutes between 1909 and 1933, interest in planting and caring for trees dates back to 1910 when a forestry committee was first established. In early 1911, the group agreed that 'an Ordinance should be introduced looking to a law governing the planting of trees &c.' in the city. The organisation conducted a study and forwarded it to the Mayor's office with the request that 'an Ordinance be drafted in accord with the substance matter of the report'. Members of the Peabody Heights Improvement Association then waged a vigorous campaign to garner support for the proposed ordinance. A 1912 report in the *Baltimore Sun* validates the effectiveness of their efforts:

> After listening to a delegation from the Peabody Heights Improvement Association at the City Hall yesterday afternoon, Mayor Preston announced that he would get behind a bill authorizing the appointment of a city forester, and giving the municipality control over the planting, care and removal of shade trees on streets within

49. K. Lewand, *North Baltimore: From Estate to Development*, Baltimore, 1989, 58.
50. Peabody Heights Improvement Association of Baltimore City, *Constitution*, Baltimore, 1908. Copy on file at the Maryland Historical Society in Baltimore, Maryland.
51. Peabody Heights Improvement Association of Baltimore City, *Constitution*.

> the city limits ... Data on the subject collected by the Peabody Association during the last 18 months was presented to the Mayor, in addition to the rough draft of a bill. It was shown that a number of large cities had assumed control over shade trees, with good results.

Preston gave final approval to Ordinance No. 154 on 17 August 1912. In the space of a year, the city established a Division of Forestry and hired R. Brooke Maxwell, a Yale-trained forester, to manage the city's trees. Over the course of the next two decades, the Peabody Heights group enthusiastically supported the work of the City Forester and his staff. Embedded throughout the Association's meeting minutes are numerous requests for tree planting and maintenance. The group also cooperated fully when it came to other sorts of beautification efforts, such as the Women's Civic League's occasional 'clean up, paint up' campaigns.[52]

Members of the Peabody Heights Association were also attentive to the needs of nearby Wyman Park. In addition to supporting park improvements and expansions, the group's membership played an instrumental role in assuring that the park was selected as the new home of the Baltimore Museum of Art. Arguing in November 1925 that 'the knoll to the North of Druid Lake in Druid Hill Park' would not be a good location for the new museum because it would destroy 'the natural beauty of the park', the Association, on 11 January 1926, endorsed Wyman Park as an ideal site for the museum. There can be little doubt that securing a site for so coveted a cultural amenity as the Museum of Art represented a major victory for the Association.[53]

A third issue that attracted the attention of the Peabody Heights Association was air pollution. As noted earlier, the first two decades of the twentieth century was a period of impressive industrial growth and change in Baltimore. One indicator of that growth was the tremendous increase in railroad traffic. According to Olson, during the period 1900–1918, the Pennsylvania Railroad was running 400 trains a day past Union Station, 'ten times the number of 1890'.[54] Although the city had anti-smoke laws on the books to protect residential areas from the pollution caused by coal-powered steam locomotives, they proved difficult to enforce. Given Peabody Height's location just a few blocks from the station, it is not surprising that a 'Smoke Committee' was formed to deal with this nuisance.[55]

The fourth issue that elicited concern among residents was 'undesirable commercial development'. The community having originally been intended as a strictly residential neighbourhood, the Peabody Heights Association was adamantly

---

52. Peabody Heights Improvement Association, *Meeting Minutes 1909–1933*. Copy on file at the Maryland Historical Society in Baltimore, Maryland.

53. Peabody Heights Improvement Association, *Meeting Minutes*.

54. Olson, *Baltimore*, 245.

55. Peabody Heights Improvement Association, *Meeting Minutes*.

opposed to businesses setting up shop, whether it was the opening of a Greek delicatessen in the basement of an apartment building or the location of a store on a busy thoroughfare. When threatened by 'invasion ... of business establishments' complaints were promptly filed with the office of the Inspector of Buildings and the problem was quickly resolved. This issue, in particular, caused members of the Association in October 1929 to 'urge upon the City Council the early passage of the new comprehensive zoning ordinance now before that body, so as to prevent the further invasion of residential neighbourhoods by business establishments and also the development of our city on modern lines', although it was several more years before such comprehensive legislation was approved.[56]

Finally, like other improvement associations in Baltimore at the time, the Peabody Heights group took steps to prevent African Americans from buying or renting in the neighbourhood, as this entry for 10 October 1910 makes abundantly clear:

> The following Resolution offered by Mr. Goldsborough, was carried; Resolved, That the Peabody Heights Improvement Association hereby advocated the enactment of such proper State or City legislation as will make it difficult or impossible for negroes, as dwellers, to invade those blocks or neighborhoods where there is a preponderance of white occupants; be it further Resolved, that a copy of this Resolution be sent to the Mayor and City Council of Baltimore.[57]

Not only did the Peabody Heights Improvement Association support Baltimore's segregation ordinance, the group sought to continue the practice of segregation long after the Supreme Court ruled such ordinances unconstitutional. Five years after the ruling, noting that 'no valid law can be enacted to effect segregation', the president of the board, on 10 April 1922, assured residents that 'property owners in any section may by contractual agreement bind themselves not to sell to a negro'. Meeting minutes from this date indicate that the association also favoured protection 'against invasion of the neighborhood by so-called 'kike' Jews as well as by negroes'.[58] The following excerpt from 11 May 1925 is valuable because it suggests that neighbourhood associations worked with one another to achieve a common goal – in this case, exclusion of African Americans:

> Under the head of new business, Mr. Harn commented on the action of the School Board in their decision to establish a negro school at 24th and Oak Streets by converting the white school at that location. This location is not within the zone of the Peabody Heights Improvement Association, but is in the area covered by the Citizens Improvement Association. Mr. Harn stated that the establishment of a negro school at this location would be the entering wedge of a negro invasion of a strictly

56. Peabody Heights Improvement Association, *Meeting Minutes.*

57. Peabody Heights Improvement Association, *Meeting Minutes.*

58. Peabody Heights Improvement Association, *Meeting Minutes.*

> white neighborhood, … Mr. Norris advised that he would discuss this matter with Dr. Watson of the Citizens Improvement Association and assure Dr. Watson that the Citizens Association would have the sympathy and cooperation of the Peabody Heights Association in any effort they care to make against this negro invasion.[59]

What is the link between segregation and the desire to attract amenities such as parks and street trees? With respect to amenities, Lawrence reminds us that 'these places are often unequally available to different social groups, and as private amenities they are usually unequally distributed within the urban landscape'.[60] Considering the power that groups like the Peabody Heights Association wielded, one cannot help but speculate whether a neighbourhood-wide strategy to improve park grounds, plant trees, and engage in other beautification activities, coupled with aggressive enforcement of segregation – formal or informal – might have contributed to an inequitable distribution of amenities, with a disproportionate share of resources and services flowing into wealthy white districts like Peabody Heights at the expense of other neighbourhoods. Evidence gathered from the records of another group, the Mount Royal Improvement Association, appears to support this assertion.

The Mount Royal Improvement Association

In 1930, the Mount Royal Improvement Association published a glossy promotional pamphlet entitled, *The Mount Royal District: Baltimore's Best Urban Section.* Richly illustrated with black and white photographs, its pages depict stately homes and apartment buildings, elegant hotels and monuments, and beautiful tree-lined streets. At first glance, one might dismiss it as a contemporary example of promotional literature, an advertisement for one of Baltimore's fashionable neighbourhoods. Closer inspection suggests another purpose. In language that is unmistakable and straightforward, it is apparent that the document was intended to serve as a reminder to prospective residents that Mount Royal was a 'protected' neighbourhood.

After describing the location of the district in very general terms, the pamphlet discusses the advantages of living there:

> The Mount Royal District is most conveniently located. The business section and railroad stations though well without the District are within easy walking distance, yet no section is so free from business encroachment. It is convenient to churches, schools, parks, theatres, and hospitals. The Lyric Theatre, where all Grand Operas

59. Peabody Heights Improvement Association, *Meeting Minutes*.

60. H. Lawrence, 'Changing Forms and Persistent Values: Historical Perspectives on the Urban Forest', in G.A. Bradley, ed. *Urban Forest Landscapes: Integrating Multidisciplinary Perspectives*, Seattle, 1995, 31.

are given, is within four blocks.[61]

Readers are then assured that this district 'is within easy walking distance of the section from which comes most of our domestic help'. Under the heading 'A Restricted Area', the document boasts that, 'The greatest achievement of the Mount Royal Improvement Association has been the subjecting of the property in its area to a restriction for white occupancy only. Already a large majority of the properties in its territory south of North Avenue have been protected by its restrictive agreement, and the protection of the entire area in the near future is well under way.'[62] Under the heading 'A Protected Area', the pamphlet goes on to state that 'The Mount Royal District is protected not only by the restrictive agreement but by the Mount Royal Improvement Association, an incorporated body whose members consist of property owners in its territory, which maintains a permanent office and is ready at all times to take any action necessary to protect the health, welfare and property rights of its members, and generally to advance the interests of the district.'[63]

Touting its history as 'the home of old and cultured Baltimore families' and the high quality of its housing stock, 'modernised and equipped with the latest improvements', the pamphlet proceeds to discuss the neighbourhood's commitment to beautification: 'The residents of Mount Royal District take great interest in their gardens and yards, and many of them have been awarded prizes in the Annual Garden Contests conducted by the Women's Civic League. They are strongly supporting the crusade to change unsightly alleys into lovely lands, and an inspection of the yards and alleys in this section reveals many beauty spots not dreamed of by the general public.' Specific reference is also made to the planting of trees: 'The Mount Royal Improvement Association is waging a systematic campaign for the extension of this work, and has appointed a lieutenant for each block to assist the residents in adopting a uniform plan of fencing and tree planting along the alleys and the beautifying of garages.'[64] The document concludes with the authors' claim that 'Lanvale Street and Lafayette Avenue ... between Mount Royal Avenue and Eutaw Place, are among the finest residential streets in Baltimore. Property values on these streets have steadily increased since they were restricted for white occupancy by the Neighborhood Corporation.'[65]

To underscore the connection that existed, at least in the minds of the members of the Mount Royal Association, between white occupancy and a beautiful

61. Mount Royal Improvement Association, *The Mount Royal District: Baltimore's Best Urban Section*, Baltimore, 1930, 3.

62. Mount Royal Improvement Association, *The Mount Royal District*, 5.

63. Mount Royal Improvement Association, *The Mount Royal District*, 3, 6.

64. Mount Royal Improvement Association, *The Mount Royal District*, 6, 12.

65. Mount Royal Improvement Association, *The Mount Royal District*, 39.

environment, the group printed the following message on a meeting announcement from c.1930:

> When the present officers of the Mount Royal Improvement Association assumed office, assurances were given that plans would be presented for the maintenance of this district as the most beautiful and most desirable urban section of Baltimore, but that this could be done only after the property owners had made the district safe for white occupancy by the execution of a sufficient number of the association's protective agreements. This condition was imposed because of the impossibility of preserving, much less improving any unrestricted section of Baltimore.[66]

Viewed in this light, it is not difficult to imagine how discriminatory housing practices and protective covenants worked in favour of neighbourhoods like Mount Royal.

Despite restrictive covenants and active improvement associations, residents of Peabody Heights and Mount Royal found it increasingly difficult to pursue their neighbourhood visions. As a growing number of white residents took advantage of the 1918 annexation to locate farther away from the downtown, the black community remained 'locked in'. In Olson's words, 'a Jim Crow space was coalescing, growing larger and more formidable ... Its space expanded grudgingly, explosively, block by block and year by year.'[67] By the late 1930s, whites were increasingly leaving the city altogether for outlying suburbs, replaced by a steady stream of African Americans migrating from the South. As these changes unfolded, both the Peabody Heights and Mount Royal Improvement Associations were forced to adjust to changing circumstances.

Shifting Demographics

In 1937, the Home Owners Loan Corporation, a Depression-era federal agency charged with refinancing homes in danger of foreclosure, conducted a survey in which it assigned 'security' grades to different Baltimore neighbourhoods based on perceived loan risk.[68] Areas considered to be high risk were outlined in red on the original maps and thus the process came to be known as 'redlining'. Among the criteria used to assess risk were occupation of residents, average annual income, predominant nationality, percentage of 'negro families' to total number of families,

66. Mount Royal Improvement Association, NOTICE OF MEETING. A special meeting of the MOUNT ROYAL IMPROVEMENT ASSOCIATION will be held at the ASSOCIATE CONGREGATIONAL CHURCH North-west corner of Maryland Avenue and Preston Street Tuesday, June 24th at 8 p.m. (Daylight Saving Time). Copy on file at the Maryland Historical Society in Baltimore, Maryland.

67. Olson, *Baltimore*, 372.

68. Home Owners Loan Corporation, Records of the Federal Home Loan Bank Board, Home Owners Loan Corporation, Records Relating to the City Survey File, 1935–1940. RG 195, Box 106, 107. National Archives II.

percentage of families on relief, and 'threat of infiltration of foreign born, negro or lower grade population'. In addition, surveyors were instructed to estimate the rate of change occurring in a given population.

According to the map and accompanying documents, Mount Washington, an old suburban area absorbed by the city after the annexation of 1918, was deemed a low-risk area and so it received a 'security grade' of 'B' (still desirable). With respect to 'favorable influences', it was described as a 'restricted, desirable residential section of moderately large houses on large lots' with 'a wooded section of good character'. No 'detrimental influences' were noted. Homeowners were 'executive, professional men' with annual incomes of $5,000 or more. Thanks to a restrictive covenant, no 'negro', foreign, or relief families were identified as living in Mount Washington at the time of the survey.

While the major portion of Peabody Heights was assigned a security grade of 'B', another part merited only a 'C' (declining) from the HOLC surveyors. As far as 'favorable influences' on the neighbourhood were concerned, it was described as a 'stable old residential section of good character' conveniently located near a commercial area. The presence of 'industrial plants' and the 'continuing obsolescence of old property', however, were viewed as detrimental influences. Although no African Americans were living in Peabody Heights at the time of the survey, a 'mixture' of foreign immigrants and a small number of 'relief' families had begun to trickle in. Mount Royal, meanwhile, received lower evaluations, with one section getting a security grade of 'C' (declining) and the other a 'D' (hazardous). While 'proximity to Druid Hill Park' and a 'central location' near employment were considered positive attributes, 'heavy obsolescence', 'negro encroachment', and 'industrial encroachment' were detrimental influences. Also contributing to the increasingly hazardous conditions was a moderate to high percentage of relief families and the presence of a mixture of Jewish and foreign families. Unlike Peabody Heights, where the population was described as 'static', Mount Royal's population was 'slowly declining'.

At the other extreme was South Baltimore. Assigned a 'D' grade (hazardous), the neighbourhood's central location 'near employment' was cited as a 'favorable' influence. 'Heavy obsolescence' of buildings and 'industrial encroachment' were listed as 'detrimental influences'. Instead of executive and professional men, inhabitants were listed as 'laborers' with an annual income of $700–1,000. Approximately twenty per cent of the population was listed as 'Negro' with the remaining eighty per cent a mixture of 'Foreign-born'. Sixty per cent of residents were identified as being on 'relief'. According to the report, an 'infiltration' of 'Negro' families and 'foreigners' made this a high risk area. Perhaps because of the aforementioned 'industrial encroachment' it was also noted that the overall population of this area was slowly declining.

How did demographic changes such as these affect the allocation of resources in the city? A letter penned by Frederick Law Olmsted, Jr to his old friend Theodore Marburg of the Municipal Art Society offers a clue. In 1939, Marburg contacted Olmsted, soliciting his advice regarding the optimal location for a major new park. Although earlier reports made it clear that Olmsted wished to provide the city with 'a roughly equitable distribution' of parks and outdoor recreation facilities 'for all its citizens', even he succumbed to the idea that some parts of town were simply too 'high risk' to warrant a substantial investment of money. In his response to Marburg he justified his selection of a site abutting an already existing park:

> The belated introduction of suitable parks might be a very important and valuable element in a sound general program of rehabilitation, intelligently and realistically directed toward salvaging such districts and reconverting them into really satisfactory and attractive residential communities ... But, in the face of wide-spread uncertainty about the future trends of a district, random and uncoordinated expenditures of a few hundred thousand dollars here and a few more there, intended to benefit a diminishing population, the future distribution of which no one now has a sound basis for estimating, is far too liable to be an extravagant and futile gesture. Certainly it is a very risky venture in which to invest trust funds when a safer one that is entirely appropriate is clearly available.[69]

Although other sections of Baltimore were arguably in greater need of a major park – neighbourhoods like South Baltimore – Olmsted recommended, and the city followed through with, the purchase of the Winans estate, also known as 'The Crimea', next to Gwynns Falls Park. Further evidence of inequity is found in The Long Range Recreation Plan of 1941, prepared by the National Recreation Association for Baltimore's Board of Public Recreation. The report concluded that while Baltimore in general lacked adequate acreage in parks, especially for children's playgrounds, the black community, in particular, was 'lacking in areas and facilities quite out of proportion to the ratio of its numbers to the total population'.[70]

An Inherited Landscape

In the early years of the twentieth century, neighbourhood associations proliferated in Baltimore. These groups, especially older established ones like the Peabody Heights and Mount Royal Improvement Associations, proved adept at attracting amenities, while discouraging unwanted land uses and other disamenities. In fulfilling their suburban vision, many of these groups supported Baltimore's segregation measure, also known as the West Ordinance. When the West Ordinance was struck down

69. F.L. Olmsted, Jr. Letter to Theodore Marburg, Chairman of the Board, Municipal Art Society of Baltimore, 12 July 1939. Copy on file at the Maryland Historical Society in Baltimore, Maryland.

70. W.W. Pangburn and F.E. Allen, *Long Range Recreation Plan, City of Baltimore, Maryland*, Baltimore, 1941, ix.

in 1917, communities attempted to 'protect' themselves against African American and foreign 'invasion' by passing restrictive covenants, thus reinforcing previously existing patterns of segregation.

During the 1920s and 1930s, as an increasing number of white residents moved to suburbs farther away from the downtown, immigrants and African Americans began to crowd into urban neighbourhoods, a trend that would accelerate after the Second World War. Deplorable practices like block busting only stepped up the pace of white out-migration. As the black population, in particular, grew, white residents and government officials became convinced that the neighbourhoods into which Blacks were moving were in decline and, further, that investments in these districts, public or private, were risky.

Ironically, 'white flight' permitted the resident black population to spread out and thereby gain access to amenities that had previously been off limits to them. Today, Baltimore's African American population enjoys a relatively high access to amenities such as parks and playgrounds but this was not, in fact, by design. Rather, African Americans are the beneficiaries of an inherited landscape, one that was left behind when whites left the city in droves during the 1950s and 1960s.

Conclusion

Neighbourhood improvement associations were key actors in the development of Baltimore's social and built geography. As private governance organisations, the improvement associations worked to serve their own interests first, with less regard for the overall welfare of the city. It should come as no surprise that Baltimore's segregation ordinance came not from the chambers of city hall but from a neighbourhood association working to 'protect' itself from outsiders.

An important consequence of the improvement associations was the uneven distribution of city services and resources. Although municipally-owned and operated, parks were strikingly sparse in African American communities in the first half of the twentieth century. This chapter demonstrates that the improvement associations lobbied very effectively to secure municipal resources for beautifying their communities at the expense of others, especially African American neighbourhoods. Other actors, including the federal Home Owners Loan Corporation and its neighbourhood ranking system, worked to reinforce disinvestment in African American communities and neighbourhoods with older housing, which further undermined the ability of marginalised districts to attract city services. 'Block busting' and 'white flight' added to the disparities between white and black spaces in Baltimore.

By any measure, the block busting, racial covenants, segregation ordinances, neighbourhood ranking system, and diversion of parks and street trees to white neighbourhoods were unjust, yet the historical examination of these tools and processes is typically ignored in the environmental justice literature. Analyses that focus solely on the present miss the important stories described in this chapter, making it difficult to answer a fundamental question of environmental justice – whether just distributions were justly achieved. The struggle for space and resources in Baltimore, described in part here, illuminates a history of environmental injustice that disadvantaged black Baltimoreans. New efforts to improve quality of life in the city, including the provision and maintenance of environmental amenities such as parks and street trees should, therefore, seek just solutions for past injustices.

Selected Bibliography

Boone, C.G. 'An Assessment and Explanation of Environmental Inequity in Baltimore', *Urban Geography*, 23, 2002, 581–95.

Boone, C.G. 'Environmental Justice as process and new avenues for research', *Environmental Justice*, 1, 2008, 149–54.

Davis, M. City of Quartz: *Excavating the Future in Los Angeles*, New York, 1992.

England, M. 'When "Good Neighbors" Go Bad: Territorial Geographies of Neighborhood Associations', *Environment and Planning A*, 40, 2008, 2879–94.

Garb, M. 'Drawing the "Color Line": Race and Real Estate in Early Twentieth-Century Chicago', *Journal of Urban History*, 32, 2006, 773–87.

Hayden, D. *Redesigning the American Dream: The Future of Housing, Work, and Family Life*, New York, 1984.

Hood Washington, S. 'Mrs. Block Beautiful: African American Women and the Birth of the Urban Conservation Movement, Chicago, Illinois 1917–1954', *Environmental Justice*, 1, 2008, 13–23.

Jackson, K. Crabgrass *Frontier: The Suburbanization of the United States*, New York, 1985.

Low, S. *Behind the Gates: Life, Security, and the Pursuit of Happiness in Fortress America*, New York, 2003.

McDougall, H. A. *Black Baltimore: A New Theory of Community*, Philadelphia, 1993.

McKenzie, E. *Privatopia: Homeowner Associations and the Rise of Residential Private Government*, New Haven, 1994.

Mohl, R.A. 'Race and Space in Miami', in A. R. Hirsch and R. A. Mohl, eds. *Urban Policy in Twentieth-Century America*, New Brunswick, 1993, 100–58.

Nightingale, C.H. 'The Transnational Contexts of Early Twentieth Century American Urban Segregation', *Journal of Social History*, 39, 2006, 667–702.

Orser, W. E. *Blockbusting in Baltimore: The Edmondson Village Story*, Lexington, 1994.

Power, G. 'Apartheid Baltimore Style: The Residential Segregation Ordinances of 1910–1913', *Maryland Law Review*, 1983, 289–328.

Pulido, L. 'Rethinking Environmental Racism: White Privilege and Urban Development in Southern California', *Annals of the Association of American Geographers*, 90, 2000, 12–40.

Purcell, M. 'Neighborhood Activism among Homeowners as a Politics of Space', *Professional Geographer*, 53, 2001, 178–94.

Schirmer, S.L. *A City Divided: The Racial Landscape of Kansas City 1900–1960*, Columbia, 2002.

Stilgoe, J. *Borderland: Origins of the American Suburb*, New Haven, 1988.

Wells, J., G.L. Buckley, and C.G. Boone, 'Separate but Equal? Desegregating Baltimore's Golf Courses', *The Geographical Review*, 98, 2008, 151–70.

Chapter Three

The Social Production of a Canadian Urban Forest

Joanna Dean[1]

The environmental justice movement has resulted in a new attentiveness to race, class and power among environmental historians. Much of the attention has been focused, however, upon the negatives of environmental racism, especially the management of toxic wastes, and little attention has been paid to the inequitable distribution of environmental benefits. This chapter is intended to bring the analysis of the environmental justice movement to bear upon an environmental amenity, the urban forest, whose unequal distribution can be correlated to the social indices of income and race. The chapter draws upon well-developed methodology for the digital analysis of canopy cover in aerial photographs; it departs from most work on the urban forest canopy cover, however, in its historical perspective and close focus upon neighbourhoods and concludes that a long temporal scale and a small spatial scale are crucial to understanding urban ecological change.

Most existing histories of the urban forest emphasise the aesthetic value of the trees. The focus of landscape historians is upon park trees and those avenues of street trees that soften the urban landscape in wealthy residential neighbourhoods; and while they are attentive to class dynamics, their emphasis is frequently upon the symbolic value of trees and the ways in which they are used to assert power.[2] My interest here is in the environmental benefits of city trees. As nineteenth century planners knew, and modern urban foresters have proven, trees improve air quality and reduce summer heat; historically street trees were known as shade trees and park trees were commonly described as the lungs of the city. Today we refer to the 'urban forest' and urban foresters emphasise the impact of urban forest canopy

1. Geomatics by Jon Pasher and Michelle Leni at the Geomatics and Landscape Ecology Research Laboratory, Carleton University, Ottawa. This research was supported by the Social Sciences and Humanities Research Council of Canada.

2. For the emphasis on aesthetics see H.W. Lawrence, *City Trees: A Historical Geography from the Renaissance through the Nineteenth Century*, Charlottesville, 2006; for the symbolic power of street trees, see T.J. Campanella, *Republic of Shade: New England and the American Elm*, New Haven, 2003; for the use of landscape to assert power see S. Pugh, *Garden, Nature, Language*, Manchester, 1988; and for a history that integrates these with the environmental amenities provided by trees, see in M. Page, *The Creative Destruction of Manhattan, 1900–1940*, Chicago, 1999, ch. 6, 'Uses of the Axe: Toward a Treeless New York'.

cover on storm water run-off, temperature fluctuations, and levels of particulate matter, ozone, sulphur dioxide, nitrogen dioxide, and carbon monoxide. They have even quantified psychological benefits: one widely cited study showed that hospital patients recover more quickly if they can see trees outside their window.[3] Some of these benefits are distributed over the city as a whole but many are localised, and the inequitable distribution of the urban forest contributes to environmental injustice.

A number of studies of urban forest change have been conducted by urban foresters, most notably by David Nowak of the American Forest Service, who has developed a sophisticated and widely used software model for the analysis of the environmental benefits of urban forest canopy cover. The most politically engaged studies are by a leading proponent of urban political ecology, Nik Heynen, who has been critical of the lack of class analysis in urban environmental history.[4] He argues: 'Marxist urban political economy more explicitly recognises that the material conditions that comprise urban environments are controlled, manipulated and serve the interests of the elite at the expense of marginalised populations.'[5] His analysis is useful for its close attention to power and class, the conceptualisation of the urban forest as a socially produced and unevenly distributed commodity, and the driving focus upon the central question of 'who produces what kind of socio-ecological configurations for whom'.[6]

This chapter describes the early stages of a pilot project applying geospatial analysis historically to the urban forest. The urban forest canopy cover is mapped in four neighbourhoods in the adjacent cities of Ottawa and Gatineau, Canada, using aerial photographs from the 1920s through to 2005. The project departs from most existing urban forest analysis by using methodology drawn from landscape ecology

3. R.S. Ulrich, 'View Through a Window May Influence Recovery from Surgery,' *Science,* 224 1984, 420–21.

4. N. Heynen, 'The Scalar Production of Injustice within the Urban Forest', *Antipode: A Journal of Radical Geography* 35:5, 2003, 980–98; N. Heynen and G. Lindsay, 'Correlates of Urban Forest Canopy Cover: Implications for Local Public Works', *Public Works Management and Policy* 8:1, 2003, 33–47; N. Heynen and H. Perkins, 'Scalar Dialectics in Green: Urban Private Property and the Contradictions of the Neo-liberalization of Nature', *Capitalism, Nature, Socialism* 16:1, 2005, 99–113; N. Heynen, 'Green Urban Political Ecologies: Toward a Better Understanding of Inner City Environmental Change', *Environment and Planning* A, 38:3, 2006, 499–516. See also G.M. Talarchuk, 'The Urban Forest of New Orleans: An Exploratory Analysis of Relationships', *Urban Geography,* 18:6, 1997, 65–86.

5. E. Swyngedouw and N. Heynen, 'Urban Political Ecology, Justice and the Politics of Scale,' *Antipode* 35, 2003, 902.

6. N. Heynen, M. Kaika, and E. Swyngedouw observe: 'to the extent that cities are produced by socio-ecological processes, attention has to be paid to the political processes through which particular socio-environmental urban conditions are made and remade. From a progressive or emancipatory position, then, urban political ecology asks questions about who produces what kind of socio-ecological configurations for whom.' 'Urban Political Ecology: Politicising the Production of Urban Natures', in N. Heynen, M. Kaika, and E. Swyngedouw, eds. *In the Nature of Cities: Urban Political Ecology and the Politics of Urban Metabolism,* London, 2006, 2,

to create maps of canopy cover for individual neighbourhoods. The neighbourhood scale of analysis allows the comparison of canopy cover in different parts of the city, and correlation of canopy cover to indices of socio-economic status and ethnicity. The maps can also be interpreted visually. For one small study area, Lowertown, geospatial analysis is combined with archival research and street-level observation of existing city trees. The geospatial analysis shows that forest cover has changed significantly over time and that the rate of change varies in different parts of the city in ways that can be correlated to socio-economic status. The close analysis of canopy cover in Lowertown suggests that the inequalities in urban forest are more complex, and more profound, than simple statistical analysis might suggest.

A Short History of the Urban Forest in Ottawa and Gatineau

The cities of Ottawa and Gatineau face each other from opposite banks of the Ottawa River, and from opposite sides of the linguistic and ethnic divide that defines Canadian identity. Ottawa, Canada's capital city, is in the mainly English speaking province of Ontario, and Gatineau is in francophone Quebec. The area was once heavily wooded: in 1613 Samuel de Champlain described rich stands of pine along the Ottawa River and in the early nineteenth century surveyors described mixed maple and beech upland forests as well as impenetrable cedar swamps. This forest disappeared with the construction of the towns of Ottawa and Gatineau (then Bytown and Hull): early paintings show only the occasional remnant tree left standing among the wooden houses, with patches of cedar in swampy areas, and scrubby second growth bush in ravines. The towns grew with the timber boom, and with the construction of the Rideau Canal in the 1830s the population of French Canadian lumbermen was augmented by Irish labourers and British military personnel. There was little interest in street trees until the middle of the nineteenth century, when the Ottawa was selected as the capital city and newly arrived members of parliament pressed for street tree planting as a key element in urban beautification. American elm, sugar and silver maple were planted along residential streets to provide shade and to grace an uneven streetscape. The federally funded Ottawa Improvement Commission laid out tree-lined parkways for visiting dignitaries and homeowners planted fruit trees and ornamental trees on their lots. Avenues of street trees became objects of municipal pride.

The trees grew quickly and by the 1920s some horticulturalists were complaining about street trees that were getting out of control. Fast growing species had been planted closely together for quick results, and street trees were competing with the expanding built environment of utility wires, sidewalks and asphalt roads. The city of Ottawa engaged in an aggressive tree trimming and tree removal campaign. City records show that 4,000 trees were removed and many thousands more trimmed or topped between 1921 and 1945. In 1962, after Dutch elm disease further reduced the tree cover, the municipal government reversed position, and recommended a programme of tree conservation and tree planting.[7]

Figure 1. Aerial Photograph of the Ottawa Area in 1923.

This early oblique aerial photograph shows the wide variations in canopy cover in the urban neighbourhoods. Ottawa, Canada's capital city, is to the south of the Ottawa River; Gatineau is to the north. The canopy study areas, Lowertown, New Edinburgh, Notre Dame and Alta Vista, are indicated on the photograph, although Alta Vista was undeveloped agricultural land in 1923.[8]

In Ottawa and Gatineau, as elsewhere, it is evident that trees follow wealth. Just how closely income and canopy cover are correlated is more difficult to establish. Precisely because trees were prestigious elements in the urban landscape, traditional sources are unreliable. City boosters exaggerated the beauty of the treed neighbourhoods and photographers positioned their cameras to accentuate the presence and height of street trees. The presence of trees in wealthy areas is often inflated and working class neighbourhoods are rarely documented. Aerial photographs provide a

7. For the history of municipal policies, see J. Dean, '"Said Tree is a Veritable Nuisance": Ottawa's Street Trees, 1869–1939', *Urban History Review: Revue d'histoire urbaine*, 34:1 2005, 46–57.

8. Source: National Air Photo Library, Ottawa, Canada. HA 10 008.

more reliable sense of relative canopy cover. In one of the earliest aerial photographs of Ottawa (Fig. 1), taken from an oblique angle in 1923, wealthy neighbourhoods can be identified by a dark canopy of trees. Sandy Hill, for example, was the home of prosperous civil servants, professionals and politicians, and was densely covered with a growing canopy of trees. The adjacent working class neighbourhood, Lowertown, has a visibly thinner canopy. The correlation of wealth and canopy is not entirely consistent: the relatively treeless area in LeBreton Flats and Hull is the product of a devastating fire in 1900 that removed most of the established trees from neighbourhoods close to the lumber yards. Vertical aerial photographs, and more recently satellite images, show the canopy in a form that can be accurately measured. Because it was a capital city, Ottawa was well photographed; vertical aerial photographs are available from the 1920s along some flight paths and from 1950 photographs captured the entire area.

Methodology: Calculating Canopy Cover

Most methods developed for urban forest analysis produce broad estimates of the canopy cover for the entire urban area.[9] For this project a methodology has been developed in conjunction with students at the Carleton University Geomatics and Landscape Ecology Laboratory to provide a detailed neighbourhood-by-neighbourhood measure of changes to the urban forest canopy over time.

First, some definitions are necessary. Urban tree cover is defined as, 'the proportion of an area, when viewed from above, occupied by tree crowns'. We measured the crown or canopy, rather than individual trees, because it is the canopy, not the number of individual trees, that provides the ecological and health benefits of the urban forest.

Second, what counts as urban forest? We included large shrubs, mainly because they are impossible to distinguish from small trees in aerial photographs, but also because the distinction between a tree and a shrub is largely an aesthetic one – trees have a single trunk and are taller than shrubs. Large shrubs, however, serve the same environmental functions as small trees.

Next, what is a neighbourhood? The areas selected for study are commonly understood to be identifiable neighbourhoods – that is, they are seen to be homogeneous and to have a particular identity within the larger city. The four neighbourhoods chosen represent different income levels, ethnicities, and dates of construction. There are a limited number of suitable aerial photographs for the

9. For an introduction to this approach see D.J. Nowak, 'Understanding the Structure of Urban Forests', *Journal of Forestry*, 92, 1994, 42–46; D.J. Nowak, R.A. Rowntree, E.G. McPherson, S.M. Sisinni, E. Kerkmann, and J.C. Stevens, 'Measuring and Analyzing Urban Tree Cover', *Landscape and Urban Planning*, 36, 1996, 49–57. For a recent discussion of methodology see J. T. Walton, D. J. Nowak, E. J. Greenfield, 'Assessing Urban Forest Canopy Cover', *Arboriculture and Urban Forestry*, 34:6, 2008, 334–40.

early years of this study, so the geographic boundaries of study areas are in most cases smaller than the neighbourhoods.

Finally, how does one measure canopy cover? Several methods have been used. The first, known as crown cover scale, involves placing a network of polygons over the photograph and then counting the number of polygons touching tree canopy.[10] The second, the transect method, involves running lines across the photograph and then measuring length of line crossing tree canopy and calculating the measure as a percentage.[11] The third, and most widely used today among urban foresters, is the dot grid method. A dot grid is laid over a photograph and the percentage of dots falling upon canopy cover is calculated. David Nowak of the United States Department of Agriculture Forest Service has developed software (UFORE – Urban Forest Effects Model) using the dot grid analysis, in conjunction with field observations of forest structure in randomly selected sample areas, to quantify urban forest structure and calculate the impact of urban forests on air quality, greenhouse gas emission and carbon sequestration.[12] American Forests has developed software, CITYgreen, using similar techniques.[13] Dot grid methodology has also been used by Heynen in his urban political ecology.

Dot grid analysis has some drawbacks, however. As the aerial photo in Figure 1 shows, urban forests are not uniform. A city-wide average will mask differentials between neighbourhoods. Nowak has acknowledged this limitation. In a study of Santiago, Chile, Nowak and Escobedo argue that 'Future studies of urban forest and their role in environmental quality should consider the ecological and socio-economic heterogeneity within the urban ecosystem.'[14] The GIS analysis in UFORE is currently being adapted to be more sensitive to spatial variation and environmental inequities.[15] Dot grid analysis also fails, as we see below, to identify

10. K.D. Moessner, 'A Crown Density Scale for Photo Interpreters', *Journal of Forestry*, 45, 1947, 434–36.

11. R.H. Canfield, 'Application of the Line Intercept Method in Sampling Range Vegetation', *Journal of Forestry*, 39, 1941, 388–94.

12. D.J. Nowak and D.E. Crane, 'The Urban Forest Effects (UFORE) Model: Quantifying Urban Forest Structure and Functions', in M. Hansen and T. Burk, eds. *Integrated Tools for Natural Resource Inventories in the 21st Century*, IUFRO Conference, St. Paul, 2000. David Nowak has applied the random dot grid analysis historically in his dissertation, 'Urban Forest Development and Structure: Analysis of Oakland, California', Ph.D. Dissertation, University of California, Berkeley and in a 1993 article: 'Historical Vegetation Change in Oakland and its Implications for Urban Forest Management', *Journal of Arboriculture*, 19, 1993, 313–19.

13. See T. Longcore, C. Li, J.P. Wilson, 'Applicability of CITYgreen Urban Ecosystem Analysis Software to a Densely Built Urban Neighbourhood', *Urban Geography*, 25:2, 2004, 173–86, for a critical review of the CITYgreen software.

14. F. Escobedo and D. Nowak, 'Spatial Heterogeneity and Air Pollution Removal by an Urban Forest', *Landscape and Urban Planning*, 90, 2009, 109.

15. The authors of the UFORE, Urban Forest Effects, model are currently developing a new GIS model that will determine environmental inequities. See the reference to the new GIS model

patterns of tree growth within neighbourhoods. The dot grid methodology has more serious limitations for a historian, because it requires a uniform set of aerial photos for the entire city. In the 1920s and 1930s aerial photographs were taken along single flight paths, rather than the grids that later covered entire cities, and so there are large gaps in the historical data.

Instead of the dot grid method, we adopted the methodology used in landscape ecology. It is an exacting process that yields precise results for a specific neighbourhood, street, or park.[16] A doctoral student, Jon Pasher, developed the process for digitally tracing the tree canopy on scanned aerial photographs. The process is described in detail in the Appendix. To summarise: firstly, photos were selected for date (we used 'leaf-on photos'), resolution, clarity, camera angle and coverage of the neighbourhood, and then scanned and geo-referenced; secondly, an area of interest was identified for analysis, incorporating the maximum area covered by all images from the different dates; thirdly, the tree canopy within the area of interest was manually digitized in order to create bitmap masks for each image; finally, the percentage canopy cover in each image was calculated. (For examples of this methodology, see http://www.carleton.ca/~jdean/urbanforest)

We tested for error using photographs of different quality and found that the resulting error rate was acceptable.[17] The error rate does not, however, include errors that were common to each reading: shadows, for example, might be read as canopy; trees photographed at an angle appear larger. All visual interpretation was done by the same interpreter/analyst for consistency in identifying tree canopy.

After the methodology was tested on one neighbourhood, New Edinburgh, it was then applied to three other residential neighbourhoods, two in Ottawa and one in Gatineau. Neighbourhoods were selected for a cross-section of income, ethnicity and date of residential construction. New Edinburgh and Lowertown are older inner city neighbourhoods, and Alta Vista and Notre Dame are suburbs. New Edinburgh and Alta Vista are upper-middle class, and anglophone, whereas Lowertown and Notre Dame are working class and more francophone.

These results were correlated with census data. In the period after the Second World War, the Canadian Bureau of Statistics compiled and published population and housing statistics by census tracts in order to 'provide a means of comparison of social and economic factors within an urban community which are often obscured in totals for a city or metropolitan area as a whole'. Census tracts were designed to

---

at http://www.ufore.org/about/04-00.html Accessed 29 August, 2010.

16. W.C. Zipperer, S.M. Sisinni, R.V. Pouyat, and T.W. Foresman, 'Urban Tree Cover: An Ecological Perspective', *Urban Ecosystems*, 1, 1997, 229–46.

17. The error rate was 0.74 per cent for a photograph that was difficult to read, 0.40 for a clearer photograph, and an average error of 0.57 per cent.

be 'fairly homogeneous with respect to economic status and living conditions'[18] and are roughly congruent with the neighbourhoods selected for analysis here. Because the sample is a small one, and because census tracts did not correspond exactly with the areas of interest defined for each neighbourhood, the findings are suggestive rather than statistically significant but they do show interesting overall trends in tree coverage and the differentials between neighbourhoods.[19] Finally, the methodology of more traditional social history as well as street-level observations of tree cover are used to bring context to the geospatial analysis in one neighbourhood, Lowertown.

Findings

The most surprising finding thus far is at the macro level. In every neighbourhood sampled the canopy cover has increased recently; in New Edinburgh it is higher today than it has been since the 1920s, and this is probably the highest it has been since the neighbourhood was built in the late nineteenth century (see Table 1). Canopy cover in new suburbs in Alta Vista went from 9.97 per cent in 1946 (when the area was a field) to 31.49 per cent in 2005. Notre Dame, a working class suburb in Gatineau, increased steadily from 6.15 per cent in 1946 to 17.63 per cent in 2005. In Lowertown, a working class francophone neighbourhood in Ottawa, the tree cover had dropped dramatically from a high of 31.78 per cent in 1948 to a low of 9.47 in 1983 but is currently rising and had reached 14.70 per cent in 2005.

The second finding, that canopy cover can be correlated to neighbourhood income, is not surprising, but the analysis of change over time suggests that this is a more complex correlation than anticipated, in part because of what we might call the temporal ecology of trees. The correlation with income is strong.[20] Lowertown's canopy cover was less than half of that of New Edinburgh; the income levels of men were about two-thirds that of men in New Edinburgh in 1951 and 1961 and women's incomes were also significantly lower. A similar correlation exists between the two suburbs: working class Notre Dame had just over half the tree cover of

18. 'Population and Housing Characteristics by Census Tracts,' Bulletin CT-5, Ninth Census of Canada, 1951, (Ottawa: Bureau of Statistics, 1953). Similar bulletins were published for 1956, 1961, 1971 and 1981.

19. Because the areas of interest were defined according to available photographs, they do not map directly to census tracts. The census tracts were also changed over time, and some of the tracts in the 1961 census do not correspond to those in the 1981 census. In the case of New Edinburgh and Notre Dame neighbourhoods, the census tracts correspond fairly closely to the areas of interest defined for mapping. The Notre Dame census tract, however, covers the pulp mill as well as the residential area. The census tract for Alta Vista is much larger than the two areas of interest, but this is a fairly homogeneous neighbourhood. The census tracts for the Lowertown neighbourhoods shift significantly and cover a much larger area of land.

20. Median household income is considered a reliable indicator of localised urban political economy. See Heynen, 'Green Urban Political Ecologies', 512.

Table 1. Canopy Cover in Four Neighbourhoods, 1928–2005 (%)

	New Edinburgh	Lowertown	Notre Dame	Alta Vista
	%	%	%	%
1928	20.51			
1946		31.78	6.15	9.97
1952				10.88
1956	18.75	21.73	8.26	8.05
1966	22.57	10.39		
1971	21.99			
1976		9.77	13.27	
1983		9.47		
1985			14.02	18.10
2005	25.86	14.70	17.63	31.49

middle class Alta Vista in 2005. In the 2001 census, 26.4 per cent of Notre Dame's residents were considered low income, and the canopy cover was 17.23 per cent. By comparison, Alta Vista, with only 2.9 per cent of its population considered low income in 2001, had a tree cover of 31.49 per cent. The difference in canopy cover between these neighbourhoods may also be due to ethnicity. Notre Dame and Lowertown are predominately francophone; Alta Vista is anglophone.[21] The dark leafy shade in Alta Vista may reflect cultural expectations as much as wealth.[22] Fraser and Kenney have reported cultural differences between the English, Italian and Chinese immigrants in Toronto.[23] Their study does not explore socio-economic differences, however, and in Ottawa/Gatineau, as in most places, ethnicity, class and income are difficult to tease apart.

The explanations for the correlation of tree cover with wealth are simple: wealthier individuals have more money and time to spend on landscaping. They are more likely to own their homes and stay in one location long enough to benefit from

21. In Lowertown in 1981 190 residents reported English as their first language and 2165 reported French. By comparison in Alta Vista 3160 spoke English and 365 spoke French.

22. The 1981 Census shows Lowertown (Census Tracts 056, 055) with median income of $22,508 and $25,589 as compared to $56,193 for Alta Vista (Census Tract 006).

23. E.D.G. Fraser, and W.A. Kenney, 'Cultural Background and Landscape History as Factors Affecting Perceptions of the Urban Forest,' *Journal of Arboriculture*, 26:2, 2000, 106–13.

tree planting and, in the suburbs, they are likely to own larger lots. They also have more political power and are more likely to benefit from tree planting initiatives.[24]

Similar observations have been made in larger, statistically significant studies. Escobedo and Nowak found the canopy cover was correlated to socio-economic status in Santiago de Chile: 'Urban forests in the high socio-economic strata had fewer public trees, greater tree cover, tree and leaf area density, and leaf area index than lower socio-economic strata.'[25] In Milwaukee, Heynen correlated canopy cover from aerial photographs taken in 2000 with census data from 223 census tracts. He concluded that 'citywide urban-forest canopy cover across all property types is positively statistically correlated with median household income ... and those tracts inhabited largely by non-Hispanic Whites'.[26] Subsequently in Indianapolis, in a study comparing data from 41 inner city census tracts characterised by low income households in 1962 and 1993, Heynen found that 'median household income was shown to significantly contribute to local metabolization in urban forest canopy cover and residential forest canopy cover. Socio-economic factors helped explain approximately fifty per cent of the canopy cover in the two models.'[27] There were complications in both of these studies, however, and in his most recent study of sixty urban areas in central Indiana income was not correlated with canopy cover, though level of education was.[28]

These were large-scale, ahistorical studies, however, and the small spatial scale and long temporal span of this Canadian project show more complex patterns: socio-economic power is important, but it is influenced by other factors, particularly ecological change over time. Trees grow slowly. Their life span can be much longer than the human life span and a maple tree reaches full canopy cover more slowly than a human reaches full height. Trees also die slowly. Tree ring analysis of bur oaks in Winnipeg, Manitoba, showed that changes in water table levels caused by residential construction in the 1940s took decades to appear as visible

24. For a discussion of the impact of home ownership and (im)mobility, see Harold A. Perkins and N. Heynen, 'Inequitable Access to Urban Reforestation: The Impact of Urban Political Economy on Housing Tenure and Urban Forests', *Cities*, 21:4, 2004, 291–99.

25. F.J. Escobedo, D.J. Nowak, J.E. Wagner, C.L. de la Maza, M. Rodriguez, D.E. Crane, and J. Hernandez, 'The Socioeconomics and Management of Santiago de Chile's Public Urban Forests'. *Urban Forestry and Urban Greening* 3–4, 2006, 105–14.

26. 'Despite a few aberrant tracts, citywide urban-forest canopy cover across all property types is positively statistically correlated with median household income (.285) and those tracts inhabited largely by non-Hispanic Whites (.138) at $\alpha = .01$. Furthermore total urban-forest canopy was negatively statistically correlated with housing vacancy (-.149) and African-American residents (-.059). N. Heynen, H.A. Perkins and P. Roy, 'The Political Ecology of Uneven Urban Green Space: The Impact of Political Economy on Race and Ethnicity in Producing Environmental Inequality in Milwaukee', *Urban Affairs Review*, 42/1, 2006, 3–25.

27. Heynen, 'Green Urban Political Ecologies', 513.

28. Heynen, 'Correlates of Urban Forest Canopy Cover', 43.

decline. Drought has been shown to have a similarly delayed impact on Norway maples.[29] This can cause some asynchrony. Tree planting or tree protection initiatives by wealthy homeowners produce results decades later, and often long after the individual owner, and even that cohort of homeowners, has moved on, or after the status or function of the neighbourhood has changed. This is particularly evident in inner city neighbourhoods where large trees persist, providing a full mature canopy cover decades after the population that fostered them has moved on. Similarly, a new suburb or a newly gentrified inner city neighbourhood will not immediately show related increases in canopy cover.[30]

For example, in 1965 the canopy cover in wealthy Alta Vista compared poorly to that of the working class neighbourhoods of Lowertown and Notre Dame. Alta Vista's canopy cover in that year was 6.05 per cent.[31] In Lowertown and Notre Dame it was higher: 13.07 per cent in Lowertown and about ten per cent in Notre Dame.[32] But this discrepancy was the product of the slow growth of trees. Alta Vista had been a field until the mid-1950s when the suburbs were built: the canopy cover of 9.97 per cent in 1946, which grew to 10.88 in 1950, was largely made up of hedgerows. When the hedgerows were removed for construction in 1965, the canopy cover dropped to 6.05 per cent but it then grew rapidly to match income levels, reaching 18.10 per cent in 1985 and 31.49 per cent twenty years

29. H.A. Catton, S. St George and W.R. Humphrey, 'An Evaluation of Bur Oak (*Quercus macrocarpa*) Decline in the Urban Forest of Winnipeg, Manitoba, Canada', *Arboriculture and Urban Forestry*, 33:1, 2007, 22–30. For the delayed impact of drought on Norway maples, see J.D. Apple and P.D. Manion, 'Increment Core Analysis of Declining Norway Maples, *Acer platanoides*,' *Urban Ecology*, 3–4, 1986, 309–21. 'Growth over the past 20 years indicates that the most severely declining group of trees exhibited reduced growth as much as 20 years ago, suggesting that serious symptoms of deterioration in the crown may not occur for many years after the onset of decline. We do not have the necessary records to determine what events preceded this change in growth but we can predict that it was similar to the drought and sidewalk renovation combination of the past 20 years. Additional conclusions from this type of study demonstrate that decline is a longterm process and further that growth reduction may precede external symptoms of decline.'(320).

30. In his Indianapolis study, Heynen makes a similar observation explaining why the percentage household income change was negative. He argued that the landscaping associated with gentrification could initially result in a loss of trees, although the long term impact would be an increase. Heynen, 'Green Urban Political Ecologies', 511.

31. Unfortunately census data for Notre Dame is not readily available for 1961. In the 1985 Canadian census, however, the percentage of the population calculated to be living on a low income in Notre Dame was 49.4%, compared to 16.69% in Alta Vista.

32. In Alta Vista the male income was $5,678 in 1961 compared to $3,219 for men in Lowertown. Women's incomes are closer, probably because many middle class women stayed home with families in the suburbs in the 1960s. The figure for Notre Dame's tree cover is a rough average between the available figures of 8.26 per cent in 1956 and 13.27 per cent in 1976.

later in 2005.[33] In Notre Dame, the canopy cover was on a slower trajectory, rising from 6.15 per cent to 8.26 in 1956; 13.27 in 1976; 14.02 in 1985; and 17.63 in 2005. In Lowertown, canopy cover hovered in the low teens through the end of the twentieth century: 8.85 in 1956; 13.07 in 1966; 12.16 in 1976; 12.16 in 1976; 11.81 in 1983; and 17.97 in 2005 (see Table 1).

Lowertown: Patterns of Canopy Cover

A close examination of historical trends in canopy cover in one neighbourhood, Lowertown, shows further variables that can be missed by large-scale geospatial analysis. Lowertown is one of the oldest parts of Ottawa. The site was originally a cedar swamp, and is lower than the rest of the city, as the name suggests. It is bounded by the more prestigious and topographically higher neighbourhoods of Sandy Hill, Parliament Hill and New Edinburgh.[34] Statistics on population, housing and families collected and published by the Canadian census show that in the post-war era, it was a cohesive, francophone, working class neighbourhood that was slowly losing population.[35] Statistics for the eastern half of Lowertown, which includes the canopy study area, show that the population halved from 1941 to 1971, dropping from 10,061 to 4,955. In 1941, 72 per cent identified themselves as French, almost 11 per cent identified themselves as British and there was a small Jewish population of almost five per cent. As for tenure, only 18 per cent were home-owners but, of the 82 per cent who rented their accommodation, two-thirds had been tenants in the area for at least five years.[36] It was not a transient population.

Incomes were low compared to the rest of city of Ottawa, and also compared to the entire metropolitan region, including Gatineau: the median male earning was $1,962, well below (83 per cent) that of the city of Ottawa ($2,368) and 85 per cent of that of the metropolitan region. At $1,305, median earnings for women were respectively 84 per cent and ninety per cent of the city and metropolitan

33. See D.M. Sharpe, F. Stearns, L.A. Leitner and J.R. Dorney, 'Fate of Natural Vegetation during Urban Development of Rural Landscapes in Southeastern Wisconsin', *Urban Ecology*, 9, 1986, 267–87.

34. See Michael Newton, *Lowertown Ottawa*, Volume 1, and *Lowertown Ottawa, 1854–1900*, Volume 2, Ottawa: National Capital Commission, 1981.

35. The statistics are from the series of bulletins produced by Census of Canada. Census tract 18, (numbered 056 in 1971) the eastern half of Lowertown, includes the canopy study area. 'Population and housing characteristics by census tracts,' Bulletin CT-5, Ninth Census of Canada, 1951. (Ottawa: Dominion Bureau of Statistics, 1953); 1956 Census of Canada, 'Population and housing characteristics by census tracts' Bulletin 4–6, (Ottawa: Dominion Bureau of Statistics, 1957); 1961 Census of Canada, 'Population and housing characteristics by census tracts' Bulletin CT-13, (Ottawa: Dominion Bureau of Statistics, 1963); 1971 Census of Canada, 'Population and housing characteristics by census tracts', (Ottawa: Dominion Bureau of Statistics, 1973).

36. In the entire metropolitan area, which included Hull, 29,895 were owners and 36,370 tenants.

averages.[37] Only a small number of the men were in managerial or professional positions, and most of the women worked in clerical or service positions. In 1961 the population had declined but the ethnic mix remained similar, with 79 per cent of French origin, 13 per cent of British origin and about 1.5 per cent Italian – Jewish residents were no longer identified as such. Tenants (76 per cent) continued to dominate and, though wages and salaries had risen in absolute terms, the median for men of $3,219 meant it had slipped to only 69 per cent of the Ottawa city median and 73 per cent of the metropolitan area median; for women the median wage of $2,147 had slipped less, to 83 per cent of city levels and 88 per cent of the metropolitan mean wage.

Tree cover in the study area through this period was also slipping: from 31.78 per cent in 1946, to 21.73 in 1956, 10.39 in 1966, 9.77 in 1976 and 9.57 in 1983. It only recovered when the neighbourhood did, rising to 14.70 per cent in 2005. New Edinburgh, the adjacent residential neighbourhood, offers a comparison. New Edinburgh had a canopy cover that hovered from 18–23 per cent in the twentieth century and then rose to 25.86 in 2005. The median earnings in New Edinburgh were much higher than in Lowertown: $2,831 for men in 1951 and $1,561 for women – that is, about 45 per cent and twenty per cent higher respectively than in Lowertown. British origin residents out-numbered those of French origin by more than 2:1.[38]

If we place the Lowertown statistics into historical context, however, a more complex story emerges. The neighbourhood was fractured in the 1960s and 1970s when it became the site of one of Canada's most expensive urban renewal projects. The census reports referred to above played a role in the process by gathering statistics on 'substandard' houses. The 1961 census showed that most of the houses (1,444) were built before 1920, with only 187 built after 1945; a quarter (323) of the 1203 dwellings were identified as 'crowded dwellings'.[39] A series of community studies and reports provided further ammunition for urban planners. One consultant observed in 1963: 'Generally all of the buildings in the area are crowded together, with little or no front yards, and narrow side yards. Many of the rear yards are completely filled with accessory buildings, most of which are in poor condition. Open space is virtually absent through the study area.'[40]

---

37. There were 56 Italians and insignificant numbers from other ethnic backgrounds. Most men worked in clerical (445), manufacturing and mechanical (409), service (305 and 153) or as labourers (272).

38. There were 318 Germans, 63 Jews, a smattering of other European ethnicities and 4 Asians.

39. A crowded dwelling was defined as a dwelling in which the number of people exceeds the number of rooms occupied.

40. *Urban Renewal Ottawa*, City of Ottawa, 1967, 108. A map of land use in Lowertown shows that the four blocks of the canopy study area were almost entirely residential with small sections of commercial property, probably corner stores, and two back sheds zoned industrial.

The reports identified the entire eastern half of Lowertown as an area in urgent need of renewal. Residents were evicted, entire blocks of housing were razed, and modern townhouses were slowly built in their place. The social impact was devastating. Between 1961 and 1971 the number of French residents dropped from 79 to 71 per cent[41] and income dropped further relative to the rest of the city: the average income was 19 per cent below the city average in 1961 and 35 per cent below in 1971.[42] Statistics, however, fail to convey the collapse of community and the sense of betrayal of residents.

The area selected for canopy study narrowly escaped renewal. It had been identified in 1967 for 'active assistance,' as it had 'a high degree of blight and deterioration' but it was just outside the initial renewal area and the municipal zeal for renewal collapsed under community protests before it was altered.[43] It was, however, adjacent to two areas that were bulldozed: eight blocks of housing to the north were demolished for access roads to a new inter-provincial bridge and multi-block sections of housing directly to the east were razed for public housing.[44] In addition, the avenue along the east side of the study area, King Edward Avenue, became the route to the new bridge. But the original houses remained and urban reformers began to value the historical character of inner city neighbourhoods. More attention was paid to heritage and a new emphasis was placed on streetscaping, including street tree planting.[45] The area is currently undergoing some gentrification.

The simple story of decline and slight recovery is complicated further if we look at patterns of tree cover within the study area. To begin with, most of the trees in the study area were on one federally landscaped avenue, King Edward Avenue, and the extent of canopy cover along King Edward skews the statistics. If we remove King Edward Avenue from the mapping, then the initial tree cover in this part of Lowertown is less dense (22.25 per cent in 1946) and the decline in canopy cover happens much earlier, in 1956, when it drops suddenly to 8.85 per cent. The tree canopy cover then hovers around 11–13 per cent for about twenty years until 2005 when it recovers to 17.97 per cent (Table 2).

---

41. *Lowertown Existing Conditions*, City of Ottawa, 1974.

42. Murray and Murray, *Lowertown East: Redevelopment Plan Preliminary Report*, City of Ottawa, January 1976, 3. See also Murray and Murray, *Development Plan – Lowertown West*, Department of Community Development, City of Ottawa, 1976; Murray and Murray, *Lowertown – King Edward, Alternatives to a Freeway*, City of Ottawa, 1974, and *Proposals for Urban Renewal Action, Part III, 'Lowertown East'*, Planning Branch, Department of Planning and Works, City of Ottawa, 1963.

43. *Urban Renewal Ottawa*, p. 123.

44. *Lowertown East: Redevelopment Plan Preliminary Report.*

45. *Lowertown East: Redevelopment Plan Preliminary Report*, 29.

Table 2. Tree canopy in Lowertown 1946–2005

	King Edward Avenue included	King Edward Avenue excluded
1946	31.78	22.25
1956	21.73	8.85
1966	10.39	13.07
1976	9.77	12.16
1983	9.47	11.81
2005	14.70	17.97

The story of the King Edward Avenue trees introduces a different set of politics. King Edward Avenue was much wider than most of the streets in the city, because it was built over the by-wash from the Rideau Canal. In the early twentieth century, the avenue was selected by the Ottawa Improvement Commission to provide a ceremonial route between the residence of the Governor-General and the Parliament buildings. The commission placed a grassy boulevard down the middle and planted it with a double row of American elm trees, creating a cathedral-like streetscape. Adapting a landscape plan by Frederick Todd, a protégé of Frederick Law Olmsted, they connected King Edward Avenue to a city-wide network of leafy avenues.[46] Most of the avenues ran through wealthy neighbourhoods but King Edward Avenue was an anomaly in working class Lowertown and was highly valued by residents, especially after 1956 when it offered respite for residents from the largely treeless streets of the neighbourhood.[47]

The avenue was so thickly planted with elm trees that the Commission's arborist recommended thinning the trees in 1946 and considered pre-emptively removing the elms in anticipation of Dutch elm disease in 1948. When the trees were finally removed in 1965, the impact on the Lowertown neighbourhood was dramatic: the canopy cover for the study area was halved from 21.73 to 10.39 per cent. The trees were never replaced. A new inter-provincial bridge was built in the early 1960s and when the proposed freeway leading up to it was defeated by neighbourhood protest in the early 1970s, King Edward Avenue, became, by default, the access route to the bridge. The central boulevard of trees was replaced by new traffic lanes and the avenue has become a barren and hotly disputed inter-provincial truck route.

46. See J. Dean, 'Ottawa's Central Park: Esthetic Forestry vs Ornamental Gardens,' in *News of Forest History*, III, 36–37 – Part 1 (2005), Proceedings of the International IUFRO Conference 'Woodlands–Cultural Heritage' 3–5 May 2004, Vienna, Austria, 21–30.

47. See, for example, D.L. Morny, *Farewell My Bluebell: A Vignette of Lowertown*, Ottawa, 1998, which opens with a description of King Edward Avenue. Heynen has also noted the impact of transportation corridors in urban tree canopy in Indianapolis. Heynen, 'Green Urban Political Ecologies', 511.

Figure 2. Aerial Photograph of Lowertown, Ottawa, 1983.[48]

By 1983 the canopy cover of the selected study area of Lowertown had dropped to 9.47%, from a high of nearly 32% in 1931. (The study area is to the west [right] of the wide central avenue, King Edward Avenue.) Although Dutch elm disease contributed, the loss was largely due to the politics of urban renewal and highway and bridge construction. The canopy cover in the adjacent neighbourhood of New Edinburgh remained steady near 20% through this period.

The comparison with a second elm-lined avenue in Ottawa is instructive. Clemow Avenue was part of the ceremonial network of avenues and was also planted thickly with American elms. Clemow Avenue, however, is in a wealthy neighbourhood. When the trees succumbed to Dutch elm disease, the avenue was replanted with street trees and through traffic was blocked off. The 2005 aerial photographs show a substantial canopy. The initial loss of trees on King Edward and Clemow was due to an ecological change, Dutch elm disease, but the long-term loss was the product of political forces: the power of one neighbourhood to limit traffic and the lack of political power in another neighbourhood to resist the imposition of a highway.

A second pattern complicating the statistical drop in canopy cover in Lowertown is the shift from public to private trees. Canopy cover in the residential blocks did not simply decline after the 1940s; the trees shifted from public to private property. The maps show that the dramatic decline in canopy cover in 1956 was in part due to the loss of large street trees. They were replaced by a greater proportion of small trees and by trees growing on private property, usually to the rear of buildings. It was not until 2005 that there was a significant return to street tree planting. Street trees were traditionally planted on city property and maintained, if not planted, by the city; yard trees are the responsibility of the homeowner. The burden of tree ownership was shifting to the private property owner.

48. Source: Roll No.: A31335 Photo No. 100. Photograph Reproduced with the permission of Natural Resources Canada 2010, courtesy of the National Air Photo Library.

Street-level observation of the existing tree cover suggests further disparities. Not only was there a shift from street tree to yard tree, but there was also a shift from ornamental tree to weed tree. It appears that many of the original street trees were silver maples. The only remaining old street trees in the study area are two massive silver maples and a third silver maple was recently removed from the area. These trees tower over the houses; their trunks crowd the small front yards. Ecologically they are well suited to the area – silver maples are also known as swamp maples, and they would have thrived in a low wet part of town. Silver maples were popular shade trees at the turn of the century because they grew very quickly into beautiful large trees, with significant capacity for cooling. Donald Peattie's *A Natural History of Trees of Eastern and Central North America* (1948) is a North American classic and his description of the silver maple is worth quoting in full:

> A tree with so many charms has naturally been planted far beyond its natural range, and everywhere within it … it is a favorite, for it cannot grow without lending grace to any spot, it makes a railroad station look like a home, and adds a century to the appearance of a village street. It is the fastest growing of all of our Maples, one of the fastest amongst all the trees suitable to our climate, be they native or exotic. It is as charming in its childhood as in age, and in its youth goes through no awkward stage.[49]

As Peattie notes, however, the silver maple had fallen from grace by 1948. The quick growth and massive size that were its virtues in the early years of street tree planting had come to be seen as flaws. 'Landscape architects have little good to say of it: they complain of the insect pests that attack it and of its comparatively short life, as well as the breakage of its brittle and too-long boughs under wind and ice damage. They urge that it be planted, if at all, in the full knowledge that its quickly achieved effects will not last long and that more permanent, if slower, plantings be started at the same time.' Peattie did not share in the condemnation: 'It may be that we should always listen to cautious and sensible people, and not allow ourselves to think too highly of a tree that will perhaps only live three times as long as we do.' But silver maples continued to be disparaged. The Lowertown giants are, by the standards of modern urban forestry, much too large for the small urban yards, even though they provide substantial environmental, not to mention aesthetic, benefits. It seems likely that Lowertown's street trees were targets of a tree-trimming programme embarked upon by the city of Ottawa in the 1940s. Although city records do not give the location of trimmed trees, the timing of the collapse in tree cover in Lowertown coincides with the programme that removed thousands of trees from downtown Ottawa.

49. D.C. Peattie, *A Natural History of Trees of Eastern and Central North America*, Boston 1948 (1999), 464–65.

The yard trees that replaced the silver maple in statistics of canopy cover were, to a large extent, the Manitoba maple, or box elder (*Acer negundo*). They probably sprang up between sheds in the neglected yards described by an urban renewal consultant in 1973. In 2009 the yards in the area were dominated by these weed trees, which have sprouted from fence lines and beside buildings.[50] These trees also thrive in a swampy environment. To quote Peattie again: 'Box Elders are the Maples' poor relations. They have little of the strength and grace, little of the autumn splendor and beautifully figured wood of their noble kin. Somewhat sprawling in growth, short lived (the top of the tree often begins to die just when full and promising stature is obtained) … No one could call them aristocratic trees.'[51]

Lowertown's Manitoba maples, it appears, matched the working class status of the neighbourhood. Did this matter? The Manitoba maple does provide many of the same environmental benefits as the silver maple. The large canopy of leaves probably captures equivalent levels of pollutants and provides a similar amount of shade. But these benefits come at a cost. As early as 1956, the Canadian government publication, *Native Trees of Canada*, identified the Manitoba maple as a problem tree: 'prolific seed production, breakage in windy or icy conditions, and shoot or sucker growth are problems in cultured landscapes.'[52] The Manitoba maple has recently been identified by the City of Toronto as an invasive species.[53] Their demand for water can also be problematic: Manitoba maples have been identified by the City of Ottawa as a cause of foundation damage, although this is a characteristic they share with the Silver Maple.[54]

The Manitoba maple in the yard, then, was a burden for the homeowner, not only by virtue of its location, but also by virtue of its habits. Heynen has similarly observed that the canopy cover in African American neighbourhoods of Milwaukee was mainly made up of weed trees and argues that, in this case, the canopy cover was a mixed blessing; weed trees posed a hazard as well as an environmental benefit to the residents.[55]

---

50. The Manitoba maple is fast growing for the first 15–20 years and typically reaches 60 but rarely 100 years, making it likely that these trees are the ones seen in the photographs of the 1950s and 1960s.

51. Peattie, *A Natural History of Trees*, 473.

52. *Native Trees of Canada*, Department of Northern Affairs and National Resources, Ottawa, Fifth Edition, 1956, 260.

53. See 'Controlling Invasive Plants', Urban Forestry Services, City of Toronto, at http://www.toronto.ca/trees/pdfs/Fact_3_Controlling_Invasive_Plants.pdf Accessed August 29th, 2010.

54. 'Trees and Foundation Damage', City of Ottawa Information Bulletin, available at http://www.ottawaforests.ca/city_trees/foundation-damage_e.htm Accessed August 29th 2010.

55. In a Milwaukee study of 223 census tracts, Heynen found that African American neighbourhoods had a relatively high canopy cover, especially as compared to the Hispanic population. His dot grid analysis did not provide the detail necessary to explain this apparent discrepancy, and he turned to interviews with arborists and politicians. Interviews revealed that the trees in African

The return to civic street tree planting visible in the 2005 canopy cover map is not a return to the status quo ante, for the massive silver maples have been replaced mainly by smaller trees. Although the map of canopy cover shows extensive planting of street trees by 2005, observation shows that many of these are small trees, such as Japanese lilac and choke cherry. These trees are preferred by the city for small lots. The latest generation of trees in the neighbourhood will not burden the city, nor will it provide extensive environmental services.

So the statistical drop in canopy cover, from 22.25 per cent to numbers hovering ten per cent or so lower underestimates the true environmental loss to the residents of the early canopy cover. It does not consider the increased burden placed on homeowners by the shift to yard trees and the shift from ornamental street tree, maintained by the city, to the weed tree in the yard, maintained, if at all, by the homeowner. This neighbourhood suffered a loss in environmental amenities, a loss that is statistically larger than that of neighbouring residential areas and more significant than the statistics suggest.

Conclusion

The geospatial analysis of urban forest tree cover pioneered by David Nowak has resulted in a rethinking of the urban forest. Because the canopy cover can be measured and its benefits assessed, the environmental services of canopy cover are now widely acknowledged. This study has argued, following Nik Heynen, that the distribution of these services is inequitable. Comparison of canopy cover in areas of four selected neighbourhoods has shown that the distribution varied by income level. This pilot project has departed from earlier work by using a methodology that permits analysis over a longer time period and also facilitates the analysis of spatial patterns of canopy cover at a neighbourhood level. It argues that a long temporal frame and small spatial scale are necessary to understand shifting inequities in the urban forest canopy cover.

A long temporal frame is necessary because trees grow slowly; they take many years to reach maturity and even environmentally stressed urban trees can persist for many decades. The methodology used in this study allowed the capture of data from the uneven patchwork of aerial photographs available from 1928, in order to track canopy cover changes in small areas over sixty or seventy years. The data shows the long trajectory of change: the slow decline and recent recovery of the urban forest in the inner city and the gradual growth in suburban areas. The

---

American neighbourhoods were often 'weed' or 'volunteer' trees: trees that grew up along fences and foundations, and existed only because of poor maintenance by landlords, and only until they posed enough of a hazard to be removed. In contrast fence lines and foundations were kept clear of saplings in Hispanic neighbourhoods. Heynen, Perkins and Roy, 'The Political Ecology of Uneven Urban Green Space'.

small spatial scale is necessary because it points to environmental inequities that are missed in larger statistical studies. In working class Lowertown, canopy cover dropped from 32 per cent in 1946 to 22 per cent in 1956 and ten per cent in 1966. The mapping shows further inequities internal to the area: most of the trees were along one federal avenue, and even this avenue was eventually sacrificed for interprovincial traffic. The mapping also shows that the burden of trees also shifted over time to the homeowner, as large silver maples planted along the streets and maintained by the city were replaced by weedy yard trees, maintained, if at all, by the homeowner. The gradual decline of canopy cover traced out in the statistics is revealed in close analysis to be a much more uneven collapse, produced by the interplay of a number of ecological and political forces.

This study was a pilot project and much work remains to be done, both in terms of mapping additional areas and further investigating the areas described here. It would be useful to correlate canopy cover with other environmental indicators. It is apparent from the initial findings here that the lack of forest canopy cover is linked to and exacerbated by other environmental inequities. The role of the car is clearly central: canopy cover is closely related to the imposition of traffic, with the ensuing deterioration in air quality. But there are other elements worth tracking. Notre Dame is subjected to the pollution of the adjacent pulp and paper mill; Alta Vista is a residential neighbourhood with little or no industry. Finally, the review of the politics underlying the statistics suggests that we might reconsider the role of aesthetics: the beauty of urban trees might be considered an environmental amenity as important as their role in the reduction of particulate matter. For the elms that once grew along King Edward Avenue have not entirely disappeared. They live on in community memory and in community politics. Elegaic photographs of the King Edward Avenue elms figure prominently on the website of a community action group, the King Edward Task Force, where they act as potent reminders of environmental injustice.[56]

Appendix: Methodology

Photos were identified at the Canadian National Air Photo Library. They were selected for date (we used 'leaf-on photos'), resolution, clarity, camera angle and coverage of the neighbourhood. The area to be studied had to be near the centre of the photo because of relief displacement.

Contact prints were made from negatives, and then scanned at 1200 dpi. Photographs were geo-referenced in GCP Works (PCI Geomatica version 9.1.7) to a previously geo-referenced 2005 digital aerial photograph set in a standard MTM Zone 9 NAD83 projection. For each newly geo-referenced image the coordinate

56. See King Edward task force web site, http://www.kingedwardavenue.com Retrieved 10 June 2010.

system was set to MTM Zone 9 NAD 83 (i.e. Other/CanQC09 in PCI Geomatica. (Modified Transverse Mercator projection.)

Each image was re-sampled to have the same arbitrary pixel size of 0.14m x 0.14m, which was the highest resolution after scanning. Each image had a different pixel size after scanning because their acquisition scales were all different.

A vector layer was created outlining the area of interest to be used for the analysis, which outlined the maximum area covered by ALL images from the different dates. This area (defined by a certain number of pixels) was then used as the overall size of the area of interest.

Each image was clipped to the minimum bounding rectangle containing the outlined area of interest created in previous step.

The trees in each image were manually digitised in order to create bitmap masks for each image. Each year's photo was delineated using a different colour for this presentation.

The number of pixels classified as trees in each image was divided by the total number of pixels present in the area of interest in order to provide a per cent canopy cover for each image.

Testing for error: for our first pilot project we tested for error by classifiying images from two dates (one very clear one and another very difficult to interpret) three times each in order to provide an idea of the error involved in the manual interpretation. Each time, the classification was done independently, in order to avoid biases by the interpreter from previous classifications. The variation of these three attempts provided an idea of the error involved in the cover estimate.

Select Bibliography

Apple, J.D. and P.D. Manion, 'Increment Core Analysis of Declining Norway Maples, Acer platanoides', *Urban Ecology*, 3:4, 1986, 309–21.

Campanella, T.J. *Republic of Shade: New England and the American Elm*, New Haven, 2003.

Catton, H.A., S. St George and W.R. Humphrey, 'An Evaluation of Bur Oak (Quercus macrocarpa). Decline in the Urban Forest of Winnipeg, Manitoba, Canada', *Arboriculture and Urban Forestry* 33:1, 2007, 22–30.

Dean, J. '"Said Tree is a Veritable Nuisance": Ottawa's Street Trees, 1869–1939', *Urban History Review: Revue d'histoire urbaine*, XXXIV:1, 2005, 46–57.

Escobedo, F. and D. Nowak, 'Spatial Heterogeneity and Air Pollution Removal by an Urban Forest', *Landscape and Urban Planning*, 90, 2009, 109.

Fraser, E.D.G. and W.A. Kenney, 'Cultural Background and Landscape History as Factors Affecting Perceptions of the Urban Forest', *Journal of Arboriculture*, 26:2, 2006, 106–13.

Heynen, N. 'The Scalar Production of Injustice within the Urban Forest', *Antipode: A Journal of Radical Geography*, 35:5, 2003, 980–98.

Heynen N. and H. Perkins, 'Scalar Dialectics in Green: Urban Private Property and the Contradictions of the Neoliberalization of Nature', *Capitalism, Nature, Socialism* 16:1, 2005, 99–113.

Heynen, N. 'Green Urban Political Ecologies: Toward a Better Understanding of Inner City Environmental Change', *Environment and Planning, A*, 38:3, 2006, 499–516.

Heynen, N., M. Kaika and E. Swyngedouw, 'Urban Political Ecology: Politicising the Production of Urban Natures.' in N. Heynen, M. Kaika and E. Swyngedouw, eds. In the *Nature of Cities: Urban Political Ecology and the Politics of Urban Metabolism*, Abingdon, 2006.

Lawrence, H.W. City *Trees: A Historical Geography from the Renaissance through the Nineteenth Century*, Charlottesville, 2006.

Nowak, D.J. and D.E. Crane, 'The Urban Forest Effects (UFORE) Model: Quantifying Urban Forest Structure and Functions', in M. Hansen and T. Burk, eds. *Integrated Tools for Natural Resource Inventories in the 21st Century*, St. Paul, 2000.

Page, M. *The Creative Destruction of Manhattan, 1900–1914*, Chicago, 1999.

Peattie, D.C. *A Natural History of Trees of Eastern and Central North America*, Boston, 1948 (1999).

Suzuki, D. and W. Grady, *The Tree: A Life Story*, Vancouver, 2004.

Swyngedouw, E. and N. Heynen, 'Urban Political Ecology, Justice and the Politics of Scale', *Antipode*, 35, 2003, 902.

Talarchuk, G.M. 'The Urban Forest of New Orleans: An Exploratory Analysis of Relationships', *Urban Geography*, 18:6, 1997, 65–86.

Ulrich, R.S. 'View through a Window May Influence Recovery from Surgery', *Science* 224, 1984, 420–21.

Walton, J.T., D.J. Nowak and E. J. Greenfield, 'Assessing Urban Forest Canopy Cover', *Arboriculture and Urban Forestry*, 34:6, 2008, 334–40.

Part II

Managing Risks

Chapter Four

Threatened by the Sea, Condemned by Man? Flood Risk and Environmental Inequalities along the North Sea Coast, 1200–1800

Tim Soens

A high flood risk is inherent to living in a coastal wetland area. From time to time people are brutally reminded of this when hit by a large-scale flooding disaster, like the Christmas 2004 Indian Ocean Tsunami or the August 2005 Hurricane Katrina in the Gulf of Mexico. Less natural however is the fact that poor people and ethnic, cultural or social minority groups are usually hit disproportionately high by these 'natural' catastrophes. At first sight, this might seem typical of flood disasters in modern societies, where most coastal wetlands are heavily populated, polluted and over-exploited.[1]

However coastal wetlands in a more distant past, long before the industrial revolution, often faced similar problems. A brief example from some seven centuries ago might illustrate this. In 1288 a catastrophic storm surge hit the southern North Sea area and inundated large parts of the coastal wetlands of southern England and the Low Countries.[2] At that time both coastal regions were densely populated after two centuries of agricultural intensification and urban industrial and commercial development. Near the Flemish village of Ossenisse, on the southern shore of the western Scheldt estuary, the damage was huge. As the dike breaches could not be restored in the immediate aftermath of the disaster, the Cistercian Abbey of Our Lady of the Dunes, fearing a further extension of the inundations into its substantial estates situated inland from Ossenisse, asked permission from its overlord, Guy of Dampierre, count of Flanders, to strengthen the embankment in its own right. In order to finance these works the abbey levied a high tax on all landowners in the area. Most of these landowners were small peasants who had already suffered heavy damage due to the inundation itself, and who probably lacked sufficient cash to pay for the re-embankment, especially as both the first campaign in the summer of 1288 and a second one a year later ended in failure. In the summer of 1292 the

1. See the chapter by C. Colten on Hurricane Katrina in this volume.
2. M. Bailey, 'Per impetum maris: Natural Disaster and Economic Decline in Eastern England, 1275–1350' in B.M.S. Campbell, ed. *Before the Black Death*, Manchester, 1991, 184–208.

re-embankment of Ossenisse finally succeeded, and the abbey was handed 499.1 hectares of land abandoned by 135 former owners, 96 of whom were small peasant landowners owning less than three hectares of land. As a result, the impact of the 1288 flood disaster was clearly totally different for the peasant smallholders living in the area and for the monks of the Cistercian Abbey, both in terms of physical and material losses. Not surprisingly, a group of peasants, at least some of them former landowners in the area, ravaged the abbey's possessions in Ossenisse, stealing horses and cows and damaging crops and farm buildings.[3]

In Ossenisse the social distribution of the flood risk was highly unequal, and many similar examples in the far or more recent past can be found.[4] However, the inequality was not always as high as was the case in Ossenisse in 1288, or in New Orleans in the aftermath of hurricane Katrina. Due to economic cycles, social policies and changes in mentality, environmental inequality, defined here as an unequal social distribution of environmental risks and profits, may increase or decrease significantly. Here a long run historical analysis can help us to understand which elements may be decisive in improving or reducing environmental inequalities. It is argued here that the evolving relationship between towns and their rural hinterland was one such element. Especially in pre-modern societies, towns existed in a symbiotic relationship with their surrounding countryside, on which they depended for their physical survival through direct food supplies and for the survival of their regional market functions. As a consequence, towns were directly affected when an environmental disaster such as storm flooding struck their hinterland. But perhaps even more importantly, many pre-modern cities built up a considerable degree of political and economic control over their surrounding countryside that fostered inequalities in the distribution of environmental costs and benefits between town and countryside.[5] This was increasingly the case in those regions of Europe that made an early transition towards capitalistic, and hence more market-oriented and town-dependent, rural economies.

In order to test this hypothesis, this chapter discusses long-term changes to the social distribution of flood risk in three coastal wetlands in different parts of the North Sea area: the Flemish Polder area, Romney Marsh in southern England and the Central Holland Peatlands. All three are particularly well documented by recent historical and interdisciplinary research, but had different degrees of urbanisation and commercialisation. The main analytical tool to explain changes

---

3. T. Soens, *De spade in de dijk? Waterbeheer en rurale samenleving in de Vlaamse kustvlakte (1280-1580)*, Ghent, 2009, 223–25.

4. See G.H. Endfield, I.F. Tejedo and S.L. O'Hara, 'Conflict and Cooperation: Water, Floods and Social Response in Colonial Guanajuato, Mexico', *Environmental History*, 9:2, 2004, 221–47; E. Guidoboni, 'Human Factors, Extreme Events and Floods in the Lower Po Plain (Northern Italy) in the 16th Century', *Environment and History*, 4:3, 1998, 279–308.

5. S.R. Epstein, ed. *Town and Country in Europe, 1300–1800*, Cambridge, 2001.

in environmental equity will be the social distribution of land rights between rural dwellers and townsmen and within the rural society itself. In pre-modern societies the social distribution of landed property and landholding strongly influenced the distribution of both environmental benefits (who profited most from the opportunities offered by a certain natural environment?) and environmental risks (who suffered most from the risks inherent in that same natural environment?) Before the fourteenth century, environmental risks and benefits in the three regions were mostly shared by the people who actually lived and worked in the area. As argued here, subsequent changes in access to land, including large-scale enclosures, often introduced a social separation of the environmental risks and benefits.

The expansion of absentee, mainly bourgeois, landownership in many coastal regions concentrated environmental profits in the hands of people who did not live in the area themselves and were relatively safe from either direct physical loss of life or substantial financial losses. This separation and social polarisation of environmental risks and benefits often coincided with decades or even centuries

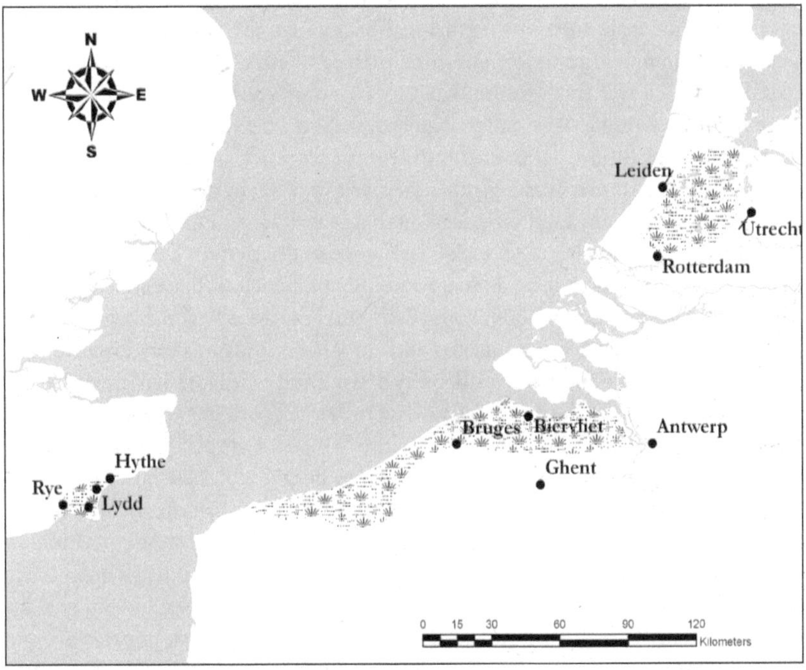

Figure 1. Map of the Study Area

Coastal Wetlands and Important Cities mentioned in the Text (Romney Marsh (UK), Flemish Coastal Plain (France-Belgium-The Netherlands) and Central Holland Peatlands (The Netherlands)

of environmental disequilibrium, which only came to an end when the interests of those who benefited most from the coastal wetland and those who supported most of the risks once again matched.

Three Coastal Wetlands: Flanders (Belgium), Romney Marsh (United Kingdom) and the Central Holland Peatlands (The Netherlands)

The medieval coastal marshlands of the North Sea area were characterised by the extreme diversity and utility of their natural resources: salt-marshes provided ideal pasture for sheep and cattle. When protected from flooding by embankment, they could be turned into high-value arable or grazing lands ideal for stock-breeding. The omnipresent peat-bogs too could be used for pasture or arable, besides hosting valuable plants and medicinal herbs. Increasingly however, they were valued as sources of fuel in a region where wood was becoming scarce. The dune barriers that protected most of the North Sea area's coastal wetlands from flooding provided the ideal starting point for fishing activities, besides offering brushwood and a variety of grasses, once again suitable for grazing. The opportunity to make a living out of a combination of all these activities fitted perfectly with a non-specialised peasant economy, which was further stimulated by two characteristics of wetland society from the High Middle Ages: a high degree of personal freedom, largely untouched by the feudal revolution of the tenth and eleventh centuries, and the increasing commercial demand for marshland products, especially wool, by the developing towns.

Of course many local variations on this general scheme existed. In the Flemish polder area, stretching from Calais in France to the western Scheldt estuary in Belgium and The Netherlands, drainage and embankment took place at an early date, starting well before the year 1000. In the Early and High Middle Ages sheep-farming on both the unembanked salt marshes and embanked lands was very important in the coastal plain. The abundant availability of wool and other agrarian surpluses was an important element in the rise of a dense urban network in the Flemish coastal plain, with the city of Bruges developing into Western Europe's premier centre of international trade from the twelfth century. The burghers of these booming towns increasingly joined the rural population in draining and embanking tidal mudflats along the estuaries of the Ijzer, the Scheldt and the Zwin. The strong position of the Count of Flanders in this area blocked the rise of independent feudal lords and guaranteed in a certain sense the high degree of personal freedom enjoyed by its population. From the twelfth to the fourteenth century, this population was predominantly one of peasant smallholders, combining arable farming and cattle breeding with important agrarian and non-agrarian seasonal activities including fishing, salt-refining, peat digging and textile production.[6]

6. In general, see Soens, *De spade in de dijk*; A. Verhulst, *Landbouw en Landschap in Middeleeuws Vlaanderen*, Brussels, 1995; E. Thoen, '"Social Agrosystems" as an Economic Concept to Explain

In the second research area, the c. 7,000 hectare Romney Marsh in southern Kent (UK), peat bogs started to develop behind a gravel coastal barrier after the last Ice Age. Some 3,000 years ago, the sea intruded again into the marsh, covering part of the peat with new layers of marine deposits and creating an environment of salt marshes and creeks. The tidal inlets created excellent conditions for harbour activity. The marsh towns of Hythe and Romney were two of the original members of the Cinque Ports, the confederation of five privileged harbours controlling most of England's Channel trade (and piracy) from the eleventh century. Embankment of the area started in the Anglo-Saxon period and continued well into the late medieval period. Once again there is evidence of individuals and groups of tenants active in reclamation in the twelfth century and before and enjoying a free juridical status. In the following centuries, Romney Marsh became one of the richest and most densely populated regions in the county of Kent, with important cereal cultivation, witnessed by the high proportion of land being manured in medieval times. Tenants in late medieval Romney Marsh practised mixed agriculture and enjoyed a free status, with little or no restriction on the inheritance and sale of customary land. Besides small-scale arable production and animal husbandry, the availability of peat, intensive salt-making and the important shipping and fishing activities of the Cinque Ports provided ample opportunities for seasonal labour and excellent conditions for the survival of a non-specialised traditional economy well into the later Middle Ages.[7]

In contrast to Flanders and Romney Marsh, peat soils did not disappear in the third research area, the Central Holland peatlands. Here the dune barrier proved to be much stronger and more stable than its Flemish counterpart and the river inundations were dammed by higher natural levees, consisting of sandy clay. Between the tenth and the thirteenth centuries however, the largest part of the peat area itself was reclaimed. Just as in Flanders, the initiative was taken by free peasant communities and was stimulated by the Count of Holland and the Prince-bishop of the neighbouring principality of Utrecht. The lord granted the reclaimers, who typically enjoyed a free juridical status, the right to settle in a confined part of the peat area in exchange for a low customary rent. Until the late medieval period, the peasant-freeholders in the newly reclaimed Holland peat-areas practised mixed agriculture, with arable farming on the peat, combined with cattle and sheep-breeding. From the fourteenth century on, this changed: drainage of the peat soils made them shrink and become increasingly unsuitable for growing

Regional Differences. An Essay Taking the Former County of Flanders as an Example (Middle Ages–19th Century)' in B.J.P. van Bavel and P. Hoppenbrouwers, eds. *Landholding and Land Transfer in the North Sea Area (Late Middle Ages–19th Century)*, Turnhout, 2004, 47–66 (CORN Publication Series 5).

7. J. Eddison, *Romney Marsh. Survival on a Frontier*, Stroud, 2000; and other studies edited by the same author (see select bibliography).

winter cereals. Between 1350 and 1500, when Holland went through a period of rapid economic and urban growth, agriculture in the peat districts underwent a radical transformation. Peasants turned towards a combination of cattle breeding, peat digging (and increasingly peat dredging) and proto-industrial activities. Thanks to this reconversion, the typical small-scale 'peat and pasture' farm would continue well into the early modern era. Just as with the Flemish coastal plain, in the medieval Dutch peat district manorialism was weak, serfdom almost absent, and the position of the Count relatively unchallenged. By contrast urban development occurred much later: by 1300 the share of town dwellers in the Dutch population did not exceed 14 per cent.[8]

Before 1300 the natural resources of the three coastal marshlands under consideration primarily benefited local peasant smallholders. Regional lords, like the counts of Flanders and Holland, and the abbeys they founded or endowed, obtained significant profits from the marshlands and so, in an indirect way, did the textile industries and tradesmen of the Flemish towns or the Cinque Ports; but in general their involvement helped rather than hampered the development of flourishing peasant societies in all three coastal wetlands during the twelfth and thirteenth centuries.

Flood Management in Traditional Wetland Economies

In the coastal wetlands the threat of inundation was never completely absent. In the older literature the ability of peasant smallholders to manage this environmental risk has often been questioned: peasants are supposed to have lacked the capital, the organisation and the technology to make efficient use of the wetland resources. Large-scale drainage or embankment projects, initiated by outsiders like Cistercian abbeys in the medieval period or bourgeois investors later on, were considered necessary in order to enhance productivity and to solve flooding problems.[9] However, when looking more closely at the management of environmental risk in the coastal wetlands of the twelfth and thirteenth century, it turns out that flood protection and drainage were primordially community based, distributing the cost of protection against natural disasters amongst those who were directly concerned with it: the peasant inhabitants. Maintenance and repair works were allotted as much as possible to individual plots of land, each tenant being responsible for the

8. B. van Bavel and J.L. van Zanden, 'The Jump-start of the Holland Economy During the Late-medieval Crisis, c.1350–c.1500', *Economic History Review*, 57:3, 2004, 503–32; P.J.E.M. van Dam, 'Sinking Peat Bogs: Environmental Change in Holland, 1350–1550', *Environmental History* 6, 2001, 32–45; W.H. TeBrake, *Medieval Frontier: Culture and Ecology in Rijnland*, College Station, 1985.

9. See H.C. Darby, *The Draining of the Fens*, Cambridge, 1956.

maintenance of his own sections of sea-walls and ditches.[10] The maintenance of larger infrastructure, such as drainage locks, could be arranged by entrusting them to large ecclesiastical landowners, as was the case with the *magna slusa* [big sluice] on the Yser near Nieuport in Flanders, transferred in 1183 to the Cistercian monks of Our Lady of the Dunes.[11] This type of organisation was perfectly adapted to the needs of a predominantly peasant society: from the twelfth to the fourteenth century, labour in the coastal plains was abundant and cheap. Peasant households could easily devote part of their (seasonal) spare labour to maintenance and repair activities on dikes and ditches. In emergency situations as well, the *dijkwere* – an extension of the old-Germanic *Lantwere* – enforced the rapid mobilisation of workers, obliging all inhabitants of a district to attend the dikes in order to counter the imminent dangers of the sea[12]. Water management in these coastal peasant societies could be very responsive to environmental challenges. For instance, in their study of investment in the marshland holdings of Canterbury Cathedral Priory in Ebony, Agony and Appledore, situated at the edge of Romney Marsh, Gross and Butcher noticed a remarkably powerful reaction to the severe storms and the flooding of 1287–88. Immediately after the flooding, investments in walling (dike maintenance and repair) were up to six or seven times higher than in normal years.[13] This highly reactive investment pattern is probably less an indication of an individual investment strategy by one single ecclesiastical landowner, than the product of a communal reaction to the inundations of 1287–88, in which Canterbury Cathedral Priory was assigned its share in proportion to the land it owned.

In the Flemish Polders a similar pattern of oscillating but highly responsive investments in flood protection and drainage could be found in the thirteenth and fourteenth centuries. The local water boards (see below) in the neighbourhood of Bruges were capable of mobilising considerable amounts of unskilled labour, which at that period was cheap and still easily available in the peasant economy of the Polders. In 1354–1355 the water board of Blankenberge, responsible for flood protection and drainage of an area of c.15,000 hectares, even hired the equivalent of 23,788 working days of unskilled labourers to counter the damage caused by a

10. G.P. Van de Ven, *Man-made Lowlands. History of Water Management and Land Reclamation in the Netherlands*, Utrecht, 2004, 58.

11. Verhulst, *Landschap*, 48.

12. A. Verhulst, 'Historische geografie van de Vlaamse kustvlakte tot omstreeks 1200', *Bijdragen voor de Geschiedenis der Nederlanden* 14, 1959, 36–37; Eddison, *Romney Marsh*, 86–87.

13. A. Gross and A. Butcher, 'Adaptation and Investment in the Age of the Great Storms: Agricultural Policy on the Manors of the Principal Lords of the Romney Marshes and the Marshland Fringe, c.1250–1320', in J. Eddison, ed. *Romney Marsh: The Debatable Ground*, Oxford, 1995, 109.

heavy storm surge. In the busiest weeks, hundreds of workers at a time were labouring literally day and night at the sluices and sea-walls of the area.[14]

In the twelfth and thirteenth centuries, environmental risk in the coastal wetlands of the North Sea area was countered by allotment systems and very large inputs of cheap labour, making the high risk inherent in a coastal wetland affordable for peasant smallholders and enabling a responsive attitude towards environmental challenges. It is, of, course difficult to measure to what extent this 'peasant' way of dealing with environmental risk was efficient. In any case, the number of flooding disasters due to storm surges was significantly lower in the twelfth and thirteenth centuries than it would be in subsequent centuries:

Table 1. Flood Disasters Following Storm Surges along the Flemish North Sea Coast and the Western Scheldt Estuary, 1000–1750[15]

	Flood disasters following storm surges	Supra-regional flood disasters following storm surges
1000–1050	2	2
1051–1100	1	0
1101–1150	2	1
1151–1200	2	1
1201–1250	3	1
1251–1300	5	3
1301–1350	7	1
1351–1400	13	4
1401–1450	17	4
1451–1500	20	2
1501–1550	21	4
1551–1600	24	2
1601–1650	17	2
1651–1700	15	3
1701–1750	5	1

Historical references to inundations following storm surges were much more numerous in the late fourteenth, the fifteenth and the sixteenth centuries, than in the twelfth or thirteenth centuries. Of course one could argue that there is a

14. T. Soens, 'Explaining Deficiencies of Water Management in the Late Medieval Flemish Coastal Plain (13th–16th centuries)', *Jaarboek voor Ecologische Geschiedenis*, 2005–06, 35–62.

15. Sources: M.K.E. Gottschalk, *Storm Surges and River Floods in the Netherlands*, Assen, 1971–77, 3 vols; J. Buisman, *Duizend jaar weer, wind en water in de Lage Landen*, Franeker, 1995–2006, 5 vols.

source problem for the earliest period, which is especially true for the local floods; and, even more importantly, that the increase in inundations reflected an increased storminess in the later medieval period, probably related to the beginning of the Little Ice Age. Both objections are relevant. Nevertheless, the table explains nothing about the number of storm surges but only about the number of inundations that followed a storm surge. Whereas the storm activity is a natural phenomenon beyond direct human influence, the impact of the storm is largely determined by human conditions: technological shortcomings, bad maintenance of infrastructure, economic, political or military disturbances that turned a storm surge into a disastrous inundation.[16] In this respect, the coastal peasant societies of the twelfth and thirteenth centuries seem to have been more successful in managing the environmental risk than the generations after them. Part of the explanation for this phenomenon could be found in the social polarisation of environmental profits and risks in the late medieval period.

The Social Polarisation of Environmental Profits

Some of the coastal wetlands studied in this paper were among the first regions in north-western Europe to make the transition towards commercialisation, not only implying the exchange of agrarian products on a market, which already took place in many estates or family-based economies of the medieval period – but also the market exchange of land, labour and capital, releasing a chain-reaction of competition, specialisation, investment and accumulation, but at the same time increasing social polarisation.[17] In the Flemish coastal plain, this transition took off in the late thirteenth century and it was first of all characterised by a radical restructuring of property relations. Statistical evidence for the decline in the number of smallholders on the coastal plain can only be found from the end of the fourteenth century, while the enclosure movement in terms of both landownership and farm sizes probably started at least half a century earlier. In the fifteenth and the sixteenth centuries however, the further expulsion of peasant smallholding gained momentum. In the rural surroundings of Oostburg for instance, the number of landowners decreased from more than 41 per hundred hectares in 1388 to only 14.3 in 1550. By the middle of the sixteenth century, 41 landowners owned 54 per cent of the land, with an average of 41.2 hectares each. Whereas in 1388 the land area of 1,228 persons did not exceed four hectares, in 1550 this remained true for only 291 people, which means that between the end of the fourteenth and the middle of the sixteenth cen-

16. A.J. de Kraker, 'Flood Events in the Southwestern Netherlands and Coastal Belgium, 1400–1953', *Hydrological Sciences*, 51:5, 2006, 913–29.

17. Thoen, 'Social Agro-systems'; B. van Bavel, 'Land, Lease and Agriculture: The Transition of the Rural Economy in the Dutch River Area from the Fourteenth to the Sixteenth Century', *Past and Present*, 172, 2001, 3–43.

tury, almost 1,000 small landowners had 'disappeared', or at least lost the property rights over their land.[18] It is not impossible that some of the peasant-smallholders who had lost their property rights continued to work the lands their ancestors had owned, but now in the capacity of leaseholder, paying a full, competitive market rent to the new owners, a large majority of whom leased out their newly acquired lands. In contrast to the declining number of customary and free holdings, the farms held in short-term lease were increasing both in number and in size. In the parish of Oostkerke, near Bruges, 72 per cent of the land was farmed by 26 large holdings, each cultivating 32.7 hectares on average by 1570.[19] In the seventeenth and eighteenth centuries this evolution continued and the Flemish Polder area became a region of '*grande agriculture*', with most of the land cultivated by farms over fifty hectares.[20]As peasant landowners lost their land, other people could of course extend their landed property. Bourgeois and, to a minor extent, ecclesiastical landownership increased in importance. In the environs of Bruges, urban dwellers owned thirty to fifty per cent of the land, with ecclesiastical landowners, most of them also situated in the city of Bruges, often owning another thirty per cent.[21] As a consequence, the natural resources of the coastal wetlands were increasingly controlled by absentee landowners living in the city.

The Flemish Polder area was not the only coastal wetland area where an early transformation of landownership and landholding took place. By the sixteenth century, Romney Marsh, once hosting one of England's highest population densities, had turned into one of the least populated regions of southern England and undergone an economic switch towards sheep breeding, supplying the textile industry of the nearby Weald of Kent. In towns like Lydd, a scattered landholding pattern had supported a dense and relatively prosperous population well beyond the disturbances of the Black Death. Small and middling traders, artisans, fishermen and mariners took full advantage of the ecological diversity of the Marsh. From the sixteenth century however, increasing acreages of marshland were enclosed by large landowners like Battle Abbey and its demesne farmers and this sometimes provoked fierce resistance from both rural and urban smallholders. In most cases, this resistance failed and most smallholders were more or less forced to sell their holdings. From 1432 to 1538 the number of farms in one marshland area, Dengmarsh, shrank from 42 to six.[22] In the towns of Romney Marsh the changing access

18. Soens, *De spade in de dijk*, 76–77.

19. Soens, *De spade in de dijk*, 87.

20. P.J. van Cruyningen, *Behoudend maar buigzaam. Boeren in West-Zeeuws-Vlaanderen, 1650–1850*, Wageningen, 2000.

21. T. Soens and E. Thoen, 'The Origins of Leasehold in the Former County of Flanders', in *The Development of Leasehold in Northwestern Europe, c.1200–1600*, Turnhout, 2008, 46–47.

22. S. Dimmock, 'English Small Towns and the Emergence of Capitalist Relations c.1450–1550', *Urban History* 28, 2001, 9, 15; and S. Dimmock, 'English Towns and the Transition c.1450–1550',

to land induced an increasing social polarisation: a small oligarchy took advantage of this evolution and accumulated land and sheep flocks to graze it, whereas the middling population faced the negative consequences of economic specialisation towards extensive sheep-breeding and the decline of traditional industries and trades. They were joined by an increasing army of landless peasants who migrated from the Marsh.[23] By the middle of the seventeenth century, half of the Level of Romney Marsh's almost 9,500 hectares were owned by just 49 individuals, each owning more than forty hectares, and largely consisting of ancient Kentish gentry families, supplemented by yeomen and minor gentry living in the shrinking towns of the Marsh.[24] In the early eighteenth century, Romney Marsh became the territory of the 'gentlemen-graziers', farming extensive marshland holdings on behalf of different landowners, and dominating local office-holding for decades. Interestingly, more than a third of them did not actually live in the marsh, but in the upland parishes, just using their marshland holdings for summer pasture. For this reason, eighteenth century Romney Marsh has been characterised as an 'absentee society', with not only the landowners, but even an important share of the landholders, living outside the area[25].

By contrast, in our third research area, the Central Holland Peatlands, direct access to land for peasant smallholders survived well into the early modern period. Whereas bourgeois landowners invested substantial amounts of money in drainage projects during Holland's Golden Seventeenth Century, their direct intervention in the peatlands remained limited.[26] The reason for the bourgeois reluctance to buy land in the peat-districts is linked to the organisation of the peat-industry itself. In the sixteenth and seventeenth centuries large-scale peat-industries owned by urban bourgeois and employing wage-labourers moved increasingly to the eastern provinces of the Netherlands where peat bogs had not yet been subject to shrinkage from centuries of drainage and intensive occupation. In central Holland, peat needed to be dredged from below the ground water level and this kind of peat-dredging became a corner-stone of the small-scale, owner-occupied peasant farm

---

Past and Present Supplement 2, 2007, 282–83; S. Sweetinburgh, 'Land Holding and the Land Market in a 15th Century Peasant Community: Appledore, 1400–1470', in A. Long, S. Hipkin and H. Clarke, eds. *Romney Marsh: Coastal and Landscape Change Through the Ages*, Oxford, 2002, 140–56.

23. See also A.F. Butcher, 'Origins of Romney Freemen 1433–1523', *Economic History Review*, 27, 1974, 16–37.

24. S. Hipkin, 'Tenant Farming and Short-term Leasing on Romney Marsh, 1587–1705', *Economic History Review* 53:4, 2000, 658–60.

25. A. Davison 'A "Particularly Convenient and Useful" Arrangement: The Symbiotic Relationship between the Agrarian Economy of Romney Marsh and the Surrounding Region in the 18th Century', in Long, Hipkin and Clarke, eds. *Romney Marsh*, 191.

26. J. De Vries, *The Dutch Rural Economy in the Golden Age 1500–1700*, London, 1974, 45–6.

in the Golden Age.[27] Both landownership and landholdings became increasingly scattered and population numbers exploded. In the peat village of Zegwaart, for instance, population increased from about 335 at the beginning of the sixteenth century to 2,099 in 1622, with the largest proportion concentrated on the higher roads and dikes.[28] In the peat-village of Ter Aar the number of holdings smaller than 4.2 hectares (five *Rijnland morgen)* increased from 16.9 to 66.7 per cent of all holdings.[29] Only after 1670, when the economic tide was turning and water levels in the peatlands became increasingly difficult to manage, would many smallholders abandon their land. By 1686 the population of the village of Zegwaart had already diminished to 1339 people.[30] By then, the abandoned and increasingly submerged peat lands were either incorporated in the area's expanding lakes, like the well-known Haarlemmermeer, or they persisted in the landscape as smaller pools, separated from each other by small ridges, witnesses of the former field system. Ultimately, most of them would be drained and devoted to agriculture again, but their former peasant landowners would not return: after drainage, large leasehold farmers, most of them concentrating on animal husbandry, made their appearance in the former Peatlands.[31]

The Social Polarisation of Environmental Risks

All three traditional coastal wetlands underwent profound social changes in the course of the late medieval and early modern period, although with a somewhat different chronology. In all three cases the changing access to land had a significant impact on the way coastal society dealt with environmental risk and the social distribution of these risks. Environmental risk was increasingly burdening the lower ranks of society. This was the result of a triple evolution: regionally, the most environmentally vulnerable areas witnessed a social decline; institutionally, smallholders increasingly lacked the instruments to deal with the environmental risk; and financially, environmental management increasingly depended on the financial strategies of absentee, urban or gentry, landowners and their capitalist farmers.

In all three coastal wetlands under consideration the environmental risk could vary substantially from one place to another and the higher risks in some areas were not always matched by higher profits. The highest risks could be found in areas with sea walls frequently threatened by the sea. Although large regions benefited from the adequate maintenance of these sea walls, their labour- and capital-intensive

27. P. van Dam and M. van Tielhof, *Waterstaat in stedenland. Het hoogheemraadschap van Rijnland voor 1857*, Utrecht, 2006, 122–33.

28. Van Dam and van Tielhof, *Waterstaat in stedenland*, 170–74.

29. De Vries, *Dutch Rural Economy*, 132.

30. Van Dam and van Tielhof, *Waterstaat in stedenland*, 314–15.

31. Van Dam and van Tielhof, *Waterstaat in stedenland*, 244.

upkeep often depended on rather small communities to whom it had traditionally been entrusted, even when environmental conditions had now become altered. Throughout the late medieval and early modern period, both in Holland and in Flanders, rural communities and smaller cities begged for financial assistance from larger regions but often met with fierce resistance, in many cases headed by the largest and politically most powerful cities and the land-owning elite living there. A good example is the Spaarndammerdijk in central Holland, the main sea wall between the cities of Haarlem and Amsterdam, protecting the Rijnland area from the water of the Ij. The maintenance of this dike fell under the responsibility of the communities – *ambachten* – adjacent to the dike. In the early sixteenth century the maintenance became increasingly costly and the dike breaches more and more frequent. As a consequence, land burdened with maintenance duties on this dike became highly unattractive, and the wealthier landowners tried to get rid of their land. It was bought by poorer people, as the only land they could afford, or, alternatively, was simply abandoned. Only after decades of problems and legal disputes was the maintenance of the Spaarndammerdijk finally handed over to the regional water board of Rijnland in 1579–81. At least financially, the environmental risk was now spread over a much larger region, but in the meantime the communities adjacent to the dike had witnessed clear impoverishment.[32]

Quite similar problems could be found in Flanders. On the isle of Biervliet in the western Scheldt area, the problem of flight of capital paralleled by a worsening of environmental conditions is particularly striking[33]. Biervliet was a small port, founded in the twelfth century in a desolate moor area, and became famous for its salt-refining industry and herring fishery. Originally Biervliet was not an island and was easily accessible by land. However, from the second half of the fourteenth century onwards, the expanding western Scheldt estuary submerged neighbouring lands and cut the town off from its hinterland. By the middle of the fifteenth century, the town had turned into a remote isle, faced with depopulation, general economic decline and bankruptcy, desperately looking for funding to maintain its harassed dikes. In the small polder of Boterzande, north of Biervliet, Saint Peter's abbey in Ghent paid an average yearly water tax of 152.6 Flemish *groten* per hectare between 1397 and 1421, whereas the abbey's income from this land seldom exceeded 100 Flemish *groten*. When a new storm surge inundated Boterzande in November 1421, the abbot and many other landowners were no longer willing to invest in dike repair and they abandoned their land. Only the town of Biervliet itself was willing to take over the lands for free. Polders like Boterzande were part of the city's outer defences against the waters of the sea and so Biervliet got into debt and paid for the repair works itself. Biervliet's economic and financial situation worsened in the

32. Van Dam and van Tielhof, *Waterstaat in stedenland*, 145–49.

33. See M.K.E. Gottschalk, *Historische geografie van Westelijk Zeeuws-Vlaanderen*, 2 vols., Assen, 1955–58, vol.2, 19–25; Soens, *De spade in de dijk*, 102–103.

course of the fifteenth century. In the spring of 1483, while civil war was raging in Flanders, a new disaster struck the town. This time a plague of rats invaded the city and started to undermine its precious dikes. For the few remaining ecclesiastical landowners in the city and its surroundings, a point of no return had been reached. They started to evacuate their possessions as soon as possible. In the autumn of 1483 a community of Augustinian nuns founded in Biervliet some decades earlier in an effort to encourage spiritual life in the languishing town, moved out. The Ghent abbey of Saint Bavo also tried to get rid of its lands in Biervliet, claiming that the gross annual rent income of this land did not exceed 440 Flemish *groten*, while in 1483 alone the abbey had to contribute 9,600 *groten* for dike repair works. After an initial period of high investment, ecclesiastical landowners like the two Ghent abbeys desperately sought to get rid of their land, preferring to abandon it rather than continue the investment in this unprofitable property, handling over the entire burden of the flood risk to the remaining local inhabitants.

In late medieval Holland there was a second evolution that contributed to an unequal geographical distribution of environmental risk, with equally important social consequences: the spread of the polder mill. Starting in 1408, when the first windmill for drainage was built near Alkmaar, the construction of polder mills enabled much more effective drainage of small areas surrounded by new purpose-built dikes. However, there were some disadvantages too: on the one hand, drainage by wind mills accelerated the shrinkage of the peat soils within the polder and thus necessitated ever more drainage; on the other hand, the surplus water drained out of the private polders provoked flooding in other parts of the peat area and contributed to the expansion of peat lakes like the Haarlemmermeer. Especially in poorer regions, which lacked the capital to build their own polder mills, flooding increased due to the introduction of the polder mill.[34]

In Romney Marsh, too, a regional polarisation occurred in the early modern period. We have already seen that an increasing number of the better-off gentlemen-graziers preferred to live outside the marsh, in the upland area. By the eighteenth century the marshlands had become so wet in winter that the over-wintering of lambs there was considered too dangerous. Many Marshland farmers concluded *'agistment'* contracts with upland farmers, permitting them to lead their lambs to the uplands in winter, for a long period of thirty weeks or more. In return, the upland farmers sent their cattle, and in particular bullocks that had to be fattened, to the Marsh in summer, the usual exchange rate being five lambs for every bullock. Although this seems a perfect example of a symbiotic relationship between upland and marshland agrarian economies, the exchange rate seems to favour the upland farmers who often earned huge money through meat production enabled

34. Van de Ven, *Man-made Lowlands*, 122–25; van Dam and van Tielhof, *Waterstaat in stedenland*, 320–21.

by Marshland grazing.[35] Furthermore, the upland farmers did not share the environmental cost of living and working in the Marshland: the '*agistment*' permitted them to use the rich summer pasture of the marsh, without even paying for the maintenance cost of drainage and sea-defence systems. They also escaped the health risk of living in a marshland area. This had to do with the flood risk, but also with the presence of malaria ('marsh fever') in the coastal wetlands.[36]

An Unequal Access to Flood Risk Management

The twelfth and thirteenth century institutional arrangements concerning flood protection and drainage perfectly fitted a peasant society with plentiful labour but little cash. In the course of the later medieval period and the early modern period, the organisation of water management underwent important changes, largely coinciding with the changes in landownership and landholding described above. As a consequence, these institutional changes occurred first in the Flemish Polders and Romney Marsh, where they became visible as early as the late thirteenth century, and only much later in Central Holland (sixteenth century). Essentially the changes concerned a centralisation and a monetarisation of water management and went hand in hand with a more restricted access to the decision-making process for peasant smallholders.

From the 1230s, landowners in the Flemish coastal plain started to organise themselves in water boards (Dutch: '*waterschap*' or '*watering*'; French: '*wateringue*'). These were associations that took over the maintenance of the drainage and flood protection system from individual landowners. From 1285, the activities of the *Blankenbergse watering* are reported in a continuous series of annual accounts. Assuming responsibility over the hydraulic infrastructure of an extensive area (c.15, 000 hectares), the board members bought materials, hired people to perform maintenance and repair works on sluices, dikes and drains, outsourced some of the maintenance, kept a separate administration, met to make minor decisions, called on the aldermen to inspect the infrastructure, and reported to a general assembly of landowners. All of their activities were financed by a uniform land tax, the *geschot* or *scot* paid in cash by the landowners proportionately to the size of their land.[37] This new type of organisation suited urban landowners perfectly – logically they preferred the payment of an amount of cash rather than labour services. For peasant smallholders such monetisation of labour duties was less favourable, in particular because of the very short-term of payment that characterised the taxes levied by the water boards and the high penalty for delayed or non-payment. In most cases,

35. Davison, 'Particularly Convenient and Useful', 191–96.

36. M. Dobson 'Marsh Fever: The Geography of Malaria in England', *Journal of Historical Geography*, 6, 1980, 357–89.

37. Soens, *Spade in de dijk*, 20–4.

taxes were due within two weeks, after which term a double or quadruple fine was levied, eventually resulting in expropriation within the year. Although by the middle of the sixteenth century almost ninety per cent of the land in the Flemish Polder area was held in short-term lease, liability for payment of the *scot* taxes and the connected access to the decision-making process of the water boards was, and is today, restricted to the owners of the land. At least in theory, leaseholders were excluded. As a consequence, there was an increasing separation of *scot* payment on the one hand and (direct) interest in adequate maintenance of the infrastructure on the other hand. In Romney Marsh the same evolution towards leasehold took place before the seventeenth century but here the responsibility for the payment of the *scot* remained tied to the *usufruct*, which meant that the leaseholder had to pay for it in normal years.[38] The largest of these leasehold-farmers profited from this situation to extend their grip on the water management organisation. By the early eighteenth century the typical *jurat* was a gentlemen-grazier like Daniel Langdon, who was not only one of the largest tenant-farmers in the area, but also a jurat of the Level and Liberty of Romney Marsh for 45 years.[39]

Both in the Flemish Polders and in Romney Marsh, peasant smallholders increasingly lost the institutional tools to intervene in the management of environmental risk. Surprisingly, this was also increasingly the case in the Dutch peatlands, where peasant smallholding flourished well into the seventeenth century. As we have seen, the essential tasks of maintaining and upgrading the flood protection and drainage system in central Holland were increasingly handed over from local communities to the regional water boards. These water boards – the *Hoogheemraadschappen* – had existed as early as the thirteenth century but in the first centuries of their existence their tasks were largely limited to the inspection of the water control system.[40] From the sixteenth century onwards, the regional water boards gradually centralised the maintenance of key infrastructure. At the same time, the decision-making process in these regional water boards was increasingly controlled by the urban oligarchy of Holland's booming towns. Cities obtained the right to delegate a representative to the regional water boards: in the important water board of Rijnland, four out of seven board members in the seventeenth century were directly appointed by cities. Furthermore, in the sixteenth century important urban landowners also started to gather with noblemen and abbeys to form colleges of 'main landowners'

38. Hipkin, 'Tenant Farming', 650; S. Hipkin, 'The Structure of Landownership and Land Occupation in the Romney Marsh Region 1646–1834', *Agricultural History Review*, 51:1, 2003, 73; D. Beck, 'The Drainage of Romney Marsh and Maintenance of the Dymchurch Wall in the Early Seventeenth Century', in. J. Eddison, ed. *Romney Marsh*, 167.

39. S. Hipkin, 'The Worlds of Daniel Langdon: Public Office and Private Enterprise in the Romney Marsh Region in the Early 18th Century', in Long, Hipkin and Clarke, eds. *Romney Marsh*, 173–89.

40. Van Dam and van Tielhof, *Waterstaat in stedenland*, 232–33.

(*Hoofdingelanden*) who could control investments in the water control system. As we have seen, these urban elites had little or no direct patrimonial interests in the peatlands: their landed property was situated in the river clay area or on the dunelands. The environmental problems faced by the peasant smallholders in the peatlands were of little or no concern to them. In the course of the early modern period, structural solutions for the increasing drainage problems in the Dutch peat areas would not be found. The major cities of Leiden, Amsterdam and Haarlem blocked every initiative for centuries. Leiden, for instance, obstructed the drainage of the ever-expanding Haarlemmermeer, in part because the city got considerable income from fishing on the lake and, above all, because it feared that, after the drainage of the lake, water would stagnate in and around the city.[41] Only after the fall of the Republic would selfish obstruction by individual cities and their regents be overruled by a henceforth stronger central state.

An Unequal Financial Spread of Environmental Risk

When peasant society in Romney Marsh and the Flemish Polders was fading away, crucial investments in drainage and flood protection became increasingly dependent on the financial strategies of large absentee landowners or commercial tenant farmers. In the fifteenth and sixteenth century Flemish Polders, the water control system was controlled by absentee landowners. By that time, this group owned and leased out most land in the area. As a consequence, financial investments in the water control system increasingly depended on the evolution of rent-income. For most of the bourgeois, noble and ecclesiastical absentee landowners, the yearly *scot* must have been an expense depressing the net profit from their leased-out landed property. Not surprisingly the amount of *scot* to be paid every year tended to follow the evolution of lease-prices: between the late fourteenth and the early sixteenth centuries those lease prices dropped from the equivalent of c.400 litres of wheat per hectare to approximately 200 litres.[42] In reaction to the depressed lease prices, *scot* taxes were equally lowered, although the number of flood problems was very high in this period (see table 1). Flooding, however, was less a worst-case scenario for absentee landowners than for peasant smallholders. On a macro-level, it could even be profitable. From the fifteenth to the eighteenth centuries the inundated Polder lands in Flanders, Holland and Zeeland offered great opportunities for re-embankment. It was precisely in these re-embankments, where the field and drainage system as well as the layout of the farms were designed for large-scale farming, that early modern 'polder capitalism' would prosper most.[43] In this respect, a temporary

41. Van Dam and van Tielhof, *Waterstaat in stedenland*, 290.

42. Soens, *Spade in de dijk*, 138.

43. P.J. van Cruyningen, 'Profits and Risks in Drainage projects in Staats-Vlaanderen, c.1590–1665', *Jaarboek voor Ecologische Geschiedenis* (2005–2006), 123–42.

increase in environmental risk due to low investment in the water control system was perfectly acceptable.

In Romney Marsh, too, a higher risk of flooding became acceptable in the course of the early modern period: agriculture in the Early Modern Marsh became radically oriented towards extensive pasture, with not only the landowners but also a considerable number of the largest farmers living outside the Marsh. The drainage of the area no longer needed to meet the high standards required for arable farming. In the landscape, the gradual abandonment of arable farming in the marsh led to the filling up of minor ditches that were no longer retained.[44] In the seventeenth and eighteenth centuries, the water control system was adapted to the needs of the gentlemen-graziers and thus to the needs of extensive sheep breeding. As a consequence, investments tended to follow wool prices. As long as the latter increased (until the 1620s) land taxes for the maintenance of the sea-walls (*wall scots)* could triple (from an average of about 10d to an average of about 30d per acre).[45] By contrast, the depression of wool prices that followed, until the 1690s, was paralleled by a stagnation of the *wall scots*, followed by a straight decline from the 1660s onwards. In the 1680s, the *wall scots* again reached the level of a century earlier.[46] We can conclude that seventeenth century *wall scots* were clearly in line with the economic fortunes of sheep breeders. A clear comparison with the late medieval period is difficult and so is an appreciation of the sustainability of the water management in the seventeenth century Marsh. Having seen that the priory of Canterbury invested 4d per acre of arable in normal years in the 1280s and 1290s, increasing to 46d in the storm year 1288, seventeenth century average investments can reluctantly be called impressive. The seventeenth and eighteenth century Marsh clearly no longer demanded the same degree or quality of drainage needed by the mixed farming economy of the late thirteenth century.

The Dutch peatlands revealed an increasing separation of the environmental problems of the peatlands on the one hand and the interests of the urban citizens who controlled the water management on the other. Here too, this was reflected in the evolution of investments in flood protection and drainage. In the regional water board of Rijnland, the scot-tax (*morgengeld*) remained surprisingly stable from the second half of the seventeenth century to the end of the eighteenth century, notwithstanding the increasing drainage problems in the peatlands.[47] Structurally, regional water boards did not react to these increased difficulties. In Rijnland the only investments that were structurally rising in the late seventeenth and the eighteenth century were the administrative costs (expenses for wages for the water

44. Eddison, *Romney Marsh,* 101.

45. The use of 'd' refers to pence, a unit of currency.

46. Hipkin, 'Tenant Farming', 669–670.

47. Van Dam and van Tielfhof, *Waterstaat in stedenland,* 258–261.

board's administration, for dinners and drinks and, of course, for the maintenance of the prestigious office – *Gemeenlandshuis* – in Leiden).[48]

In all three coastal wetlands the collapse of peasant society not only corresponded with a period of increased environmental problems, but also with a clear social polarisation of environmental risk at the expense of the remaining peasant smallholders. In the long term however, this rather 'unsustainable' situation would not persist. Often a new environmental equilibrium was found. In the Flemish Polders for instance, the uncontested hegemony of the large farmers and absentee landowners in the eighteenth century would be paralleled by a decrease in environmental problems in the area (see Table 1). Landowners were increasingly interested in stable long-term relationships with their farmers and the latter tended to become landowners in the area themselves, which also gave them access to the decision making process in the water management system.[49] Probably for the first time since the thirteenth century, the interests of most of the people who actually lived in the Polder area and those who owned the land and controlled the decision-making process in the water management bodies were in alignment with each other. In eighteenth century Romney Marsh the same coalition between landowners and farmers could be found. With regard to environmental justice however, two remarks must be taken into account: this new sustainability was achieved through the destruction of a previous, perhaps also sustainable, peasant society capable of providing a living to considerably more people; and, second, the environmental benefits of this new environmental equilibrium were first and foremost enjoyed by a limited group of small farmers in cooperation with mostly absentee landowners.

Conclusion: Environmental Inequalities and Flood Risk along the North Sea Coast

Throughout history natural disasters always seem to hit the weakest groups in society hardest. Compared to middle class and elite groups their adaptation and mitigation strategies often seem more limited. Nevertheless there is nothing 'natural' about this situation. Environmental inequalities are always produced by specific social and economic conditions and these conditions can change over time. In this article we have tried to examine long-term changes in environmental inequalities in three coastal wetland areas that were particularly prone to flooding, focusing on the interaction between coastal towns and their surrounding countryside. In the Flemish Coastal Plain, in Romney Marsh and in the Dutch Peatlands inhabitants had to cope with a permanently high flood risk. Their ability to do so changed over time and so did the social distribution of flood risk among the different groups in

48. Van Dam and van Tielhof, *Waterstaat*, 224.

49. Van Cruyningen, 'Profits and Risks'.

society. In the medieval period all three coastal wetlands supported a dense and relatively prosperous population, often with a smaller degree of social polarisation than was the case in 'dryland' communities.[50] In order to maintain the environmental equilibrium in these vulnerable environments, communities had to invest heavily in the water control system. As long as labour was cheap and abundant, all three regions managed to do so. Nevertheless in two of the three areas – the Flemish Polder area and Romney Marsh – the traditional smallholding economies of both rural and urban communities in the coastal marshes did not survive. Their formerly secure property rights to land were either contested by enclosing landlords and their farmers, as was the case in Romney Marsh, or had to be given up due to economic pressure. In both regions landownership became concentrated in the hands of an oligarchy of urban or gentry landowners roughly between the fourteenth and the seventeenth centuries, while, at the same time, farm sizes increased significantly. At the same time, many small towns in the marsh also faced social problems, due to the decline of traditional industries, loss of access to the marsh and the in-migration of landless peasants. The result was a social polarisation of environmental risk. Geographically, the areas and cities most prone to inundation and flooding witnessed a downward social evolution; institutionally, smallholders were disadvantaged by the increasing monetarisation and centralisation of a water management organisation more and more closed to them. This was even the case in Golden Age Holland, where the political supremacy of the large towns cut off peasant communities in the peatlands from access to vital control over drainage systems and flood protection. Finally, investments in flood protection and drainage were no longer determined by direct environmental challenges but by abstract calculations of profit and loss by large landowners and tenant farmers and thus increasingly depended on market cycles, not on the environmental needs of the inhabitants of the area. In this respect, the separation of interests between those who owned the land and controlled investments in the vital water control system and those who actually lived in the area tended to increase both environmental risk and environmental inequalities in the coastal wetlands. This unsustainable situation only came to an end when the interests of those who benefited most from the coastal wetland and those who took most of the risks were realigned.

50. A. Wareham, 'Water Management and the Economic Environment in Eastern England, the Low Countries and China, c.960–1650: Comparisons and Consequences', *Jaarboek voor Ecologische Geschiedenis*, 2005–2006, 21.

Select Bibliography

Van Bavel, B. and J.L. van Zanden, 'The Jump-start of the Holland Economy during the Late-Medieval Crisis, c.1350–c.1500', *Economic History Review*, 57:3, 2004, 503–32.

Van Dam, P. and M. van Tielhof, *Waterstaat in stedenland. Het hoogheemraadschap van Rijnland voor 1857*, Utrecht, 2006.

Eddison, J. *Romney Marsh. Survival on a Frontier*, Stroud, 2000.

Eddison, J. and C. Greene, eds. *Romney Marsh. Evolution, Occupation, Reclamation*, Oxford, 1988.

Eddison, J. ed. *Romney Marsh: the Debatable Ground*, Oxford, 1995.

Eddison, J., M. Gardiner and A. Long, eds. *Romney Marsh: Environmental Change and Human Occupation in a Coastal Lowland*, Oxford, 1998.

Hipkin, S. 'Tenant Farming and Short-term Leasing on Romney Marsh, 1587–1705', *Economic History Review*, 53:4, 2000, 646–76.

de Kraker, A. 'Flood events in the Southwestern Netherlands and Coastal Belgium, 1400–1953', *Hydrological Sciences*, 51:5, 2006, 913–29.

Long, A.J., S. Hipkin and H. Clarke, eds. *Romney Marsh: Coastal and Landscape Change Through the Ages*, Oxford, 2002.

Soens, T. 'Explaining Deficiencies of Water Management in the Late Medieval Flemish Coastal Plain (13th–16th centuries)', *Jaarboek voor Ecologische Geschiedenis*, 2005–06, 35–62.

Soens, T. *De spade in de dijk? Waterbeheer en rurale samenleving in de Vlaamse kustvlakte (1280–1580)*, Gent, 2009.

Stabel, P. *Dwarfs among Giants. The Flemish Urban Network in the Late Middle Ages*, Leuven–Apeldoorn, 1997.

Thoen, E. '"Social Agrosystems" as an Economic Concept to Explain Regional Differences. An Essay Taking the Former county of Flanders as an Example (Middle Ages–19th Century)', in B.J.P. van Bavel and P. Hoppenbrouwers, eds. *Landholding and Land Transfer in the North Sea Area (Late Middle Ages–19th Century)*, Turnhout, 2004, 47–66 (CORN Publication Series 5).

Tys, D. 'Landscape, Settlement and Dike Building in Coastal Flanders in Relation to the Political Strategy of the Counts of Flanders, 900–1200' in M. Fansa, ed. *Kulturlandschaft Marsch–Natur, Geschichte, Gegenwart*, Oldenburg, 2005, 106–26.

Van de Ven, G.P. *Man-made Lowlands. History of Water Management and Land Reclamation in the Netherlands*, Utrecht, 2004.

Verhulst, A. *Landbouw en Landschap in Middeleeuws Vlaanderen*, Brussels, 1995.

Chapter Five

Floods and Inequitable Responses: New Orleans before Hurricane Katrina

Craig E. Colten

Hazard events are not mindful of society. Yet there are inevitable inequities in their impacts that arise from social and economic circumstances. In recent years, vulnerability has shaped our understanding of exposures to and abilities to respond to hazard events. Through that lens we see that biophysical events do not recognise class or race and that social and economic restrictions have a powerful influence on the immediate and long-term impacts of hazard events. This is particularly true for rural, agricultural populations.[1]

Recently, the orientation for examining human exposure has begun to shift from vulnerability to a perspective framed by the concept of resilience. Based on ecological theories, resilience repositions the focus from social and economic barriers to the ability of groups to learn from past experiences and prepare to face extreme events with minimal disruption.[2] This suggests that effective planning can overcome the inequities that populations face and shifts the focus from post-hazard responses to advance preparation. Vulnerability still plays a role, since not all have equal access to the planning process or the means to boot-strap themselves into a more resilient position.

Scholarship on environmental justice, which has not always been aligned with hazards research, initially focused on the inordinate burden placed on minority and, later, low-income populations. While not presented in terms of vulnerability, it effectively paralleled that concept. The poor and unempowered were less able

1. D. Liverman, 'Drought Impacts in Mexico: Climate, Agriculture, Technology, and Land Tenure in Pueblo and Sonora Mexico', *Annals of the Association of American Geographers*, 80, 1990, 49–72; B. Wisner, P. Blaikie, T. Canon and I. Davis, *At Risk: Natural Hazards, People's Vulnerability, and Disasters*, 2nd ed. New York, 2004; G.H. Endfield, I.F. Tejedo, and S.L. O'Hara, 'Drought and Disputes, Deluge and Death: Climatic Variability and Human Response in Colonial Oaxaca, Mexico', *Journal of Historical Geography*, 30, 2004, 249–76; G.H. Endfield, I.F. Tejeco, and S.L. O'Hara, 'Conflict and Cooperation: Water, Floods, and Social Response in Colonial Guanajuato, Mexico', *Environmental History*, 9:2, 2004, 221–47.

2. B.L. Turner III, *et al.* 'A Framework for Vulnerability Analysis in Sustainability Science', *Proceedings of the National Academy of Sciences* 100, no.14, 2003, 8074–79; W. N. Adger, 'Social-ecological Resilience to Coastal Disasters', *Science* 309, 2005, 1036–39.

to deflect the imposition of unwanted land uses on their neighbourhoods. At the root of early environmental racism studies was the notion that society deliberately imposed burdens on the less powerful.[3] As research into environmental equity probed deeper historically, it became evident that those who made decisions about where to place currently unwanted land uses had often done so when an entirely different population resided at the location enduring environmental degradation.[4] Pulido offered the concept of white privilege to reframe the discussion in terms of broader institutional processes that produced inequities and to move away from the narrow charges of environmental racism, while making the same point.[5] On another front, historical geographers began to trace the creation of inequities by looking at the delivery of urban amenities and the relationship of municipal services to specific racial and low-income groups.[6]

This chapter takes that discussion a step further. In a city that has faced numerous floods from the Mississippi River and tropical hurricanes, officials have made repeated adjustments to their flood protection policies. The basic question to be addressed here is: have the public flood protection policies and practices been equitable? Have minorities and the poor been provided for equitably or have they suffered disproportionately? Certainly Hurricane Katrina left the impression that urban Blacks and poor residents shouldered an inordinate burden.[7] But how have municipal and federal programmes shaped the social geography and flawed infrastructure that the 2005 hurricane so dramatically overwhelmed?

Floods and Equity

A consistently prominent element in discussions about Mississippi River flooding has been that minority and low-income populations have shouldered the burden. In many respects, the initial inequity fell on enslaved labourers. French and Spanish law mandated that landowners build and maintain levees along their riverfront properties and enslaved African workers toiled to complete construction projects. Their efforts, no matter how gruelling and heroic, were no match for the mighty river and, when the river broke through the flimsy levees, the ensuing calamities

3. R. Bullard, *Dumping in Dixie: Race, Class, and Environmental Quality*, Boulder, 1990.

4. A. Hurley, *Environmental Inequalities: Class, Race, and Industrial Pollution in Gary, Indiana, 1945–1980*, Chapel Hill, 1995.

5. L. Pulido, 'Rethinking Environmental Racism: White Privilege and Urban Development in Southern California', *Annals of the Association of American Geographers* 90:1, 2000, 12–40.

6. C.E. Colten, 'Basin Street Blues: Drainage and Environmental Equity in New Orleans, 1890–1930', *Journal of Historical Geography*, 28, 2002, 237–57; J.E. Wells, G.L. Buckley, and C.G. Boone, 'Separate but Equal? Desegregating Baltimore's Golf Courses', *Geographical Review*, 98, 2008, 151–70.

7. Social Science Research Council, *Understanding Katrina: Perspectives from the Social Sciences*, http://understandingkatrina.ssrc.org/ 2005 Accessed May 2008.

fell on the urban poor. Nineteenth century accounts commonly portray suffering and distress among the poor.[8]

During the flood of 1927, rural planters imposed inequities on rural African Americans,[9] while New Orleans' elite redirected floodwaters onto Isleno trappers in St. Bernard Parish.[10] Wealthy decision makers opted to protect their interests while allowing the less powerful to suffer. Following the 1927 flood, as the Corps of Engineers shifted its 'levees only' policy to one of 'levees and outlets', it retained a commitment to protect wealthy planters and urban financiers. When it built a massive floodway through the Atchafalaya Basin, it disrupted the economic base of many rural Acadian trappers and fishermen.[11]

The other important flood hazard, hurricanes, adversely affected marginal populations, although not always African Americans. One can argue that the most marginalised in Louisiana occupied coastal locations during the nineteenth and early twentieth century. Houma Indians, displaced from their more northerly homeland, ranged down the bayous of Terrebone Parish (Fig. 1).[12] Acadian trappers and fisher families lived in the swamps and in the coastal marshes along the bayous south of New Orleans.[13] Isleño (Canary Island) trappers occupied the flanks of the lower river and earned a living from the marshlands south-east of New Orleans.[14] By the early twentieth century, a small community of Filipino shrimpers lived in Manila Village on Barataria Bay (Fig. 1) and, by the 1980s, a sizable number of Vietnamese shrimpers along the lower river added to the mix of minority populations.[15] All were highly exposed to hurricanes and subject to repeated impacts. They learned to evacuate the area or ride out storms in modest stilt houses and then get on with their lives as soon as the winds died down and the storm surge recede.

8. For example, see E. Fenner, 'On the Inundation of 1816', *Southern Medical Reports* 1, 1849, 56–62; C. Forshay, 'Crevasse of 1849: A Chapter of the Hydrography of the Mississippi River', *Southern Medical Reports*, 1, 1849, 63–70; C.E. Colten, *An Unnatural Metropolis: Wresting New Orleans from Nature*, Baton Rouge, 2005.

9. J. M. Barry, *Rising Tide: The Great Mississippi Flood of 1927 and How it Changed America*, Boulder, 1997.

10. G. Gomez, 'Perspective, Power, and Priorities: New Orleans and the Mississippi River Flood of 1927', in C.E. Colten ed. *Transforming New Orleans and Its Environs: Centuries of Change*, Pittsburgh, 2000, 109–21.

11. M.L. Comeaux, *Atchafalaya Swamp Life: Settlement and Folk Occupations*, Baton Rouge, 1972; M. Reuss, *Designing the Bayous: the Control of Water in the Atchafalaya Basin, 1800-1995*, Alexandria, Va. 1998.

12. F. B. Kniffen, *The Historic Indian Tribes of Louisiana from 1542 to the Present*, Baton Rouge, 1994.

13. Comeaux, *Atchafalaya Swamp Life*; L. Estaville, 'The Louisiana French Homeland', *Journal of Cultural Geography*, 13, 1993, 31–45.

14. G. Din, *The Canary Islanders of Louisiana*, Baton Rouge, 1988.

15. C. Bankston III, 'Southeast Asians in Louisiana', in C.A. Brasseaux, ed. *A Refuge for All Ages: Immigration in Louisiana History*, Lafayette, 1996, 661–77.

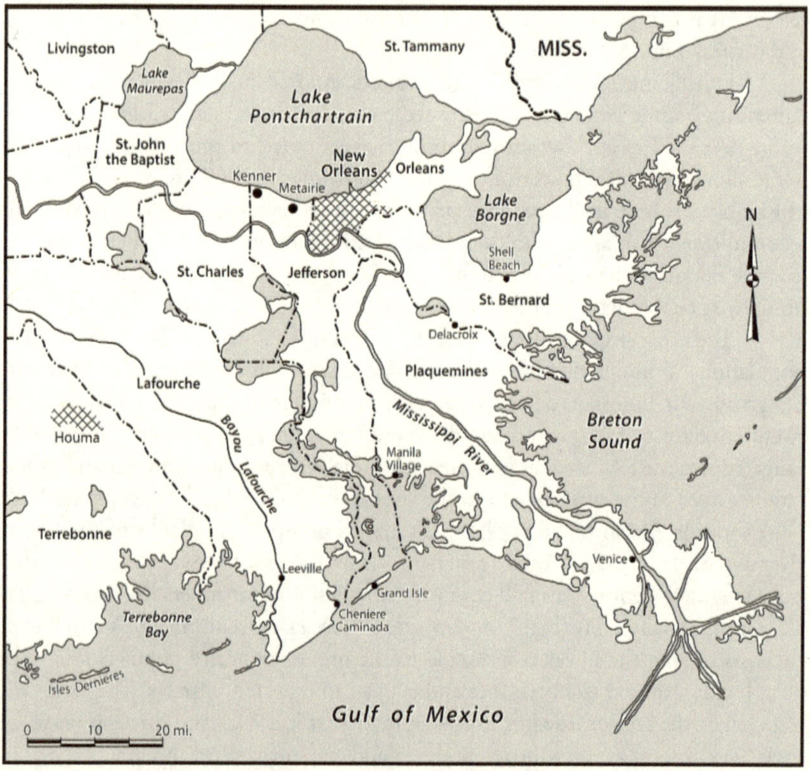

Figure 1. South-east Louisiana[16]

Their encounters with hurricanes are poorly chronicled, as is typical of the history of other marginalised populations. New Orleans journalists reported that the hurricane of 1905 inundated the houses of the 'poor people' in Bayou Lacompte on the north shore of Lake Pontchartrain, a community of mixed African Americans and Creoles. In 1909 the coastal communities in Terrebone and St. Bernard parishes fled before the storm.[17] As previous storms had done, the 1915 hurricane swept across the St. Bernard wetlands 'inflicting loss of life, destruction of property, and suffering on many of the poorest people'.[18] Fatalities were regularly

16. Source: cartography by Clifford Duplechin.

17. *New Orleans Times Picayune* (hereafter *NOTP*), 'Swept by Tidal Wave', 22 September 1909, 1.

18. Din, *Canary Islanders*, 144.

high among those in St. Bernard, Plaquemines, and Terrebone parishes in the early twentieth century.[19]

Despite the repeated destruction to coastal fisher and trapper families, the most notorious nineteenth century hurricanes destroyed retreats for the urban elite. In 1856 a storm obliterated a resort on Isle Derniere, a barrier island south and west of New Orleans. Wind and wave killed over 300 people trapped there by the storm.[20] In 1893, an even more devastating tropical storm demolished the hotel on Grand Isle, another barrier Island resort, and destroyed the Acadian fishing village on adjacent Cheniere Caminada. It killed over 1,100 holidaymakers and local residents.[21] Granted, the resort guests likely travelled with a retinue including African Americans and some resort employees were likely Acadian, but many fatalities were wealthy. These two storms prompted the decline of island resorts in Louisiana, which meant that those with options withdrew from exposed positions.[22] Yet the Islenos, Houma Indians, Acadians, and Filipinos who depended on marine and marsh wildlife for livelihoods could not retreat from the coast. Consequently, they continued to suffer from repeated landfalls. Subsequent efforts to construct flood protection structures in south Louisiana tended to focus where expenditures were most economically efficient – around concentrated urban populations. It is in the overlapping processes of government involvement in hurricane prediction, evacuation, structural protection and emergency response in the twentieth century that we begin to explore the inequities.

Hurricanes: Preparations and Response in New Orleans

Several conflicting tendencies put New Orleans area residents in more a vulnerable situation in terms of their physical exposure to hurricanes in the early twentieth century than had been the case the century before. The city was nearing the limits of expansion on the relatively safe natural levee, due to a burgeoning population forcing its growth toward the hurricane exposed Lake Pontchartrain shore. Beginning in 1900, local officials commenced a coordinated public works project that gradually extended water, sewerage, and drainage services to the entire city and made lakefront development feasible for the first time. Assisted by newly drained real

19. *NOTP*, 'Over One Hundred Lives Lost in Terrebone', 23 September, 1909, 1; *NOTP*, 'Sheriff Reports Forty Dead in Parish Below', 31 September, 1915, 2; *NOTP*, 'Known Dead from Storm', 30 September 1915, 1; *NOTP*, 'Fifteen Negroes Reported Dead below City', 1 October, 1915, 15; *NOTP*, 'Death Toll of Storm Placed at 350', 2 October, 1915, 1; Din, *Canary Islanders*, 146.

20. U.S. Army Corps of Engineers (hereafter USACE), New Orleans District, *History of Hurricane Occurrence along Coastal Louisiana*, New Orleans, 1972, 15.

21. USACE, *History of Hurricane Occurrence*, 18.

22. K. Meyer-Arendt, 'Resort Development along the Gulf of Mexico Littoral', Louisiana State University, PhD. dissertation, 1987.

estate, the economic boom of the 1920s accelerated residential sprawl northward across what had been a sea-level wetland but was becoming drained neighbourhoods behind a modest lakefront levee.[23]

The city's black population had grown during the late nineteenth century with the movement of freed slaves to the region's leading metropolis. From about 25,000 free and enslaved Blacks in 1860, the number tripled to over 77,000 in 1900 and continued upwards to 129,000 in 1930, although, as a percentage, the African American population remained between 26 and 28 per cent during the immediate post-bellum period. With limited residential options and few resources in a rapidly expanding city, these newcomers, as previous immigrants to the city, found housing away from the natural levee on the wetland fringe known as 'back of town'. By 1881, there was a clear pattern emerging, similar to other southern cities: African Americans occupied the least desirable real estate in low areas subject to flooding.[24] With the expansion of the city's public works, whites vacated older less-desirable neighbourhoods and moved into newly developed tracts near the lake. Deed covenants, by restricting the purchase of certain real estate to whites, reinforced racial separation and the city became rigidly segregated during the early twentieth century.[25] The movement of poor African Americans into low neighbourhoods, at the same time that exclusive white residential areas developed along the lakefront, increased exposure for both groups.

Beyond the city limits, high fur prices during the 1920s compelled St. Bernard Parish Islenos to remain in exposed locations where they could trap marsh-dwelling muskrats. Likewise, with the availability of manufactured ice and rail transport, shrimping and crabbing expanded to serve a growing urban market during the late nineteenth century.[26] This trend continued into the twentieth century, as shrimp landings more than doubled between 1918 and 1927, and the oyster catch overcame a late nineteenth century decline and edged upwards following implementation of improved conservation practices.[27] Coastal and near-coast communities of marginalised minorities that relied on marine and wetland resources in Plaquemines, Lafourche, and Terrebone parishes had ample economic justification to remain in

---

23. Colten, *Unnatural Metropolis*, 77–107.

24. R. Campanella, *Geographies of New Orleans: Urban Fabrics before the Storm*, Lafayette, 2006; J. Kellog, 'Negro Urban Clusters in the Post-bellum South', *Geographical Review*, 67, 1977, 310–21; J. Ueland and B. Warf, 'Racialized Topographies: Altitude and Race in Southern Cities', *Geographical Review*, 96, 2006, 50–78.

25. Colten, *Unnatural Metropolis*, 77–107.

26. H. Padgett, 'Sea Fisheries of the Southern States: Retrospect and Prospect', *Geographical Review*, 53, 1963, 36.

27. U.S. Bureau of Fisheries, *Fishery Industries of the United States*, Washington, 1920; U.S Bureau of Fisheries, *Fishery Industries of the United States*, Washington, 1929.

the path of extreme, although irregular, weather. Their labours fed urban diners and earned them a decent livelihood.

Tropical cyclones visited the south-east Louisiana coast with some frequency in the early twentieth century. Strong storms produced damage and flooding in 1901, 1905, 1909 and 1915. Lesser storms brought wind and rain in 1900 and 1906.[28] Areas that were most susceptible to flooding were the extensive low-lying marshlands south and east of New Orleans and the city's growing lakefront district. Even with modest winds, storm surge could wash across the marshlands where resource gathering communities existed and push up the bayous and canals into the lower sections of the densely developed city.

The 1915 hurricane proved to be the most devastating of the early twentieth century storms. Over eight inches of rain fell on the city within a day, while winds pushed storm surge into low-lying districts where flood waters rose as high as eight feet and damaged an estimated 25,000 structures. The local newspaper reported a considerable number of black fatalities and injuries. Of those identified by race 49 whites perished and 26 Blacks, along with another 132 unidentified victims, totals roughly in accord with the racial demography of the area.[29]

Surge raised water levels to over eleven feet across the lower delta and St. Bernard Parish wetlands. South of the city, the surge exceeded nine feet. The impact was most severe in low-lying areas, both across the coastal wetlands and in the urbanised areas. Waves and surge wiped out the Filipino community at Manila Village.[30] Leeville, an Acadian community on lower Bayou Lafourche, endured the destruction of all but one house.[31] Although most whites evacuated the small oyster-gathering community of Yscloskey, fifteen Blacks who remained behind died.[32] Plaquemines Parish, home mainly to Islenos, suffered extensive property damage and loss of life.[33]

Among the public services available to reduce the impact of storms were weather forecasts, disseminated primarily through newspapers at the time and therefore providing more immediate assistance to more literate citizens. The *Times Picayune* reported on the approaching storm three days before landfall and on the day of landfall stated that the hurricane 'will probably strike the coast near or immediately east of the mouth of the Mississippi River'.[34] Poor schools had limited reading abilities among urban African Americans. Islenos, Acadians and Houma

28. USACE, *History of Hurricane Occurrences.*
29. *NOTP*, 'Known Dead', 1.
30. *NOTP*, 'Fifteen Negroes', 15.
31. USACE, *History of Hurricane Occurrences*, 23.
32. *NOTP*, 'Fifteen Negroes', 15.
33. *NOTP*, 'Death Toll', 1.
34. *NOTP*, 'Tropical Storm Expected', 29 Sep, 1915, 1.

Indians used Spanish or French as their first language and also enjoyed limited benefits of public education. In 1910, illiteracy rates in Lafourche (42 per cent) and Terrebone (38 per cent) parishes were well above the Orleans percentage (29 per cent). Thus, newspaper accounts had less influence on these communities. The principal beneficiaries of printed weather forecasts were large shipping companies and railroads that undertook preparations before storms made landfall. Acknowledging that print reports were not equally accessible or entirely efficient, the weather service posted hurricane warning flags in coastal areas and also sent word to exposed communities by telegraph, telephone and even by messengers.[35] These warnings were disseminated equitably to the most exposed locations, although there were no formal evacuation plans at the community level and travel was a personal responsibility that required adequate financial resources and a notion that other locations offered greater security.

Following the 1915 hurricane, the Orleans Levee Board began a protracted effort to erect hurricane protection along the city's exposed lakefront. This represented a public programme to protect all exposed citizens. The local drainage authority reported that the 1915 storm surge had overwhelmed the six-foot protection levees near the lakefront and along the drainage canals[36] and, as early as 1916, the Orleans Levee Board received legislative approval to construct a higher and more substantial sea wall though it did not begin work until 1926. Work suffered disruption when the Levee Board diverted its attention to riverfront repairs following the 1927 flood and then the Great Depression prompted additional fiscal delays. With financial assistance from federal Depression-era programs in the 1930s, the levee board erected a 9.5-foot high sea-wall and created 6,000 acres of raised land with dredged lakebed sediments repositioned behind the flood protection structure.[37] Completed in 1934, the barrier protected a mixed population of Blacks living in low areas behind the natural levee and a growing suburban white population closer to the lake.[38]

Even at a time of explicitly racist Jim Crow policies and when denial of many basic services to Blacks was common, rational engineering and the urge to defend the larger city and its economy provided a basis for protecting both European and African American citizens of New Orleans. In 1930, whites constituted the city's majority (67 per cent), so expenditure on structural defences benefited more whites, but did not exclude Blacks. Thus, within New Orleans, it was virtually impossible to protect the economic and social elite without also protecting minority and low-income residents. Rural coastal parishes were unable to take on comparable

35. *NOTP*, 'Tropical Storm Expected', 1.

36. *Report of George G. Earl to Sewerage and Water Board*, New Orleans, 1915.

37. New Orleans Levee Board, *Building a Great City*, New Orleans, 1954.

38. Campanella, *Geographies of New Orleans*; Colten, *Unnatural Metropolis*; S. Carter, *A Report of Metropolitan New Orleans Land Use, Real Property, and Low Income Housing*, New Orleans, 1941.

structural protection projects due to their huge costs and Isleno, Acadian, Native American, and Filipino populations remained highly susceptible to storms.

By the time of the next major tropical cyclone in south-east Louisiana, hurricane tracking and warning capabilities had improved greatly. Early use of radar and also aircraft flights into storms enabled more precise tracking and improved forecasting of the 1947 storm. A full day before the 19 September landfall, military, municipal and other organisations set their hurricane preparation procedures into motion. These preparations included mobilising workers and equipment needed for rescue and repairs and also securing buildings and equipment subject to storm damage. Reported by both the print and radio media, warnings had become much more accessible to both the illiterate and to those whose first language was not English. The National Weather Service had also developed standardised plans that encouraged those in low areas to evacuate and those in safer locations to board up their windows and stockpile food and water.[39] For New Orleans, the bureau advised lakefront residents to move to higher and safer sites. Some residents evacuated the marsh community of Delacroix in St. Bernard Parish, while there was a more complete response from communities in Plaquemines Parish.[40] Overall some 23,000 evacuees, both urban and rural, flocked into the city's shelters, as did about 3,500 to 4,000 from downriver communities.[41] News accounts reported about 3,000 'French speaking trappers, fishermen, and truck farmers' crowded into the city's Municipal Auditorium.[42] Thus warnings, evacuation procedures, and shelters were available to the marginalised coastal dwellers.

Flooding caused by the storm struck both the vulnerable coastal populations, and also more affluent city dwellers. A storm surge of more than eleven feet washed over the rural wetland communities and destroyed twenty homes in Delacroix.[43] Wind-blown waves breached the sea-wall and caused flooding in urban lakefront neighbourhoods.[44] An editorial lamented the fact that the Lake Pontchartrain sea-wall 'proved insufficient to contain the fury of southerly whipped waters and that a large section of Lakeview and Gentilly was consequently flooded'.[45] Urban

39. I. Tannehill, *Hurricanes: Their Nature and History*, Princeton, 1938, 137–38.

40. *NOTP*, 'Take Precautions, Quit Lakeshore for Safer Spot', 19 Sep 1947, 1; *NOTP*, 'Delacroix Island Village Battens Down for Storm', 19 Sep 1947, 6; *NOTP*, 'Residents of Lower Coast Swarm into City for Safety', 19 Sep 1947, 9.

41. *NOTP*, 'Toll of Hurricane Surprisingly Low', 20 Sep 1947, 1; *NOTP*, 'Red Cross Maps Rehabilitation', 20 Sep 1947, 11.

42. *NOTP*, 'City's Hurricane Loss Light', 20 Sep 1947, 2.

43. USACE, *History of Hurricane Occurrences*; *NOTP*, 'Red Cross Maps', 11.

44. *NOTP*, 'City's Hurricane Loss Light', 2.

45. *NOTP*, 'Delacroix Island, 6.

flooding affected these two middle-class suburbs but also forced the evacuation of working class public housing projects near the Industrial Canal.[46]

In addition, extensive flooding covered middle class suburbs in adjacent Jefferson Parish in the 1947 event. During the storm, waves broke over the raised highway that doubled as a levee and, once within the raised perimeter, lake water could not be drained out. Water stood several feet high in homes in Metairie and covered the runways of the municipal airport.[47] An estimated 15,000 suburban residents endured up to two weeks of flooding.[48]

The effects spanned the social and economic spectrum and created broad-based support for more protection works.[49] Authorisation and construction of lakefront levees in Jefferson Parish protected a more exclusively white and middle class population – about 85 per cent white in 1950.[50] There were no comparable federal projects started in the wake of the 1947 storm to protect the coastal wetland dwellers. Granted, physical conditions in the coastal wetlands presented contrasting engineering situations and there were significantly lower population densities, but the post-1947 levees protected a narrow segment of the population while also encouraging expansion of post-war, white-flight residential areas. Indeed, the cost–benefit analysis required the 'potential' value of anticipated development to justify construction.[51] As federal dollars underwrote new suburban structural protection, the Orleans Levee Board announced plans to finance higher levees in working class eastern New Orleans to 9.5 feet and basically to the same standard as the lakefront levees. Thus, before the 1947 storm, there were inequities in the protection. Eastern New Orleans, with several public housing projects, had seven-foot levees, while the predominately middle-class and white lakefront had 9.5-foot levees.[52] Rural coastal communities had no comparable structural protections.

Preparations for and Responses to Hurricane Betsy

The next major storm to sweep over south-east Louisiana was Hurricane Betsy in September 1965. It exposed shortcomings in hurricane protection structures and accelerated an ongoing effort to enlarge and strengthen the defensive perimeter

46. *NOTP,* 'Red Cross Maps', 11.
47. *NOTP*, 'New Flood Perils Arise', 21 September 1947, 1, 4.
48. *NOTP*, 'Drastic Steps to Relieve Metairie Flood Planned', 22 September 1947, 1; USACE, *Letter from the Secretary of the Army: Lake Pontchartrain, Louisiana*. Senate Document, 81st Cong., 2nd sess., 1950.
49. *NOTP*, 'Aftermath and Future', 25 Sep 1947, 10 and USACE, *Letter from the Secretary*, 5.
50. USACE, *Letter from the Secretary*, 5.
51. C. Colten, *Perilous Place, Powerful Storms: Hurricane Protection in Coastal Louisiana*, Jackson, 2009.
52. *NOTP*, 'Levees will be Raised', 1 Oct 1947, 1.

around the region. Leading up to the storm, the region had both structural and procedural protections in place.

By 1960, the city had its 1930s lakefront sea-wall and improved levees around much of the eastern portion of the city. The Jefferson Parish's lakefront had an eight foot high lakefront levee; St. Bernard Parish had modest 'back levees' that stood five feet high, protecting the urbanised portions of the parish along the natural levee; and similar 'back levees' offered modest protection in Plaquemines Parish. Levees of this height provided sound protection against high tides and modest gales, but not from hurricanes. Far-flung coastal communities relied on stilt construction to minimise the impacts of hurricane-induced flooding. Thus, ethnic minorities like Islenos, Acadians, Native Americans, and Filipinos lived in the most exposed locations and with the least structural protection.

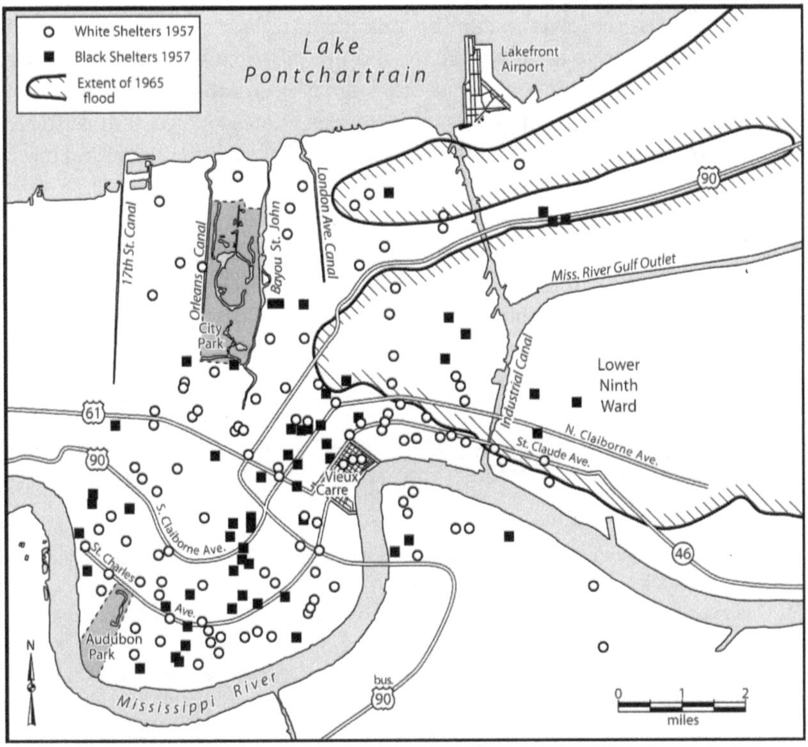

Figure 2. Hurricane Evacuation Shelters and Flooded Zones, 1957[53]

53. Source: Louisiana Civil Defense, 1957 and USACE, 1965 (cartography by Clifford Duplechin).

As part of post-Second World War civil defence preparations and also growing federal attention to hurricane preparations,[54] the city had prepared a plan for sheltering those who had to evacuate during a storm. The 1957 plan listed 187 schools that could be pressed into use as shelters.[55] It designated 123 of the facilities as shelters for whites and 64 for Blacks. In a city noted for significant differences in quality of its white and black schools, there was obvious potential for inequity in terms of the structural and sanitary conditions of the buildings.[56] But in terms of exposure to danger, eighteen white and eight black shelters suffered inundation due to Betsy (Fig. 2). The majority of both black and white shelters escaped high water and thus effectively served evacuees of both races. Since the area of inundation could not be predicted, another measure of anticipated risk affords some insight into equity. Twenty-seven of the 64 Black shelters (42 per cent) were lakeward of Claiborne Avenue, a corridor that approximately follows the one foot above sea level contour (Fig. 2). By contrast, 47 of the 123 white shelters (38 per cent) were distributed across the lower terrain. By a narrow margin, a higher percentage of shelters for Blacks were in safer locations and this relative equity between shelter locations is quite surprising given the economic pressures that restricted African Americans and the poor to low areas. Yet the expansion of whites into desirable but low-lying lakefront neighbourhoods offset potential inequities and placed more white schools in susceptible locations.

There were evacuation plans that provided procedures for those who lived in the coastal lowlands to retreat before the storm. Army engineers reported that 250,000 people, many from the coastal parishes, evacuated to storm shelters.[57] Yet, while most fled Grand Isle, many stayed behind at Shell Beach, where the storm destroyed approximately 95 per cent of the homes. Authorities reported that over three quarters of low-lying St. Bernard Parish dwellers heeded warnings and evacuated.[58] A small number of residents rode out the storm in the Plaquemines Parish courthouse, while an undetermined number stayed in their homes and some fled to higher ground.[59] The overall fatality count of 81 suggests evacuation was effective.

---

54. U.S. Weather Bureau, *A Model Hurricane Plan for a Coastal Community*, National Hurricane Research Project Report No. 28, Washington, 1959.

55. Louisiana Civil Defense Agency, *Louisiana Survival Plan: Reception and Care Study: Resources Study*, Baton Rouge, 1957.

56. See A. Fairclough, *Race and Democracy: The Civil Rights Struggle in Louisiana, 1915–1972*, Athens, GA, 1995; J. Landphair. 'Sewerage, Sidewalks, and Schools: New Orleans Ninth Ward and Public School Desegregation', *Louisiana History*, 40:1, 1999, 36–62.

57. USACE, *New Orleans District, Report on Hurricane Betsy*, New Orleans, 1965.

58. *NOTP*, 'Southern Louisiana Begins Survey of Storm Damage', 11 Oct 1965, 8–9; *NOTP*, 'Grand Isle but a Wasteland Now', 12 Oct 1965, 1; and *NOTP*, 'Deaths Rising in City, State', 12 Oct 1965, 1, 6.

59. *NOTP*, 'Pointe a la Hache, Phoenix are Reported Wiped Out', 12 Oct 1965, 11.

Following Betsy, Congress authorised a major hurricane protection project that would considerably expand the territory encircled by levees. It called for a new levee system around eastern New Orleans, improved levees around the more densely settled sections of St. Bernard and Plaquemines parishes, as well as projects that enclosed a portion of the west bank of the Mississippi River and highly exposed areas in Lafourche Parish and at Grand Isle. The plan also included improvements to the lakefront levees in Orleans and Jefferson parishes.[60]

At the time, New Orleans still had a majority white population (63 per cent in 1960), but that was changing. Construction of interstate highways, improved hurricane protection in Jefferson and St. Bernard parishes and the overall tendency of those with means to depart central cities for the suburbs contributed visibly to white flight from New Orleans proper. Its white population fell to 57 per cent in 1970 and 45 per cent in 1980. Meanwhile its suburban neighbours became increasingly white (Table 1).

Table 1. Parish Populations and Federal Levee Expenditures, 1970–2000 [61]

Year	Parish	Population	%Black	Estimated $m*	Completed %
1970	Jefferson	208,769	15	148.9	8
	Orleans	627,525	37	148.9	8
	Plaquemines	22,545	28	nd	nd
	St. Bernard	32,186	4	148.9	8
1980	Jefferson	337,568	12	394.0	25
	Orleans	593,471	45	394.0	25
	Plaquemines	25,225	23	110.0	53
	St. Bernard	51,185	5	394.0	25
1990	Jefferson	454,592	14	501.0†	80†
	Orleans	557,515	55	501.0	80
	Plaquemines	26,049	21	168.0	62
	St. Bernard	64,097	4	501.0	80
2000	Jefferson	455,466	23	527.0	88
	Orleans	484,674	67	527.0	88
	Plaquemines	26,757	23	173.0	79
	St. Bernard	67,229	8	527.0	88

60. Colten, *Perilous Place*, ch. 4.

61. Source: U.S. Census and USACE, Annual Reports.

† data for 1992

* Estimated dollars and per cent completed for the Lake Pontchartrain and Vicinity Hurricane Protection Project included all or portions of Jefferson, Orleans, and St. Bernard parishes. Hence the same values for these parishes appear in the table. These values do not represent comparable amounts for each parish but one sum shared by the single project that encompassed all three parishes.

Attributing environmental injustices to the bodies that funded, designed, and built the hurricane protection system is extremely problematic. The Corps of Engineers' assessment of Hurricane Betsy's damage did not even acknowledge race or income.[62] Subsequent planning documents reported populations but, with typical engineering efficiency, did not indicate race, ethnicity or poverty levels.[63] Furthermore, populations were in flux during the protracted construction period and changed dramatically in several areas between 1966 and 2005. The project lagged behind schedule and its budget ballooned, thereby distorting any connection between design and equitable protection. Any attempts to tie dollars expended to population protected is complicated by the general reporting of estimated costs in large units that do not correspond to census territories (Table 1). What is obvious is that the pace of construction, while slower than the original projections, did come closer to completion for the more urbanised portions of the project areas by 2000. The more densely populated sections of Jefferson, Orleans, and St. Bernard parishes enjoyed 88 per cent completion by the beginning of 2000. The lower delta communities of fishers and farmers had less complete protection (Table 1).

As the hurricane protection construction inched forward during the 1990s, local demographics were changing. New Orleans' population reached an overwhelming African American majority of 67 per cent by 2000. The newly protected eastern New Orleans area had become a largely Black middle-class residential area; most census tracts had over seventy per cent African Americans. In addition, a sizable Vietnamese community lived within the far eastern reaches of the city's leveed territory. Thus, urbanised areas closest to the Gulf of Mexico and potential storm surge had grown in absolute numbers and also in terms of the percentage of African Americans and other minorities. Jefferson Parish, too, saw a sizeable increase in its African American population. Most notably, the Black population remained low on the safer east bank but increased to 37 per cent on the west bank. This area had endured lengthy delays in construction of the hurricane levees and was only 32

62. USACE, *Report on Hurricane Betsy*.

63. See Secretary of the Army, *Letter from the Secretary of the Army: Lake Pontchartrain and Vicinity*, Louisiana, Washington, 1965; and USACE, *Interim Survey Report: Lake Pontchartrain Louisiana and Vicinity*, New Orleans, 1962.

per cent complete in 2004.[64] Thus, the most exposed urbanised population was an African American population that had experienced rapid growth but did not constitute a majority.

Plans for evacuation in the post-Betsy period underwent a fundamental change. With new levees that would potentially impound more flood water, along with the construction of interstate highways, using numerous small shelters within the city was no longer a viable solution. Total evacuation by automobiles on the new expressways became the objective, though an evacuation investigation noted the problems presented by certain segments of the population. In general, it reported that, with aggressive warnings, a high percentage would evacuate from areas near the open coast. It suggested that Hispanics would tend to evacuate at the same rate as the general population but concluded that Vietnamese would leave at about half the rate of the general population. The report also indicated that about fifteen per cent of New Orleans residents were without personal vehicles and would have to rely on public transport.[65] Yet the final plan developed by local agencies relied heavily on self-evacuation, with only contingent provisions for the immobile, impoverished, illiterate, infirm or incarcerated. Despite a 1994 executive order to take environmental justice into account for all federal projects, the agencies charged with managing levee construction and evacuation did not explicitly address this mandate.[66] Additionally, the Corps' Risk and Reliability study since Hurricane Katrina fundamentally ignores equity.[67] At a time when the USEPA was making a concerted effort to showcase its environmental justice initiative in New Orleans and even the federal Minerals Management Service was investigating potential impacts of offshore oil extraction on minority and low income populations, hurricane preparation included no comparable considerations for equity.[68] Warnings, with rapid dissemination via television and radio, were equitable.

Conclusions

Residents of south-east Louisiana had come to recognise the seasonal risks of hurricanes during the colonial period, but risk did not outweigh the opportunities

64. USACE, *West Bank and Vicinity Hurricane Protection Project*, http://www.mvn.usace.army.mil/hps2/ Accessed August 2006.

65. Federal Emergency Management Agency (FEMA) and USACE, *Southeast Louisiana Hurricane Preparedness Study*, New Orleans, 1994, 4–6.

66. See C.E. Colten, 'Environmental Justice in a Landscape of Tragedy', *Technology in Society*, 29:2, 2007, 173–9.

67. USACE, *Risk and Reliability Report*, [http://nolarisk.usace.army.mil/] 2008, accessed May 2008. http://www.mvn.usace.army.mil/hps/risk_depth_map.html

68. C. Colten, 'Environmental Justice in the Big Easy? The Agriculture Street Landfill Tragedy', *Environmental Practice* 3:1, 2001, 19–26; S. Hemmerling and C. Colten, *Environmental Justice: A Comparative Perspective in Southeast Louisiana*, New Orleans (forthcoming).

provided by the region. In the nineteenth century, a pair of powerful storms revealed the impacts of tropical storms to both elite and lower class citizens and prompted the closure of coastal resorts. Permanent retreat was not a viable option for fisher and trapper families who remained tied to their livelihoods in the exposed coastal and wetland communities. Wealth and permanent inland homes enabled some to make locational adjustments and reduce their risks before 1900.

By the early twentieth century, while large corporations such as shipping companies and railroads, were able to take precautionary actions, public agencies provided warnings to coastal residents. The warnings also benefited the residents of coastal fishing communities and some evacuated and found shelter in urban setting. Nonetheless, fatalities tended to be higher in those communities where minority ethnic populations were predominant and Isleno, Acadian, Native American and Filipino communities remained in high-risk locations due to their economic dependence on coastal resources.

Structural protections erected around portions of New Orleans after the 1915 hurricane protected both poor black and wealthy white residents. Yet, before the 1947 hurricane, the working class eastern neighbourhoods and suburban St. Bernard Parish offered inferior protection to that of the middle class neighbourhoods near the lakefront and the coastal wetlands. Despite structural inequalities, the 1947 flood overwhelmed levees across the region and impacted all classes and races. Structural improvements after 1947, by design, provided a more consistent level of protection to the entire urbanised area but not to down-river parishes and the minority communities there. Urban evacuation plans, while strictly segregated in the late 1950s and early 1960s, offered surprisingly equitable protection to whites and Blacks. Rural, coastal minorities received warnings and could find shelter in the city.

Adjustments in both structural and procedural preparations following Betsy in 1965 encouraged development in high-risk locations. This follows a recognisable pattern of development associated with structural improvements in flood protection. In the process, planners gave no explicit attention to matters of race or class, even though race was a prominent issue locally and drove much of the urban population redistribution. Nonetheless, Blacks moved into the eastern New Orleans neighbourhoods, while whites shifted into St. Bernard and into lakefront areas in Jefferson Parish. Expenditure on hurricane structures inevitably provided limited protection for those of all races, classes and minority ethnic communities. Yet, risk increased for all as well. As some federal agencies responded to a 1994 Executive Order to address environmental justice, no comparable efforts emerged in hurricane preparations. And, tragically, this remains true in the post-Katrina period.

Select Bibliography

Bankston, C. , III, 'Southeast Asians in Louisiana', in C. A. Brasseaux, ed. *A Refuge for All Ages: Immigration in Louisiana History*, Lafayette, 1996, 661–77.

Campanella, R. *Geographies of New Orleans: Urban Fabrics before the Storm*, Lafayette, 2006.

Colten, C.E. *Perilous Place, Powerful Storms: Hurricane Protection in Coastal Louisiana*, Jackson, 2009.

Colten, C.E. *An Unnatural Metropolis: Wresting New Orleans from Nature*, Baton Rouge, 2005.

Comeaux, M.L. *Atchafalaya Swamp Life: Settlement and Folk Occupations*, Baton Rouge, 1972.

Din, G. *The Canary Islanders of Louisiana*, Baton Rouge, 1988.

Fairclough, A. *Race and Democracy: The Civil Rights Struggle in Louisiana, 1915–1972*, Athens, GA, 1995.

Gomez, G. 'Perspective, Power, and Priorities: New Orleans and the Mississippi

River Flood of 1927', in C. E. Colten, ed. *Transforming New Orleans and Its Environs: Centuries of Change*, Pittsburgh, 2000, 109–21.

Kniffen, F.B. *The Historic Indian Tribes of Louisiana from 1542 to the Present*, Baton Rouge, 1994.

Landphair, J. 'Sewerage, Sidewalks, and Schools: New Orleans Ninth Ward and Public School Desegregation', *Louisiana History* 40:1, 1999, 36–62.

Ueland, J. and B. Warf, 'Racialized Topographies: Altitude and Race in Southern Cities', *Geographical Review*, 96:1, 2006, 50–78.

U.S. Army Corps of Engineers (USACE), New Orleans District. *History of Hurricane Occurrence along Coastal Louisiana*, New Orleans, 1972.

Wisner, B., P. Blaikie, T. Canon, and I. Davis, *At Risk: Natural Hazards, People's Vulnerability, and Disasters*, 2nd ed. New York, 2004.

PART III

Water-related Inequalities

Chapter Six

Urban or Suburban Water? Working Class Suburbs, Technological Systems and Environmental Justice in Swedish Cities in the Late Nineteenth Century[1]

Jonas Hallström

By 1900, some Swedish working class suburbs had gained a nationwide reputation due to the poor sanitary and moral conditions that were said to prevail there.[2] New working class suburbs were the result of city growth and industrialisation and with them came questions of whether or not their inhabitants should enjoy the same technological amenities as people in the city, for instance in terms of water supply, sewerage and other technologies used to improve the urban environment. For their part, city governments were uncertain about whether to include these new peripheral areas and their inhabitants on the same basis as those for which they had formal responsibility.[3] These working class suburbs, or shanty towns (*kåkstäder*) as they were often called, developed neither from previous urban nor rural settlements. They represented something entirely new in late nineteenth century Sweden. Surprisingly little has been written about them and their relationship to the cities.[4]

---

1. See also J. Hallström, 'The Growing Pains of the Pipe-bound City: The Extension of Water and Sewerage to Suburban Areas in Norrköping, Sweden, 1860–1890', *Public Works Management & Policy* 6:3, 2002, 186–99; J. Hallström, *Constructing a Pipe-Bound City: A History of Water Supply, Sewerage, and Excreta Removal in Norrköping and Linköping, Sweden, 1860–1910*, Linköping, 2002, part II; J. Hallström, 'Technology, Social Space and Environmental Justice in Swedish Cities: Water Distribution to Suburban Norrköping and Linköping, 1860–90', *Urban History* 32:3, 2005, 413–33.

2. See, for example, 'För dagen: Fakta tala', *Svensk Läraretidning*, 19:29, 1900, 481–82. Retrieved 23 February 2009 from http://www.runeberg.org

3. H. Meller, *European Cities 1890–1930s: History, Culture and the Built Environment*, Chichester, 2001, 2–3; Hallström, *Constructing a Pipe-Bound City*, 13–38, 168–228; A-M. Thagaard, *Backarna. Liv och historia i en förstad*, Malmö, 1992, 53–108.

4. I. Johansson, *Stor-Stockholms bebyggelsehistoria. Markpolitik, planering och byggande under sju sekler*, Hedemora, 1991, 236–7. My own studies of the academic literature on working class suburbs in Sweden confirm the view of Johansson; virtually nothing has been written since 1991. Johansson, *Stor-Stockholms bebyggelsehistoria*; Hallström, *Constructing a Pipe-Bound City*; and Hallström, 'Technology, Social Space and Environmental Justice' are exceptions.

This chapter considers how the Swedish cities of Norrköping and Linköping treated new working class suburbs just outside the jurisdictional areas of their administrative control during the period from c. 1860 to 1890. Attitudes to the suburbs were expressed in discussions relating to piped water supply and sewerage and primary sources about these two cities will also be used to compare secondary material concerning a wider Swedish urban context, namely the cities of Stockholm and Malmö, during roughly the same period.

The issue of extended amenities in contiguous urban areas relates to the relatively new concept of environmental justice or equity and has inspired several historical studies within the field of urban history in the past ten years or so.[5] In this chapter environmental justice is considered from the point of view of Henri Lefebvre's and David Harvey's notions of the social character of space. They assume that space, which is sometimes seen as something neutral outside the social context, is in fact an implement of social control. Lefebvre has coined the term 'social space' to denote this and Harvey argues, in line with Lefebvre, that 'spatial practices' – that is, appropriation, production, control or representation of space – are inextricably linked to social relations in the city.[6] The environments and spaces members of social groups inhabit and the environmental technologies they spatially extend or withhold therefore tell us something about social relations, environmental justice and the evolution of technology in the city.

5. A brief selection of such works includes R. Guha and J. Martinez-Alier, *Varieties of Environmentalism: Essays North and South*, London, 1997; J.R. McNeill, *Something New Under the Sun: An Environmental History of the Twentieth-Century World*, New York, 2000; M.A. Flanagan, 'Environmental Justice in the City: A Theme for Urban Environmental History', *Environmental History* 5, 2000,159–64; C.E. Colten, 'Basin Street Blues: Drainage and Environmental Equity in New Orleans, 1890–1930', *Journal of Historical Geography* 28:2, 2002, 237–57; J.A. Tarr, 'The Metabolism of the Industrial City: The Case of Pittsburgh', *Journal of Urban History* 28:5, 2002, 511–45; B. Luckin, 'Environmental Justice, History and the City: The United States and Britain, 1970–2000', in D. Schott, B. Luckin and G. Massard-Guilbaud, eds. *Resources of the City: Contributions to an Environmental History of Modern Europe*, Aldershot, 2005, 230–45; J.F.M. Clark, '"The Incineration of Refuse is Beautiful": Torquay and the Introduction of Municipal Refuse Destructors', *Urban History* 34:2, 2007, 255–77; H. L. Platt, 'From Hygeia to the Garden City: Bodies, Houses, and the Rediscovery of the Slum in Manchester, 1875–1910', *Journal of Urban History* 33, 2007, 756–72.

6. H. Lefebvre, *The Production of Space*, Oxford, 1991, 401–23; D. Harvey, *The Condition of Postmodernity: An Enquiry into the Origins of Cultural Change*, Cambridge, 1990, 201–25. See also S. Gunn, 'The Spatial Turn: Changing Histories of Space and Place', in S. Gunn and R.J. Morris, eds. *Identities in Space: Contested Terrains in the Western City Since 1850*, Aldershot, 2001; E.W. Soja, 'Reassertions: Towards a Spatialized Ontology', in J. Agnew, D.N. Livingstone and A. J. Rogers eds. *Human Geography: An Essential Anthology*, Oxford, 1996, 623–5, 633–5.

The Cities of Norrköping and Linköping

The cities of Norrköping and Linköping were very small compared to the larger and mid-sized cities on the continent of Europe and in Great Britain. Nonetheless they were considered to be cities administratively, financially and culturally, and were both experiencing urbanisation during the period after 1860 (see Table 1).[7] Norrköping's population expanded from 17,000 to 46,000 and Linköping's from 5,000 to 18,000 in the period 1850 to 1910. By comparison, Stockholm grew from 93,000 to over 342,000 in the same period; Göteborg, from 26,000 to 168,000 or half of the size of Stockholm; and Malmö from 13,000 to 83,000 between 1860 and 1910. The respective growth rates are presented in Table 1.

Table 1. Population in Five Swedish Cities, 1850–1910 (1850 = 100) [8]

	1850	1860	1870	1880	1890	1900	1910
Stockholm	100	121	146	181	264	323	368
Göteborg	100	142	216	293	401	500	643
Malmö	100	145	196	291	370	465	637
Norrköping	100	118	141	158	194	242	274
Linköping	100	117	138	167	241	278	346

Norrköping is situated on the river Motala Ström in the eastern part of Östergötland County in south-eastern Sweden, just as the river reaches the bay Bråviken and the Baltic Sea. Louis de Geer made Norrköping the centre of his manufacturing empire in the early seventeenth century, when he established the first paper and brass mills there, followed a few years later by the textile manufactory of Drags. By the mid-nineteenth century Norrköping was dominated by several smaller woollen manufactories.[9] In the 1840s and 1850s the popularity of finer cloth declined, while the demand among workers and farmers for cheap fabrics increased. The textile manufacturers consequently mechanised production in order

7. Swedish cities were defined as such primarily on administrative grounds, not on population. This meant that market places and new urban areas that were not designated cities – *köping* and municipality (*municipalsamhälle*) – could in some cases be larger as regarded population: J-O. Drangert, M. C. Nelson and H. Nilsson, 'Why Did they Become Pipe-bound Cities? Early Water and Sewerage Alternatives in Swedish Cities', *Public Works Management and Policy* 6:3, 2002, 173.

8. Source: *Historisk statistik för Sverige*, vol. I: Befolkning 1720–1967, Stockholm, 1969, 61–5.

9. B. Horgby, *Surbullestan*, Stockholm, 1989, 38; T. Söderberg, 'Norrköpings ekonomiska och sociala historia 1719–1870', in B. Helmfrid and S. Kraft, eds. *Norrköpings historia IV: Tiden 1719–1870*, Stockholm 1968, 17–73; C. Hugerth, *Holmen: A Swedish Industrial History*, Norrköping, 1996, 2; L. Schön, *En modern svensk ekonomisk historia. Tillväxt och omvandling under två sekel*, Stockholm, 2000, 97–8.

to mass produce this cloth. The dominance of the woollen industry within the city was considerable: in 1870, 52 per cent of the industrial workers of the city were employed in this sector and Norrköping's woollen industry was the primary centre of production in Sweden. Unsurprisingly, in the 1860s Norrköping was one of the most successful industrial cities in Sweden, second only to the capital, Stockholm.[10]

Linköping lies on the river Stångån just south of Lake Roxen, around forty kilometres south-west of Norrköping. It had been an ecclesiastical and educational centre for centuries and the seat of the county government since 1634. Linköping was also the agricultural centre of the surrounding fertile plain, the so-called *Östgötaslätten*, and some of the landed gentry owned houses in the city. Prior to receiving the right to direct international trade in 1874, local trade, handicraft and small businesses in Linköping thrived by virtue of their close proximity to the farming district.[11] In contrast to Norrköping, Linköping was not an industrial city, at least not judged by the economic activities within the actual city boundary. If the industries on the other side of the river, in the St. Lars rural district (*landskommun*), are also taken into account, the city stands out as a little more industrialised but it was still far behind Norrköping. The few industries that did exist within the city were mostly situated along the river. Asklund's tobacco factory, which was the largest one, had around fifty workers in the 1860s. On the eastern bank of Stångån, in Ladugårdsbacke, lay L. T. Brogren's brewery, as well as mills, textile factories and a distillery at Tannefors and Nykvarn.[12]

The Norrköping and Linköping Suburbs

One consequence of industrialisation in Sweden was the appearance of suburban shanty towns. Housing shortages, with accompanying high rents and increased land prices, drove the working class in particular beyond the jurisdictional reachof

10. Horgby, *Surbullestan*, 38–41; Schön, *En modern svensk ekonomisk historia*, 98–100; J. Svensson, S. Godlund and K. Godlund, 'Norrköpings ekonomiska och sociala historia 1870–1914', in B. Helmfrid and S. Kraft eds. *Norrköpings historia V. Tiden 1870–1914*, Stockholm, 1972, 2, 10–11, 78–80, 95–7, 180–1, 239–245; K. Nyberg, 'Brittisk teknik, svensk överföring och finländskt mottagande? Textilindustrin i Sverige och Finland 1809-1870. Exemplet ylleindustrin', *Historisk tidskrift* 120:4, 2000, 623–4.

11. S. Andersson, 'Politik och demokrati', in H. Nilsson, ed. *Norrköpings historia. 1900-talet*, Linköping, 2000, 298–9; G. Bagge, *Ungdomsminnen från östgötabygder*, Linköping, 1922, 50–60; S. Hellström, 'Befolkning och social struktur', in S. Hellström, ed. *Linköpings historia 4. Tiden 1863–1910*, Linköping, 1978, 11–13; R. Marks von Würtemberg, *Min barndoms Linköping*, Linköping, 1955, 81–100; D. Torbrand, 'Kommunikationernas förändring', in Hellström, ed. *Linköpings historia 4*, 92–4.

12. P. Almroth and S. Kolsgård, 'Näringsliv', in Hellström, ed. *Linköpings historia 4*, 111–43; Hellström, 'Befolkning och social struktur', 11–13.

the city's borders or administrative areas.[13] Since construction in these areas was not restricted by national building codes, fire-protection and health laws it was accordingly less expensive.[14]

Norrköping's northern suburb, which belonged to the Östra Eneby rural district north of the city, was seen by many as a shanty town. The area was marshy and unsuitable for farming, so the property owners divided it into building sites which were then sold at prices lower than those in the city. The suburb had its origins around 1860 and was situated along the main eastern railway line that was under construction from 1863 to 1866; in 1860 the suburb had about sixty inhabitants and did not grow much during the 1860s.[15] The transformation into a suburb, sufficiently crowded with mostly workers and poor people to be regarded as a problem area by local and regional authorities, occurred during the economic boom of the 1870s, when population increased very rapidly.[16]

If the housing and sanitary conditions in Swedish industrial towns at this time were generally poor, in Norrköping these conditions were noticeably worse, particularly in its suburbs.[17] In the northern suburb, houses were often crowded and built of shoddy materials; even stables were sometimes turned into apartments.[18] Sanitary measures of any kind were lacking, water supply was rudimentary and the suburban sewers ended in open trenches. These conditions differed within the northern suburb, however. In certain areas the standard of accommodation was better.[19]

---

13. The city's planned area, as defined by the city plan, constituted the actual city in administrative and judicial terms and the national urban general regulations, building codes, health and fire laws were only applicable there. Areas outside the planned area were considered to be the countryside, despite the fact that they sometimes lay within the city border. Swedish cities could thus include land that was administered by the city but was not formally urban. See Y. Larsson, *Inkorporeringsproblemet. Stadsområdets förändringar med särskild hänsyn till svensk förvaltningspraxis, del I & II*, Stockholm, 1913, 373–6.

14. Larsson, *Inkorporeringsproblemet*, 13, 105.

15. M. Arvidsson, 'Från förstad till stadsdel. En studie av inkorporeringen av Norra förstaden med Norrköpings stad 1886–1916', unpublished Masters thesis, Linköping, 1995, 10–11; R. Myrdal, *Styrelse, förvaltning, politik 1863-1919, Norrköpings historia 11*, Uppsala, 1972, 301–3, 396; B. Gejvall-Seger, 'Stadsplanering och bebyggelseutveckling i Norrköping 1719–1970', in B. Helmfrid and S. Kraft, eds. *Norrköpings historia VI: Tiden 1914–1970*, Stockholm, 1976, 81, 90.

16. Horgby, *Surbullestan*, 31–3; Larsson, *Inkorporeringsproblemet*, 105; Arvidsson, 'Från förstad till stadsdel', 11.

17. Svensson, Godlund and Godlund, 'Norrköpings ekonomiska och sociala historia', 259–67.

18. Horgby, *Surbullestan*, 31–3.

19. Norrköping City Archives (NSA), Norrköping City Council Archives, supplement No. 15, 1893; Svensson, Godlund and Godlund, 'Norrköpings ekonomiska och sociala historia', 267. The northern suburb consisted of three distinct areas. *Marielund* was the western part, *Fredriksdal* the eastern part, and *Sandby* the northern part (see Fig. 1).

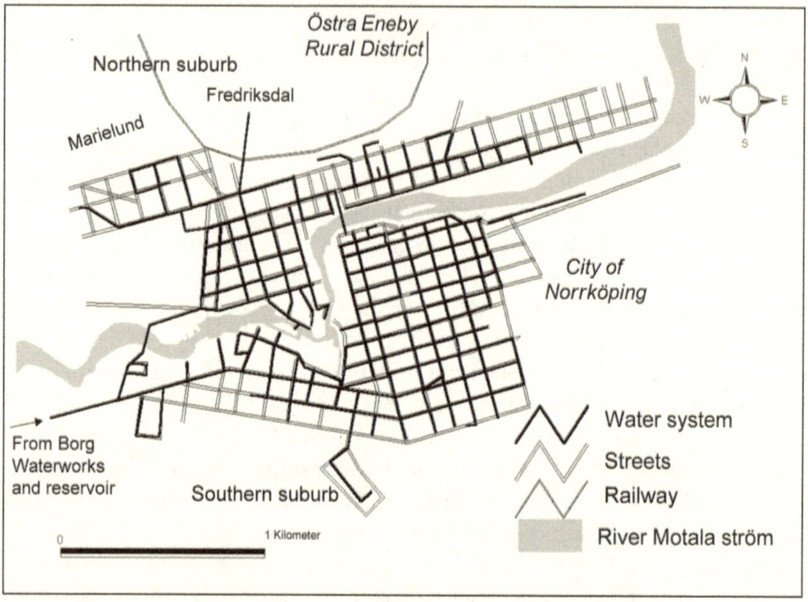

Figure 1. The Norrköping Water System in 1896[20]

The network of streets roughly corresponds to the planned area of the city of Norrköping.

In Linköping the area east of the river Stångån was not under the city's jurisdiction. It belonged to the St. Lars rural district, which surrounded the city and one of Linköping's several suburbs, Ladugårdsbacke, was situated there. Like Norrköping's northern suburb, it was considered to be a shanty town; the area was partly marshy and sanitation was very poor. It was also a working class area, in close proximity to the riverside industrial districts of Nykvarn and Tannefors.[21]

---

20. Sources: Hallström, 'Technology, Social space and Environmental Justice in Swedish Cities', 424; T. Schmid Neset and J. Hallström, 'The Pipe-bound City in Time and Space: Applying GIS to the Historical Development of Two Cities', in R. Coopey and T. Tvedt, eds. *A History of Water Volume 2: The Political Economy of Water*, London, 2006.

21. Myrdal, *Styrelse, förvaltning, politik*, 283; Almroth and Kolsgård, 'Näringsliv', 111–24; D. Torbrand, 'Stadsplanering', in Hellström, ed. *Linköpings historia 4*, 25-32; H. Nilsson, *Mot bättre hälsa. Dödlighet och hälsoarbete i Linköping 1860-1894*, Linköping, 1994, 140–1. Ladugårdsbacke consisted of several smaller areas such as *Ladugårdsbacke*, *Ladugårdsgärde* and *Tomteboda*, but the former will hereafter be used as a common name for the whole area (it is the most frequently used in the literature).

City versus Suburb: The Extension of Water Systems in Norrköping and Linköping

Modern water supply and sewer systems were constructed in Norrköping between 1872 and 1875. According to customary practices in Norrköping, water pipes outside the planned area had to be paid for by the building owners themselves and this meant that few in the northern suburb could afford a connection to the city's piped water supply. A company just north of the city border, the engineering workshop Norrköpings Mekaniska Verkstad, was granted the right to distribute water to the inhabitants of the eastern part of the suburb, Fredriksdal, particularly to those living in close proximity to its premises. In this way, the City Council could attend to the needs of the poor suburban district and still make some money, since the company paid for both the infrastructure and the water.[22]

In the 1880s, the number of inhabitants in the northern suburb expanded quickly, reaching some 3,000, and, in 1885, it became a municipality (*municipalsamhälle*).[23] It was really a part of the rural district Östra Eneby, adjacent to the city of Norrköping, but by being designated a municipality the suburb came under the national urban laws while remaining a part of the rural district administratively, judicially and ecclesiastically.[24] For instance, the building of houses and privies came under urban building codes, although it is difficult to say how much effect the regulations had in practice.[25]

In the mid-1880s, the inhabitants of the northern suburb began to demand help from the city with essential infrastructure and services. Both private and common water taps, for instance, were too few. In 1886, a committee from Fredriksdal wrote to Norrköping Waterworks Board to ask if a water pipe could be extended from the engineering workshop along one of the streets.[26] The Waterworks Board was not in favour of any new extensions of the water supply. Water consumption was increasing exponentially and the pumps of the Norrköping waterworks were

22. NSA, Norrköping City Council Archives, AI a:12 and AI a:13, minutes 26 November 1874, section 6 and 9 September 1875, section 6; Svensson, Godlund and Godlund, 'Norrköpings ekonomiska och sociala historia', 178–9; Hallström, *Constructing a Pipe-Bound City*, 170–203, 336–7. Although the sources are not clear on this point, it seems likely that the engineering workshop in its turn charged the inhabitants for the water.

23. Arvidsson, 'Från förstad till stadsdel', 11, 20.

24. Arvidsson, 'Från förstad till stadsdel', 12; L. Nilsson, *Den urbana transitionen. Tätorterna i svensk samhällsomvandling 1800–1980*, Stockholm, 1989, 53–6; Larsson, *Inkorporeringsproblemet*, 107; Gejvall-Seger, 'Stadsplanering och bebyggelseutveckling', 82; NSA, Norrköping City Council Archives, supplement No. 43, 1893 and No. 1, 1894.

25. 'Byggnadsordning för Norrköpings Norra Förstäders område', *Kongl. Maj:ts Befallningshafvandes i Östergötlands län allmänna kungörelser*, Serien B, No. 43, 1885, Norrköping, 1922.

26. The Waterworks Board was the city's administrative unit for the water and sewer systems. It was placed under the City Council, but was also financially responsible to the city's Financial Department.

already functioning at maximum capacity, so expansion outside the planned area could not be permitted. Furthermore, if the suburb was brought within the orbit of Norrköping then it would be very difficult to deny other suburbs similar access to water infrastructure. The Board was of the opinion that more water would increase rather than decrease sanitary problems, since drainage was so poor in the suburb.[27]

Not only was the Waterworks Board divided on this question, so was the City Council, and this became apparent at a meeting in late 1886 when a protracted discussion took place and several proposals were put to a vote. A detailed account of the discussion, which was published in the local newspaper *Norrköpings Tidningar* the day after the meeting, reveals deep disagreement among the members of the Norrköping City Council on this question.[28]

The proponents of the water extension started out by contradicting the Waterworks Board's technical argument, namely that the pumps were running at their maximum rate. Fredrik Blombergh was of the opinion that the pumps were capable of working either a little faster or a little longer per day without actually straining them and thereby both the city and the northern suburb could be supplied with water. The opponents, who basically supported the Waterworks Board, did not share this optimistic view of the capacity of the city's water supply and were afraid that the city itself soon would encounter a shortfall in its water supply.[29]

Issues of quality, quantity and availability of the water were also debated, and the assumption was that these aspects influenced the sanitary conditions in the suburb one way or another. The proponents thought that the northern suburb needed not only a pure and ample supply of water, which was necessary from a sanitary point of view, but also a readily available and convenient supply. That is, pipes had to be extended, if not to each household, at least in close proximity to them. The suburb had been under the national building code since 1885 but most buildings were still not in any condition to have supply pipes installed.[30]

The opponents, on the other hand, regarded the water tap at the premises of the engineering workshop as sufficient, since it delivered pure city water. An extension to the system of piped water would certainly be convenient for the people in the area but was not at all necessary for sanitary purposes. The single water

27. NSA, Norrköping City Council Archives, minutes 21 October 1886, section 4; Norrköping Waterworks Board Archives, AI:2, minutes 22 September 1886, section 2 and 6 November 1886, section 3.

28. NSA, Norrköping City Council Archives, minutes 18 November 1886, section 16; Norrköping Waterworks Board Archives, AI:2, minutes 19 March 1887, section 3; Account of the City Council meeting 18 November 1886, *Norrköpings tidningar* (*NT*), 19 November 1886.

29. *NT*, 19 November 1886.

30. *NT*, 19 November 1886; 'Byggnadsordning för Norrköpings Norra Förstäders område'.

standpipe was sufficient; if they wanted more water they could dig their own wells, which would also be cheaper for them.[31]

Another sanitary aspect was the fear of epidemics, especially cholera. This fear was widespread among both opponents and proponents of the extended water supply proposals, particularly since, in their view, it was very likely that an epidemic would spread to Norrköping as well: '[Epidemics] hang . . . because of the suburb, like a sword of Damocles over the city of Norrköping', as Axel Molin said.[32] The proponents were of the opinion that it would be an advantage for the city itself if the northern suburb had piped water, though it would need improved sewerage first, while the opponents argued that the extension of water in such a situation would only mean an even bigger threat to the city, because of the poor drainage in the suburb.[33]

In addition, humanitarian arguments were advanced by the proponents. Since so many inhabitants of the northern suburb worked in Norrköping, it was 'the duty and obligation of the city to extend a helping hand to the inhabitants of the suburb', said Fredrik Blombergh.[34] The opponents, on the other hand, wondered where to draw the line if other surrounding areas, notably working class areas, demanded the right to water on the same grounds.[35]

At this time, the City Council was aware that the northern suburb intended to seek incorporation within the city of Norrköping but both groups favoured the separation between the suburb and the city in the long run. For those who sought to extend the water supply to the suburb, it was a means to ensure the continued separation of city and suburb by assisting the independence of the suburb. Opponents feared that expanding the water system would either draw the suburb and the city closer to each other – since not only water but also other public services would then be integrated – or the suburb would grow even faster and intensify the environmental problems of both areas.[36]

When discussing such humanitarian issues, both proponents and opponents of the water extension saw the northern suburb as a 'growth':[37] 'As it is now, the

31. *NT*, 19 November 1886.

32. *NT*, 19 November 1886. Epidemics probably lived on in the memories of the city's inhabitants as the city had been hard hit by cholera in 1853 and 1866: E. Gullberg, 'Norrköpings kommunalstyrelse 1719–1862', in B. Helmfrid and S. Kraft, eds. *Norrköpings historia IV. Tiden 1719–1870*, 114–5; Svensson, Godlund and Godlund, 'Norrköpings ekonomiska och sociala historia', 268.

33. *NT*, 19 November 1886; 'Byggnadsordning för Norrköpings Norra Förstäders område'.

34. *NT*, 19 November 1886.

35. *NT*, 19 November 1886.

36. *NT*, 19 November 1886.

37. In Swedish *utväxt*, implying a tumour, something sick or unnatural that has grown out of something else.

northern suburb is undeniably a serious growth on the city of Norrköping; sooner or later it will be necessary to operate on it, and this operation will not be easy', said Fredrik Blombergh, one of the proponents.[38] The suburb was also referred to as a weed, which should be denied water in order to avoid its spreading.[39] As a matter of fact, representatives of the suburb itself also noted that it had become a 'sewer for ... the worst and basest elements' of the city.[40] The biological imagery was as powerful as the environmental issues at stake.

The content of this 'sewer' was mostly workers and poor people, as well as some criminal 'elements'. At times, it seems to have been the city's tacit policy to rent dwellings for its poor people in the northern suburb, so that the rural district Östra Eneby was eventually obliged to support them.[41] Other sources also show that during roughly the last quarter of the nineteenth century the city's workers sometimes felt driven out of the city. A letter to the editor of *Norrköpings Tidningar* from a worker, for example, complained about the building of new workers' dwellings on a plot on the outskirts of the city: '[I]s the spot most suitable for workers' dwellings in order to separate the worker from the city as much as possible? So it seems ... since everyone knows that there are other spacious, healthy and good building sites in the city itself.'[42]

Eventually, there was a small majority in favour of a proposal that on the whole resembled the suburb committee's original one, though with some additional requirements. The water pipe was to be about 200 metres long and it would supply those living adjacent to it with city water for a fee (see Fig. 1). The pipe was to be supervised by Norrköping's waterworks engineer, B. L. Hellström, and the committee had to see to it that proper drainage was arranged. The proponents of the extension of water were therefore ultimately successful and arguably a humanitarian strand in the Norrköping City Council was victorious. In the long run, however, infrastructural improvements in the northern suburb were not prioritised by the council, since no further extensions occurred until 1909 and 1910.[43]

38. *NT*, 19 November 1886.

39. *NT*, 19 November 1886.

40. Larsson, *Inkorporeringsproblemet*, 555.

41. Horgby, *Surbullestan*, 32. The costs for poor relief should be paid for by the administrative authority of the area in which the poor lived, in this case Östra Eneby: Larsson, *Inkorporeringsproblemet*, 106.

42. *NT*, 28 May 1872.

43. Hallström, *Constructing a Pipe-Bound City*, 63–103, 170–6, 184–5, 203.

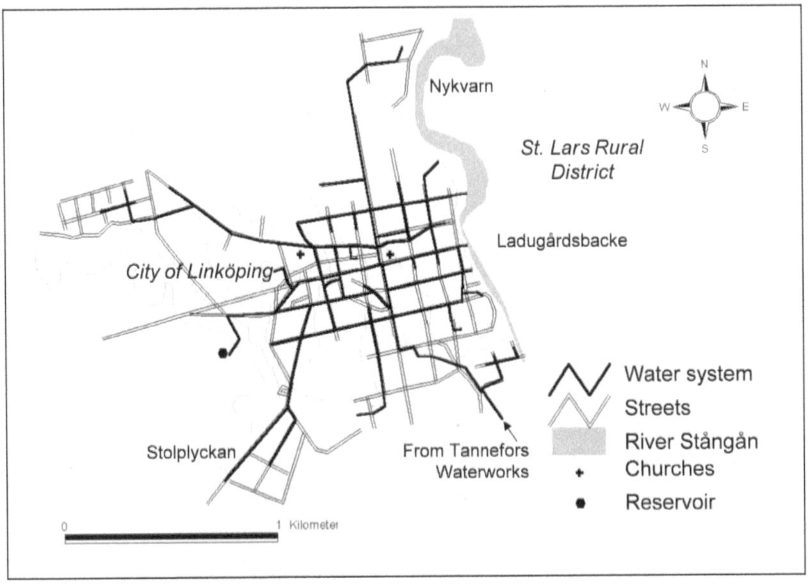

Figure 2. The Linköping Water System in 1900 [44]

The street network roughly corresponds to the planned area of the city of Linköping.

Between 1874 and 1876 a modern waterworks and a sewer system were built in Linköping. The first suburb to attract the attention of the city authorities as regards the extension of water was Ladugårdsbacke in 1881. The brewery owner Ludvig Theodor Brogren and sixteen home owners in Ladugårdsbacke wanted to connect to the city's water supply, and expected to be given water on the same conditions as people in the city. The Linköping Water Company, which owned and administered the water supply with the city as a twenty per cent joint owner, acted quickly. Within two days their engineer, C. J. Stenmark, had drawn up a plan and made an estimate of the costs. It showed that the expected income from water fees would by far surpass ten per cent of the costs of construction, according to the so-called 'ten per cent rule'.[45] Brogren's brewery was a large consumer and alone would pay 400 Swedish kronor a year.[46]

44. Sources: Hallström, 'Technology, Social space and Environmental Justice in Swedish Cities', 430; Schmid Neset and Hallström, 'The Pipe-bound City in Time and Space'.

45. This rule was set up by the Linköping Water Company in 1875 to ensure that every new extension had paid within ten years. See Hallström, *Constructing a Pipe-Bound City*, 143. This was a common way of financing the extension of technological systems to new areas in other Swedish cities as well: Larsson, *Inkorporeringsproblemet*, 354.

46. Linköping City Archives (LiSA), Linköping City Council Archives, AI:19, minutes 22 June 1881, section 73; Linköping Water Company Archives, AI:1, minutes of the shareholders' meeting 31

With the influential Brogren spearheading this matter, the issue of water supply might well have been settled quite easily, especially since he was a member of the City Council and a shareholder in the Water Company. However, the city was strongly opposed to extending a water pipe to Ladugårdsbacke. This was because, firstly, the city's Drafting Committee[47] maintained that Ladugårdsbacke lay outside the city border and, secondly, because the waterworks pumping station was already working at or near full capacity and an extension would hasten the installation, and thus cost, of a new pump at the Tannefors waterworks. The view of the Drafting Committee thus differed from that of the Water Company, which should have had the most accurate knowledge of the capacity of the waterworks. Thirdly, the committee feared that other areas around the city would ask for water pipes if this case was approved.[48]

During this City Council meeting, Brogren argued for his and the suburb's request. There had been no objections to their proposal among the shareholders of the Water Company and the pipe would be of great importance from a sanitary point of view and in the event of a fire.[49] Nils Östling expanded on his view of the subject. He believed that the proposed extension would not repay the cost, since there were no guarantees concerning the level of water consumption in Ladugårdsbacke. Besides, Östling claimed, the contract with the Water Company did not allow for extensions outside the city.[50] He also said that the advantages of piped water in case of epidemics or fire were not so significant, since there was easy and ample access to river water. However, this argument was somewhat dubious, since there was growing concern about the deteriorating quality of the river water.[51]

The ensuing City Council decision on this matter was not unanimous but the proposal to extend a water pipe to Ladugårdsbacke was rejected.[52] And so the suburb was denied access to the city's piped water supply until 1921, in spite of its incorporation into the city in 1911.

May 1881, section 8; Account of the City Council meetings 22 June and 26 July 1881, *Östgöta Correspondenten* (*ÖC*), 25 June and 28 July 1881.

47. The Drafting Committee was a part of the City Council that prepared important political questions.

48. LiSA, Linköping City Council Archives, AI:19, minutes 26 July 1881, section 81; Linköping Water Company Archives, AI:1, minutes of the shareholders' meeting 30 May 1882, section 7; Account of the City Council meeting 26 July 1881, *ÖC*, 28 July 1881.

49. *ÖC*, 28 July 1881. There can be no doubt that Brogren and the Water Company also had economic interests in the extension.

50. This was a dubious statement. The contract between the city and the Water Company allowed for extensions outside the city's planned area, provided the case in question was submitted to the City Council, but it said nothing about expansion outside the city boundary: Hallström, *Constructing a Pipe-Bound City*, 204, 211–17.

51. *ÖC*, 28 July 1881.

52. LiSA, Linköping City Council Archives, AI:19, minutes 26 July 1881, section 81.

Figure 3. North-western Norrköping (right), the Railroad (middle) and Part of the Northern Suburb, Fredriksdal (left), 1893 [53]

Water Distribution to Workers' Suburbs in Stockholm and Malmö

Was the situation in Norrköping and Linköping unique or part of a larger picture of tension between city authorities and burgeoning suburbs? Little has been written about working class suburbs in other Swedish cities during the nineteenth and early twentieth century but the outline for Norrköping and Linköping may serve as a context for the development of water supplies elsewhere. Outside Stockholm, the Swedish capital, there emerged several working class suburbs in the late nineteenth century, many of which resembled the northern suburb and Ladugårdsbacke. Two examples will be examined, Årstadal to the south and Hagalund north of the city. Årstadal was an area near the bay of Årstaviken and the railroad and was very marshy in part. Count Erik Sparre and the engineer Charles J. Smith started to sell building lots there in the mid-1870s on conditions very unfavourable to the residents. Owing to the housing shortage in the city, however, and despite the initial terms of the sales, by 1882 a number of wooden houses accommodated about 1,000

53. Source: Courtesy of Norrköping City Archives. Photographer unknown.

mostly working class inhabitants in Årstadal.[54] There was no building code for the area and sanitary conditions were very unsatisfactory; the water supply consisted of defective well, drains and sewers were lacking and privies were overflowing. The results were several serious fever epidemics emanating from the suburb. Neither Sparre, the Brännkyrka district, nor the city of Stockholm would have anything to do with the 'alien element' that the suburb constituted and water and sewerage were not extended to Årstadal during the nineteenth century, in spite of its close proximity to the reservoirs that supplied the Stockholm water system.[55]

Hagalund was located closer to the city, had better topography and the overall conditions were also more favourable, although some parts were considered to be slum areas of the worst sort. Hagalund had been developed in the same pragmatic and unregulated way, and by the same developer, as Årstadal. In 1890 Charles J. Smith was commissioned by the landowner, Captain Albert Amundson of the Royal Corps of Engineers, to start selling property in Hagalund and this was done in the same speculative manner as in Årstadal. Soon workers, artisans and the poor began moving to the suburb and, after a few years, the municipality made a considerable financial effort to build a rudimentary sewer system in the suburb, even though, in the early 1900s, the water supply was still obtained from wells.[56]

The workers' suburbs of Stockholm had a very low reputation in the city and the city feared that both epidemic disease and working class revolts might start there and hit the city. The suburbs lay in neighbouring rural districts and had emerged as the result of speculative activities by wealthy landowners, who were seen by city residents and taxpayers as responsible for the essential services and infrastructural provisions in the suburbs. Consequently, the city of Stockholm consistently denied all its suburbs access to the urban water and sewer systems until the early twentieth century.[57]

Outside the city of Malmö, in the very south of Sweden, a working class suburb called Kirseberg developed from a rural district east of the city in the 1860s. During the 1870s it grew very rapidly as a result of the economic boom and the establishment of new factories in the city. Droves of migrants moved in from the countryside and, since the city itself was short of housing, many of them settled in Kirseberg – that is, outside the city's administrative area, where building regulations were lacking and houses could be built rapidly and with inexpensive materials. In Kirseberg, as in Hagalund and Årstadal, a landowner knew how to take advantage

54. Johansson, *Stor-Stockholms bebyggelsehistoria*, 237–42.

55. Johansson, *Stor-Stockholms bebyggelsehistoria*, 237–58; E. Isgård, *I vattumannens tecken. Svensk VA-teknik från trärör till kväverening*, Örebro, 1998.

56. O. Roth, *Hagalund – en lifskraftig förstad. Kort historik rörande detta municipalsamhälles tillkomst, utveckling och uppsving*, Stockholm, 1905; Johansson, *Stor-Stockholms bebyggelsehistoria*, 254–8.

57. Johansson, *Stor-Stockholms bebyggelsehistoria*, 237–65; Larsson, *Inkorporeringsproblemet*, 355.

of the situation: Carl Petter Andersson bought building lots in Kirseberg and re-sold them at a much higher price, though still lower than in the city of Malmö itself.[58]

Powerful interests in Malmö wanted to prevent poor Kirseberg settlers from entering the city. In addition to workers and the poor, Kirseberg was also known to harbour criminal 'elements' and was the locus of prostitution. The merchant L. Osberg, a representative of the city's poor-relief organisation, wanted to exclude workers and artisans by demanding a six-month rent in advance for new tenants. By this means only a respectable class of workers, a labour aristocracy, would be admissible to the city. Though the proposal was stillborn, it is nonetheless indicative of the attitudes prevailing towards workers and the poor from suburbs like Kirseberg.[59] Conversely, deportation to the Kirseberg suburb was a threat held over those whose behaviour was considered unacceptable, as in the case of an elementary school teacher found to be unsuitable for teaching in the city schools. Status, salary and access to the city were conditional, therefore, for adults, as for poor children who were excluded from the city and its superior educational provision.[60] In this sense labelling the suburb and stigmatising its residents were closely aligned.

Sanitation and drainage in Kirseberg were inadequate and water was provided by wells. In 1879 Malmö's second waterworks was built in Bulltofta, south-east of Kirseberg, and the water tower was placed in the suburb. Some newly built houses were demolished to make way for the large conduit between the waterworks and the water tower but the suburb inhabitants were not allowed to connect to the new facility. Not until the early twentieth century were there any connections to the water system from Kirseberg and it was only in the 1920s that the Malmö water and sewer systems were fully integrated to include city and suburb.[61] Two generations of suburbans endured environmental inequalities as a result.

Swedish Cities, Working Class Suburbs and Water Distribution in Perspective

By Swedish standards Norrköping and Linköping were innovators in introducing modern water supply and sewerage in the 1870s. As Melosi points out, using examples from the USA, investment in a water system in the modern era has often indicated a commitment to city growth. Financiers and administrators of large technological

58. Thagaard, *Backarna*, 53–61.

59. Thagaard, *Backarna*, 66, 81–4.

60. 'Skolschismen i Malmö', *Svensk Läraretidning* 13:42, 1894, 536. Retrieved 23 February 2009 from http://www.runeberg.org

61. Isgård, *I vattumannens tecken*, 20–1; Thagaard, *Backarna*, 74, 83–5; A-M. Thagaard and A.K. Ribbing, *Livet på Backarna – att växa upp i en förstad i början på 1900-talet*, Malmö, 1985, 168–9; G. Olsson, 'The Struggle for a Cleaner Urban Environment: Water Pollution in Malmö 1850–1911', *Ambio: A Journal of the Human Environment* 30:4–5, 2001, 290–1. A modern sewer system for the city of Malmö was completed in 1908.

systems such as water and sewerage, which require considerable initial investment, have therefore tended to favour boundary extensions as a way of financing the whole system, for example, by amortising loans or paying investors.[62] In both Linköping and Norrköping the water and sewer systems were expanded appreciably by the Water Company and the Waterworks Board up to the 1890s, but always to areas that were deemed financially viable, were considered necessary for future economic or industrial activities or were otherwise uncontroversial.[63]

The working class suburbs that emerged around both cities, on the other hand, presented new problems regarding their legal status, property ownership, topography, sanitation and other factors that differed from areas within the city. Since these working class suburbs were new phenomena in the 1860s and 1870s, there were few precedents on which to draw regarding how to handle the problem of water distribution and management. Several other Swedish cities, particularly industrial ones, faced similar problems, though cooperation between cities seems to have begun only after 1900.[64]

The working class suburbs of Norrköping, Linköping, Stockholm and Malmö shared some common characteristics. These areas were located outside the cities' planned areas and often beyond the administrative borders, where building, fire or public health codes did not apply. The property on which the suburbs were built was not suitable for farming or commercial activities, resulting in land speculation for non-agricultural purposes. Property owners used the housing shortage in the cities to make money on suburban building lots with little alternative value. They did little or nothing to provide water or drainage in these suburbs, so the environment was marshy and unhealthy and epidemics often originated in the crowded housing areas.

These alarming conditions in close proximity to the principal urban centres, and the fact that the suburbs were expanding rapidly, forced the cities to deal with them. By looking at the evidence from Norrköping and Linköping it is possible to uncover contemporary attitudes towards the working class suburbs, which actors and factors were involved and what the outcome was in terms of water management systems. Many arguments were advanced by different actors in both cities, for and against the extension of water to the suburbs, but these need to be analysed as part of a larger picture where many complex factors interacted.

---

62. M.V. Melosi, *The Sanitary City: Urban Infrastructure in America from Colonial Times to the Present*, Baltimore, 2000, 119–21. Cf. A. Kaijser, *I fädrens spår. Den svenska infrastrukturens historiska utveckling och framtida utmaningar*, Stockholm, 1994, 57–63.

63. Hallström, *Constructing a Pipe-Bound City*, 168–228. A middle-class Norrköping suburb – the southern suburb – received water and sewerage as early as 1876 and 1877, for example (see Fig. 1).

64. Hallström, *Constructing a Pipe-Bound City*, 200–1, 336–7.

First of all the geographical location and topography of each suburb played an important part. The fact that the areas were marshy made them not only unsuitable for farming and commercial activities but also for the building of properly constructed houses, which is probably one reason why cities did not favour growth in this direction. The Norrköping case also shows that the marshy environment and poor drainage made many, though not all, council members oppose an extension of the water system as they feared that this would exacerbate the already waterlogged conditions in the suburbs. There is considerable evidence that water supply was less of a problem than water disposal and thus Ladugårdsbacke and other suburbs were geographically disadvantaged in being sited in marshy areas and on the opposite side of the river Stångån to the main area of Linköping (see Fig. 2). Despite the fact that the working class suburbs in Norrköping, Stockholm and Malmö were situated in close proximity to the areas for which the cities were administratively responsible and to communication systems such as roads and railways, they were unattractive areas for urban expansion due to their geographical and environmental disadvantages. The city of Göteborg, on the other hand, included the working class suburbs Masthugget and Majorna into its planned area and original water system as early as 1872. The main reason for this was that these areas were considered necessary for the future commercial and industrial growth of the city: they both lay near the harbour, which was essential for the city's commercial life.[65]

Secondly, there were technical considerations involved in the extension of a water supply. The capacity of the existing water system was crucial but, because it was difficult to assess the exact extent of future population expansion and thus demand for water in both the city and its suburbs, there was room for different interpretations of need. The Linköping Water Company and the City Council thus made very different estimates of the capacity of the waterworks in 1881, as did the council members during debates in Norrköping in 1886. It is very difficult to say whose judgement was accurate. Water consumption in both cities had been soaring for some time but technical measures could be taken to increase water production within the parameters of the existing system. At any rate, what seemed feasible from a technical standpoint was not always consistent within a political and social context.

Thirdly, economic factors were always very important when deciding whether to extend a water system. In the case of working class suburbs, how far the respective cities could afford to incur further debts and operating costs was unclear. Since mostly poor people and workers lived in the suburbs, their ability to pay taxes was concomitantly low.[66] In Linköping the city was not as optimistic about the future economic sustainability of the extension as the Water Company

65. H. Bjur, *Vattenbyggnadskonst i Göteborg under 200 år*, Göteborg, 1988, 37–61.

66. In the northern suburb there existed illegal water taps, which also indicates that many inhabitants could not afford their own service pipes and the water fees. See Hallström, *Constructing a Pipe-Bound City*, 188–89.

was. Agencies in both Linköping and Norrköping also feared either that the suburbs would demand further urban services or that similar areas around the cities would request piped water, should Ladugårdsbacke and the northern suburb receive water pipes. Conservative accounting was, therefore, a key factor behind restrictive policies concerning water supply to the suburbs even in Malmö and Stockholm.

Fourthly, sanitary aspects of a water extension were considered in both Linköping and Norrköping. In Linköping, for instance, council member Nils Östling thought that piped water would not benefit Ladugårdsbacke from a sanitary viewpoint. On the whole, there were few concerns about epidemics in Ladugårdsbacke, maybe because the river separated it from the city and because the suburb of Stolplyckan to the south-west caused greater sanitary concern. Part of its drainage flowed into the city (see Fig. 2). In Norrköping some council members were of the opinion that water would even worsen sanitary nuisances, due to the poor drainage in the northern suburb. Yet the fear of epidemics originating in the northern suburb was very widespread in the Norrköping City Council, since many of the suburb's inhabitants worked in the city. This was probably one reason why the suburb eventually received a water pipe, though on condition that drainage was improved first.

Finally, from an ideological point of view, a bourgeois definition of the urban space was also crucial. During the period from 1860 to the 1890s an urban working class was forming in Sweden and, just as on the continent and in Great Britain, there was a fear of revolts, immorality and disease emanating from working class districts.[67] In Norrköping and Linköping, as well as in Stockholm and Malmö, this fear also played an important part in the treatment of the suburbs and thus the respective city governments sometimes displayed a very negative attitude to workers and their housing areas. Denying water and other services to working class suburbs may have been a way of handling this fear, by keeping the workers and the problems associated with them out of the city and preserving the environmental quality of the urban space. In Norrköping there was a condescending, yet also very ambivalent, attitude to the workers, as in the metaphors of 'growth' and 'weed' used to describe the poor. The bourgeois city fathers of Norrköping had a relationship to the workers and, therefore, in the opinion of several council members, a sense of moral responsibility, which was why the northern suburb eventually received the short water pipe that was debated in 1886.[68] The aristocratic and bourgeois elite in Linköping did not have such a complex relationship with the working class as the bourgeoisie in Norrköping did. To the extent that such a class existed in

67. C. Hamlin, *Public Health and Social Justice in the Age of Chadwick: Britain, 1800–1854*, Cambridge, 1998; C. Prendergast, *Paris and the Nineteenth Century*, Oxford, 1992; Johansson, *Stor-Stockholms bebyggelsehistoria*; O. Wetterberg, 'Renhållning', *Bebyggelsehistorisk tidskrift*, 44, 2002, 5–6.

68. Hallström, *Constructing a Pipe-Bound City*, 170–203.

Linköping, it lived and worked mainly outside the town and the city was thus not as dependent on workers as Norrköping. Ironically, this may have made it easier to deny Ladugårdsbacke water so consistently, by referring to the legal power of the city border.

Conclusions

In this chapter the extension of water to working class suburbs has been studied for two Swedish cities, Norrköping and Linköping, and the primary material been complemented with secondary material from Malmö and Stockholm. When looking at why working class suburbs were most often denied – and on one occasion reluctantly given – piped water some factors stand out as particularly important. The disadvantaged geographical location of these suburbs was a drawback when extensions to the cities' piped water systems were discussed. Defective drainage, for example, was used as a direct argument against the extension of water, because the increased input of water could not be efficiently drained. Limited technical capacity of the entire system was also often seen by city governments as an obstacle to extension. However, the view of the capacity could differ considerably, as there was no absolute way of measuring either technical constraints or future loads on the system.

For the Swedish cities in this study, the extension of infrastructure to a suburb also potentially meant some kind of responsibility for the new area. This was an economic burden that they often would not accept, especially since the future of the suburbs was uncertain as regards, for example, tax and water-fee-paying capacity. In this sense, the existing economic geography and social topography were commonly confirmed through non-provision of an urban sanitary infrastructure; those who could not afford it had to do without. One notable exception was Norrköping, whose Waterworks Board finally built a 200-metre pipe after a long council debate in 1886. In this case it was probably the fear of epidemic disease that determined the issue, since the suburb lay close by and its inhabitants worked in the city. Furthermore, a majority of the bourgeois elite of Norrköping felt a sense of moral responsibility towards the working class inhabitants.

The perspective on urban social relations introduced by Lefebvre stresses that the underpinning of social practices is spatial and thus that space has social connotations.[69] Each of these contributing factors – geography, technology, economy, sanitation and ideology – had a spatial dimension to it and they inter-related in space. The working class suburbs were located in geographically unfavourable places to which extensions of water could be technically demanding, at least in the city governments' rhetoric, which also affected the cities' economic considerations. Since the suburbs often lacked fiscal resources, it was considered risky for the cities to

69. Lefebvre, *The Production of Space*, 404.

finance the extensions and the prevailing economic geography was consequently reinforced. Fear of epidemics prompted city governments to prioritise the wellbeing of the suburbs only when the cities were threatened.

From Lefebvre's perspective, urban, bourgeois ideology and practices thus resulted in a spatial exclusion of the suburbs, since water distribution during this period of Swedish urban history seems, to a great extent, to have been aimed at keeping particular cohorts of workers out of the cities and prioritising existing city residents at the suburbs' expense. In this sense, the difference between urban and suburban spaces was confirmed through water technology, since areas with an already poor environment missed out on improved environmental services. Decisions not to extend water to poor, marshy, unproductive suburban land implied that environmental services were primarily built for an existing group of city centre residents, thereby underlining existing environmental and spatial injustices between city and suburb.

Most city governments' actions, therefore, seem to contradict a common concept in the history and sociology of technology – 'momentum', coined by Hughes, which indicates that it is difficult to stop the growth of technological systems due to an 'inertia of motion' caused by technical, financial, organisational and other interests in the system. This study shows that, in the case of working class suburbs, the bourgeois elite often worked various factors in favour of *not* extending the systems, or, if one re-interprets Hughes, diminishing the momentum.[70]

Select Bibliography

Colten, C.E. 'Basin Street Blues: Drainage and Environmental Equity in New Orleans, 1890–1930', *Journal of Historical Geography* 28:2, 2002, 237–57.

Drangert, J-O., M.C. Nelson and H. Nilsson, 'Why did They Become Pipe-bound Cities? Early Water and Sewerage Alternatives in Swedish Cities', *Public Works Management & Policy* 6:3, 2002, 173.

Flanagan, M.A. 'Environmental Justice in the City: A Theme for Urban Environmental History', *Environmental History* 5:2, 2000, 159–64.

Hallström, J. *Constructing a Pipe-Bound City: A History of Water Supply, Sewerage, and Excreta Removal in Norrköping and Linköping, Sweden, 1860–1910*, Linköping, 2002.

Hallström, J. 'Systemteori och teknik. En introduktion till stora tekniska system', in P. Gyberg and J. Hallström, eds. *Världens gång – teknikens utveckling. Om samspelet mellan teknik, människa och samhälle*, Lund, 2009.

70. T.P. Hughes, 'The Evolution of Large Technological Systems', in W.E. Bijker, T.P. Hughes and T.J. Pinch, eds. *The Social Construction of Technological Systems: New Directions in the Sociology and History of Technology*, London, 1987; J. Hallström, 'Systemteori och teknik. En introduktion till stora tekniska system', in P. Gyberg and J. Hallström, eds. *Världens gång – teknikens utveckling. Om samspelet mellan teknik, människa och samhälle*, Lund, 2009.

Hamlin, C. *Public Health and Social Justice in the Age of Chadwick: Britain, 1800–1854*, Cambridge, 1998.

Harvey, D. *The Condition of Postmodernity: An Enquiry into the Origins of Cultural Change*, Cambridge, 1990.

Hughes, T.P. 'The Evolution of Large Technological Systems', in W.E. Bijker, T.P. Hughes and T.J. Pinch, eds. *The Social Construction of Technological Systems: New Directions in the Sociology and History of Technology*, London, 1987.

Johansson, I. *Stor-Stockholms bebyggelsehistoria. Markpolitik, planering och byggande under sju sekler*, Hedemora, 1991.

Larsson, Y. *Inkorporeringsproblemet. Stadsområdets förändringar med särskild hänsyn till svensk förvaltningspraxis, del I & II*, Stockholm, 1913.

Lefebvre, H. *The Production of Space*, Oxford, 1991.

Luckin, B. 'Environmental Justice, History and the City: The United States and Britain, 1970–2000', in D. Schott, B. Luckin and G. Massard-Guilbaud eds. *Resources of the City: Contributions to an Environmental History of Modern Europe*, Aldershot, 2005.

Melosi, M.V. *The Sanitary City: Urban Infrastructure in America from Colonial Times to the Present*, Baltimore, 2000, 119–21.

Prendergast, C. *Paris and the Nineteenth Century*, Oxford, 1992.

Tarr, J.A. 'The Metabolism of the Industrial City: The Case of Pittsburgh', *Journal of Urban History* 28:5, 2002, 511–45.

Wetterberg, O. 'Renhållning', *Bebyggelsehistorisk tidskrift*, 44, 2002, 5–6.

Chapter Seven

At the Limits of the European Sanitary City: Water-related Environmental Inequalities in Berlin–Brandenburg, 1900–1939

Christoph Bernhardt

In recent years the scholarly debate on social inequalities has, broadly speaking, moved from structure based approaches like class analysis towards culture-focused approaches to problems of gender and social capacity.[1] Opportunities to acquire social goods are receiving more attention than the mere status of social groups.[2] This chapter considers environmental inequalities from a point of view that concentrates on institutional and spatial perspectives and the effects of legal rules and public policies on environmental segregation. Patterns of social and environmental segregation in the field of water provision and sewerage will be analysed in terms of the example of Berlin and the surrounding state of Brandenburg, from the beginning of the twentieth century to the 1930s.

Two major questions are relevant in this context. Firstly, it is important to critically reconstruct how the 'sanitary city' that had been established in the late nineteenth and early twentieth centuries was transformed in the era of the welfare state after World War I. Here recent research has underlined that, in welfare systems, the state was a key actor in the production of partly new patterns of social injustice.[3] The public politics of hygiene, for example, not only improved living conditions for many urban inhabitants in environmental terms, but was also a means of repression

1. I want to thank Richard Rodger and Geneviève Massard-Guilbaud for their helpful remarks on the first draft of this article and Robert Bernsee for his help in compiling some of the empirical materials. Some aspects of the problems discussed here have also been analysed in Christoph Bernhardt, 'Die Grenzen der sanitären Moderne – Aufstieg und Krise der Wasserpolitik in Berlin-Brandenburg 1900–1937', in C. Bernhardt, H. Kilper and T. Moss, eds. *Im Interesse des Gemeinwohls. Regionale Gemeinschaftsgüter in Geschichte, Politik und Planung*, Frankfurt, 2009, 85–114. On the state of debate on social inequality in Germany with special regard to methodological questions see S. Hradil, *Soziale Ungleichheit in Deutschland*, 8th edition, München, 2006, 28, 43.

2. W. Merkel, 'Lebenschancen. Soziale Gerechtigkeit im 21. Jahrhundert', in *WZB Mitteilungen* 120, 2008, 68. Aufl., 10.

3. See, for example, F. Ewald, *L'Etat providence*, Paris, 1986.

and social control. The second major question is, how far can certain institutional and technological structures and practices change their environmental status in different historical contexts? In the history of water management, for example, the ideal of good environmental standards is very closely linked to the existence of centralised networks for drinking water and sewage, mostly run by public administrations. But this might not be valid to the same extent for the period after the First World War, in a context of newly emerging class and settlement structures. As will be demonstrated, the publicly produced sanitary city not only caused new environmental segregation and injustice on a regional scale; it also became an 'ossified' system of legal institutions that was attacked by collective practices of a mass movement of 'wild settlers'.

Urban Sanitary Welfare in Liberal Berlin: Three Major Interventions to Build the Egalitarian European Sanitary City, 1870–1914

The implementation of centralised systems of public water and waste-water networks in the greater Berlin region from the 1870s was a very complex political and economical process, not least because these systems were very costly and could only be implemented in a context of major political and cultural change. In political terms there was a fundamental ambivalence in the project that has not yet been clearly analysed. In fact the bourgeoisie in the German capital, as represented by their chief planner James Hobrecht (1825–1902), managed ideologically to integrate two very contradictory political concepts. Parallel to an extremely liberal ideology of town planning, as exemplified by a street-plan dating from 1862 – the so-called Hobrechtplan[4] – a strongly interventionist form of water network planning was realised from the early 1870s onwards.[5]

In the field of water policies, several historic decisions were made in the 1870s and 1880s. Firstly, the old system of public wells in the streets and cesspools in the backyards was replaced by a centralised 'egalitarian' public drinking water and sewage system. In this period the emerging social–liberal political faction, led by the medical professor and politician Rudolf Virchow,[6] succeeded in convincing the majority of municipal bodies to construct a centralised underground waste-water pipe system. As this project could not make a financial profit, it was pre-financed by the city. Secondly, the town council decided to take over the private waterworks and run them under the auspices of public administration. Here the

4. J.F. Geist and K. Kürvers, *Das Berliner Mietshaus 1740–1862*, München, 1980.

5. I have discussed these 'two faces of James Hobrecht' more extensively in my article 'Die Vertreibung des Wassers aus der Stadt und aus der Planung: Zur Hygenisierung der öffentlichen Räume im 19. Jahrhundert am Beispiel Berlins', in C. Bernhardt *et al.* eds. *Geschichte der Planung des öffentlichen Raums*, Dortmund, 2005, 71–84.

6. H. Schipperges, *Rudolf Virchow*, Reinbeck, 1994.

cherry-picking strategy of the previously private enterprise had caused serious public criticism, since the rich quarters had been privileged and the poorer quarters had been insufficiently connected.[7] Even more important in terms of an egalitarian sanitary policy was the decision to introduce a so-called 'compulsory connection and use' legislation (Anschluss- und Benutzungszwang), established step by step from 1874. Starting with a police recommendation that called for a plan for sewage discharge and the prohibition of cesspools, the connection of every parcel of land within the municipal boundaries to water and sewage networks was legally imposed. At the same time the exclusive use of the public networks was decreed.[8] This policy helped rapidly to increase the number of water customers, improve sanitary conditions and raise the revenues of the water and sewerage works. It was very thoroughly executed by the authoritarian Prussian building bureaucracy and proved to be quite successful. It was much more efficient than similar attempts in other cities, for example, in Greater Paris.[9]

In the three decades before the First World War large parts of the municipal area and the overwhelming majority of inhabitants of Berlin, as well as those of neighbouring cities, were connected, both horizontally, by including the poorer quarters, and vertically, up to the highest and poorest storeys under the roofs of the Berliner Mietskaserne – tall tenements of mass housing.[10] This meant a strong convergence in urban living conditions across social divisions. One of the surprising aspects of this process was that both house-owners and tenants accepted the high additional costs that both had to pay for the new infrastructure.

Individual legal actions from house-owners, like that of entrepreneur Wilhelm Riedel, started in the 1888 but did not find support from the high administrative court. Nor could the critical municipal councillor Schultz organise a majority in the local parliament against the new systems.[11] In fact the different social groups concerned did not resist contributing to the financing of the costly new system, as was the case in other European cities and Istanbul.[12] In practical terms, the comfort of water toilets and individual water taps seems to have interacted with strong pressure from the police authorities and the municipal administration. On the basis of

7. See S. Mohajeri, *Hundert Jahre Berliner Wasserversorgung und Abwasserentsorgung 1840–1940*, Stuttgart, 2005, 67–8.

8. H. Tepasse, *Stadttechnik im Städtebau Berlins. 19. Jahrhundert.* Berlin, 2001, 107.

9. See C. Bernhardt, 'Zwei Wege der Suburbanisierung im frühen 20. Jahrhundert. Die Stadtregionen Paris und Groß-Berlin 1900–1930', in C. Zimmermann, ed. *Zentralität und Raumgefüge der Großstädte im 20. Jahrhundert*, Stuttgart, 2006, 41–60.

10. Mohajeri, *Hundert Jahre*, 68–72.

11. Tepasse, *Stadttechnik*, 107. For the arguments of the municipal councillor Schultz see Mohajeri, *Hundert Jahre* 105–06.

12. On Istanbul see N. Dinckal, *Istanbul und das Wasser. Zur Geschichte der Wasserversorgung und Abwasserentsorgung von der Mitte des 19. Jahrhunderts bis 1966*, München, 2004.

a fundamental political agreement, sophisticated financial concepts for public and private investment helped to establish a cultural hegemony that favoured a new egalitarian underground water infrastructure without causing major social conflicts.[13]

/ohnung Nr. 12 (1904): Stube und Küche einer 2-Raumwohnung im Quergebäude

Figure 1. The Egalitarian Face of the Modern Water System: Water Tap in the Flat of a Poor Berlin Tailor, 1902[14]

The most important effect of this egalitarian public policy was the decline in mortality from cholera, typhoid fever and other water-borne diseases. This was, at least, the position of the municipal water authorities, which undertook a massive propaganda campaign with the help of extensive statistical data to demonstrate the value of the system for public health and to legitimise its own existence as a public water body (see Fig. 2).

Recent research has argued that this remarkable reduction in water-related social inequalities regarding disease and death cannot exclusively be explained by the implementation of the water and sewage systems. As Vögele and others have

13. Tepasse underlines that 'the strong logic of a large scale technical system' ("Sachzwang" großtechnischer Anlagen) could not be stopped by opponents. Tepasse, *Stadttechnik*, 107.

14. Source: *Unsere Wohnungs-Untersuchungen in den Jahren 1902–1903. Im Auftrag des Vorstandes der Ortskrankenkasse für den Gewerbebetrieb der Kaufleute, Handelsleute und Apotheker*, bearb. von Albert Kohn, Berlin, 1903.

underlined, a more complex set of factors was involved, including a general change in urban lifestyles and also better living conditions for workers.[15] But it cannot be denied that the implementation of centralised water networks contributed significantly to the reduction of traditional water-related diseases and social inequalities. Good drinking water at affordable prices and central sewerage became the technological and cultural standard for most of the inhabitants of Berlin and a key element in the vision of the European sanitary city.

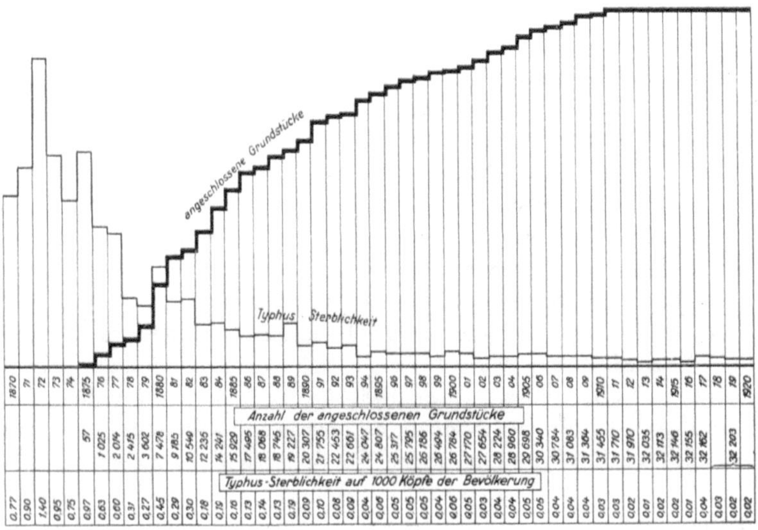

Abb. 21. Graphische Darstellung der Zahl der an die Kanalisation angeschlossenen Grundstücke in den Jahren von 1875 bis 1920 und der Typhussterblichkeit von 1870 bis 1920.

Figure 2. Connections to the Public Sewage System and Typhoid Mortality in Berlin 1870–1920[16]

The Socio-spatial Limits of the Sanitary City in Berlin–Brandenburg

A closer examination of this seemingly universal system shows that it had some clear spatial limits. Firstly, the new infrastructure systems were implemented fairly rapidly in major cities like London, Paris and Berlin; medium-sized German cities like Leipzig, Heidelberg, and Strasbourg often introduced only partial systems, whether for financial, geological or cultural reasons. Consequently a broad variety

15. See J. Vögele and U. Koppitz, 'Sanitäre Reformen und der epidemiologische Übergang in Deutschland (1850–1920)', in S. Frank and M. Gandy, eds. *Hydropolis. Wasser und die Stadt der Moderne*. Frankfurt, 2006, 75–93.

16. Source: H. Hahn/F. Langbein, *50 Jahre Berliner Stadtentwässerung*, Berlin 1928.

of technological and institutional pathways emerged.[17] At the periphery of Berlin several forms of decentralised technological solutions also survived. Amongst these the discharge of waste-water and faeces on agricultural fields and gardens by private landowners was the most important. Not surprisingly, this system was primarily to be found in the suburban parts of Greater Berlin that were marked by a low population density. But this spatial differentiation in waste-water technology represented the key element in two powerful and competing institutional and cultural systems: the European sanitary city on the one hand, providing centralised public networks of water supply and discharge with compulsory connections and run by public enterprises making limited profit; and, on the other hand, decentralised, less organised and cheaper private discharge of waste-water, which showed some problematic, or at least uncontrolled, environmental and hygienic effects.

In fact, centralised systems were not as dominant as public opinion and most historical research suggests, as a result of the focus on major cities. Not only did nearly every medium-sized city leave considerable parts of its territory outside the central water system planning, traditional systems of discharge also remained in operation. This is especially true for the towns in Brandenburg, the region around Berlin, with a relatively low population density. A quantitative analysis of the 35 cities in Brandenburg with more than 5,000 inhabitants has revealed a broad variety of decentralised concepts and even shown a certain empirical dominance of these systems.[18] Case studies of small towns like Zehdenick to the north of Berlin indicate that the farmers insisted on the use of the urban waste-water as fertiliser.[19] These differences between the metropolis and the small towns in the introduction of the new technologies cannot be explained by financial restrictions, lower population density or the supposed backwardness of these towns. On the contrary: their concepts showed a remarkable technological and institutional flexibility. So some of the municipalities introduced new systems only for part of their territory and did not legally establish a compulsory system of connection and use.[20] In some way they resisted campaigns led by public health experts in the Royal Institute for Water Hygiene who seriously claimed that 'nobody should be allowed to exclude himself from the water networks'.[21]

17. See, for example, F. Brüggemeier, *Tschernobyl, 26. April 1986. Die ökologische Herausforderung*, München, 1998, 74–83.

18. This research was made on the basis of several statistical surveys. For details see C. Bernhardt, 'Die Grenzen', 96.

19. *Verwaltungsbericht der Stadt Zehdenick*, Zehdenick, 1900, 19.

20. O. Sprung, 'Die Wasserversorgung in den brandenburgischen Städten', in E. Stein, ed. *Brandenburg. Kultur und Arbeit der Städte in der Provinz Brandenburg*, Berlin, 1929, 44–8, here 47.

21. Quoted from the statement by Prof. Thiesing at the conference of the German Municipalities Association (Deutscher Städtetag) in Berlin, 1913. *Verhandlungen der IV. Mitglieder-Versammlung des Reichsverbandes deutscher Städte am 19. und 20. Mai 1913 im Rathause zu Berlin*, Berlin 1913, 45.

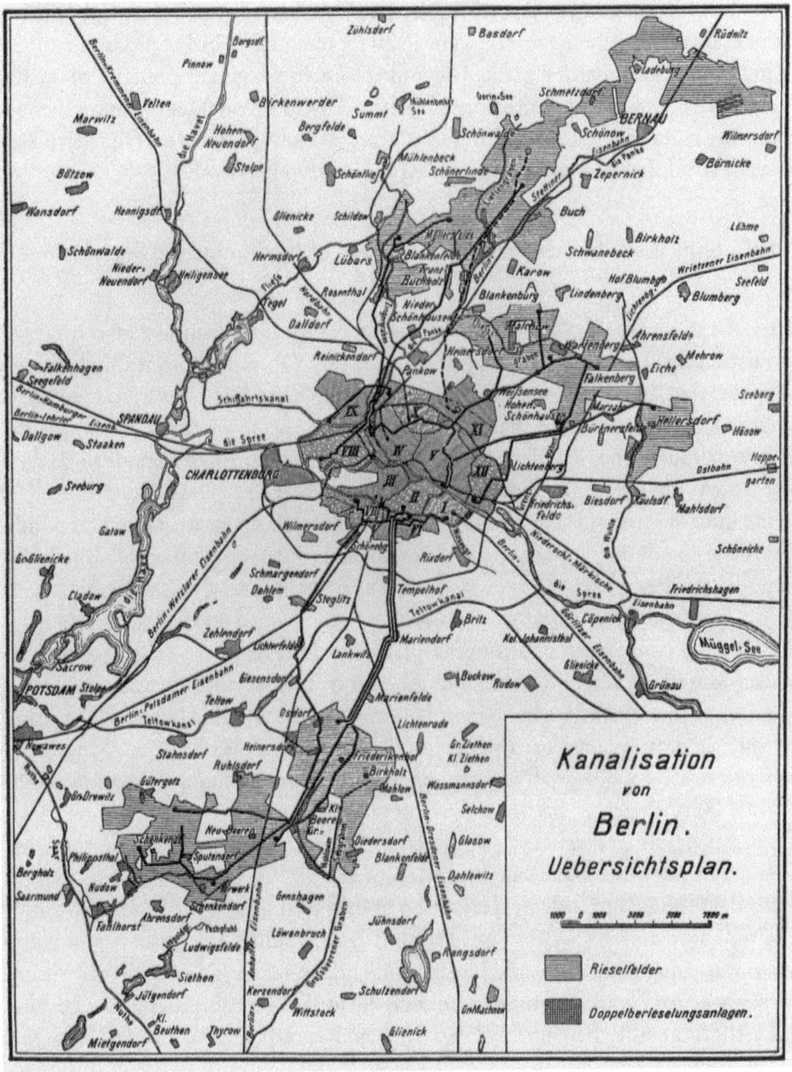

Die Kanalisation
von Berlin,
Übersichtsplan,
1906

Figure 3. The 'Aqua-ecological Footprint': Berlin Sewage Farms in the Region of Brandenburg, c.1900[22]

22. Source: H. Hahn/F. Langbein, *50 Jahre Berliner Stadtentwässerung*, Berlin 1928.

Seen from a regional perspective, the German capital and the neighbouring cities that were legally incorporated in 1920 represented a kind of modern sanitary island in a traditionally regulated region. From this regional point of view, the metropolitan sanitary system imposed its large environmental footprint on the surrounding regions, especially through the extended sewage farms and the privileged areas of the different waterworks serving the metropolis (see Fig. 3)

Pollution, Water Scarcity and New Patterns of Environmental Segregation, 1918–39

After the First World War the prospects for continued expansion of centralised infrastructural systems were not very promising. On reflection, leading experts admitted that, in the pre-war years, municipalities had sometimes introduced new technologies in areas were there had been no real need for them: 'Amongst the municipalities in the rich Germany it had become a question of honour to provide one's town with a centralised water network even where there was no serious need.'[23] In the inter-war years, serious financial crises seemed to signal an end to that policy. But the vision of the sanitary city had developed a very strong presence. The Berlin water authorities continued to expand the network in the newly established Greater Berlin, created in 1920, trying to connect all unconnected areas of the urban map. In 1924 the connection of a number of former independent villages to the network was started, 'even if the small number of consumers in these villages did not promise to prove efficient in economic terms'.[24] In the following year, the water authorities proudly presented to the public the expansion of the drinking water network to eleven villages. It was argued that the main egalitarian intention was to provide every citizen of Berlin with water of good quality[25] but, at the same time, this policy of sanitary welfare also protected the jobs and budgets of the authorities.

On the other hand, in the course of the 1920s, the new sanitary system showed its limits, not only in spatial, but also in technological and social regards. The rapidly increasing water consumption rates resulted in constantly growing waste-water quantities. If, as the Berlin municipal water administration calculated, the sewage farms could clean a maximum daily load of 45 cubic metres per hectare, this critical quantity had already been surpassed as early as 1924. In 1927 the real

23. M. Beninde (President of the Prussian State Institute for the hygiene of water, soil and air), 'Die voraussichtliche Entwicklung der Wasserversorgung in Deutschland in den nächsten Jahren und die hygienische Einstellung hierzu', in *Mitteilungen aus der Landesanstalt für Wasser-, Boden- und Lufthygiene*, H. 29, 1925, 1–8.

24. H. Bärthel, *Wasser für Berlin*, Berlin, 1997, 159.

25. Bärthel, *Wasser.*

load stood at 62 cubic metres[26] and so groundwater pollution increased dramatically, not to mention the growing soil pollution from heavy metals and other chemical substances. To continue the established practices would have meant expanding the sewage farm system by investing large amounts of money. But buying greater areas and constructing new farms would not have been efficient, as the municipal water authorities found out in a detailed survey in 1928, because of the diseconomies of scale.[27] Consequently, the new technology of biological cleaning was tested and introduced.[28]

Another problem resulting from the success of the sanitary city concept was the over-exploitation of water resources. While daily water consumption increased in Berlin from 92 litres per capita in 1912 to 148 litres daily in 1929, and in Potsdam from 85 to 107 litres between 1913 and 1927, the Berlin region suffered increasingly from local water shortages. Some new institutions and technologies, like the introduction of domestic water metering,[29] helped to slow the rate of increase of water consumption but could not solve the problem. In 1921 the water flow of the Spree fell to a critical rate of five to six cubic meters per second, far below the officially defined point of environmental damage (Schadensgrenze) of fifteen cubic meters per second.[30] Such problems had previously been unknown in the Berlin region, which generally benefited from a wealth of natural water resources. As regional water resources fell short, projects for new waterworks had to be developed at an increasing distance from the capital. The municipal waterworks increasingly expanded into the region (see Fig. 4). Areas that, as far as waste-water discharge was concerned, appeared to be spaces of danger became spaces of potential financial profit.

In a process of growing economic concentration, some major municipal and county waterworks competed in the exploitation of regional water resources with around 1,000 private water producing enterprises, like chemical factories and breweries, and with a number of public organisations like hospitals and military sites. To an estimated regional consumption of 260 million cubic metres per year these organisations contributed about forty million cubic metres.[31] In order to meet the growing demand, municipal water authorities presented plans in 1928 to redirect large water resources from the Oder river catchment to Berlin.[32]

---

26. S. Hahn, *Denkschrift über die Erhöhung der Leistungsfähigkeit der Rieselfelder durch den Bau von Belebt-Schlammanlagen*, Berlin, 1928, 4.
27. Hahn, *Denkschrift*, 12.
28. *100 Jahre Berliner Entwässerungswerke 1878–1978*, Berlin 1978, 22.
29. Sprung, *Die Wasserversorgung*, 47.
30. Hahn, *Denkschrift*, 4.
31. *7 Jahre Landesplanungsverband Brandenburg-Mitte*, Berlin 1937, 35.
32. See the criticism of A. von Feilitzsch, 'Die Zukunft der Berliner Wasserversorgung', in *Gesundheits-Ingenieur*, H. 49, 1928, 785–87.

Figure 4. Public Water Enterprises in the Berlin Region (1930s) [33]

So a type of large-scale spatial inequality emerged between the water-consuming and waste-water-exporting metropolis on one side, and an underprivileged and exploited suburban and rural periphery on the other. Increasingly, lakes were under threat of drying up and the farmers around Berlin suffered from the lower groundwater table and a decline in their agricultural revenues.[34] In addition they came under growing pressure from the expansion of the sewage farms, for which the authorities acquired ever more land. By the mid-1930s the capital's sewage farms covered an area of around 12,000 hectares.[35] The ecological footprint of Berlin seemed set to develop in the long run in a way similar to that of American, and especially Californian, cities like Los Angeles.[36] However, this prospect did not become reality, as water use and pollution were in decline from around 1930 as a consequence of the economic crisis.

In the course of the 1920s, the Berlin sanitary system was confronted with a third trend that forced increasing numbers of poor people to leave the inner

33. Source: *7 Jahre Landesplanungsverband Brandenburg-Mitte*, Berlin, 1937.

34. *7 Jahre Landesplanungsverband*, 34.

35. *7 Jahre Landesplanungsverband*, 37.

36. See M. Reisner, *Cadillac Desert, The American West and Its Disappearing Water*, New York, 1986.

city. The key problem was the high prices and rents for modern, well-equipped, sanitary flats in times of serious social crisis and mass unemployment. On the one hand, the political promises of a 'clean warm home' for everybody had become a key element of municipal policy, and especially of the Social Democratic Party as a strong partner in the coalition that governed Berlin. Production of publicly funded new houses reached numbers higher than anywhere in Europe. So-called 'smokeless settlements' in Berlin-Steglitz and in other German cities like Frankfurt represented, as was argued, new standards in environmental and hygienic terms.[37] What was more important in terms of social and environmental justice was the fact that these smokeless flats, as well as the numerous newly built modern apartments in general, were only affordable by well-paid workers. By contrast, the numbers of the unemployed and homeless urban poor increased rapidly. In a way the municipal welfare policy, in the context of steadily growing social polarisation, strengthened, or at least contributed to, social and environmental segregation. In the late 1920s a paradox arose, caused by the world economic crisis: some 3,000 publicly financed, comfortable social housing flats stood empty, while more than 100,000 poor people settled, partly illegally, at the periphery, where they occupied large areas of land.[38] This settler movement was driven, as will be shown, not least by the costly sanitary standards that had been fixed in Berlin building legislation.

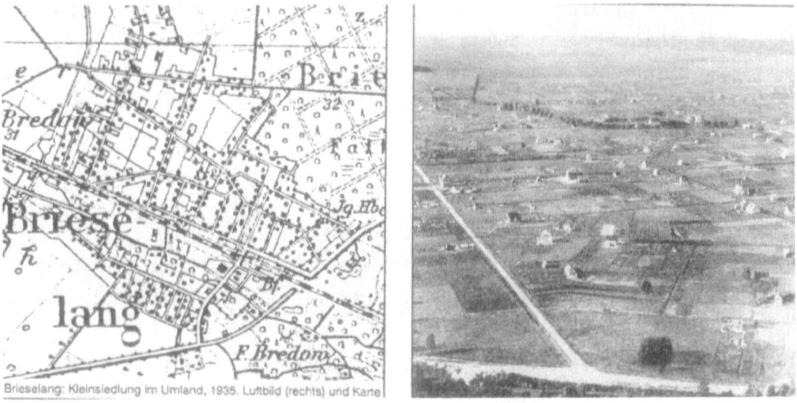

Figure 5. Settlement of Brieselang in the West of Berlin, around 1930 [39]

37. See C. Bernhardt, 'Aufstieg und Krise der öffentlichen Wohnungsbauförderung in Berlin 1900–1945', *Wohnen in Berlin. 100 Jahre Wohnungsbau in Berlin*, Berlin, 1999, pp. 45-87 and G. Kuhn, '*Menschen müssen sich selbst entwerfen.' Rationalisierung der Wohnkultur und kommunale Wohnungspolitik in Frankfurt am Main (1900–1930)*, Bonn, 1998.

38. C. Engeli, *Landesplanung in Berlin–Brandenburg*, Stuttgart, 1986.

39. Source: *7 Jahre Landesplanungsverband Brandenburg-Mitte*, Berlin, 1937, 23.

The 'Wild-Settlers': A Mass Movement and the Contest against the European Sanitary City

Along the boundaries of Berlin, as in other major European cities, the movement of so-called 'wild-settlers (*wilde Siedler*) grew up during the 1920s and occupied an estimated 120,000 parcels of land in 1933. Some 70,000 of these parcels were situated within the Berlin boundaries and a considerable number were illegally squatted.[40] The movement was partly inspired by the powerful vision of living in single family houses and garden cities that had become very popular after the First World War. At the same time it represented a kind of sub-proletarian opposition to the high sanitary standards in Berlin. Even if this opposition did not, as the communist party KPD did, publicly criticise the housing policy of the municipal administration[41] these poor people did, in fact, resist the public housing system and preferred to, or were forced to, live at the periphery. From around 1900 this movement had steadily expanded despite many 'obstacles and difficulties erected by legislation and authorities', as its spokespeople complained.[42]

What made settling at the periphery – 'within biking distance of the final station' – so attractive was the lower prices of land and construction. The parcels of land, with an average size of 750–1,200 square metres, were affordable by many poor people, since savings could be made for street building and sewerage. In contrast, waste-water was discharged in back yards or used as fertiliser for vegetables and the wealth of natural resources and the collective labour and energy of the settlers were other important resources in the struggle to survive. Beyond the municipal boundaries of Berlin, the authorities reacted by reducing the standards prescribed by the building legislation, though, within Berlin city itself, the municipal bodies were not ready for similar reforms, even where some projects received public funding. It was only in 1932 that building standards were reduced to some extent.[43]

The decline in regional water resources, with increasing shortages and pollution, together with uncontrolled settlement, were amongst the driving forces that propelled the introduction of regional planning authorities for the Greater Berlin region (Landesplanungsverband Brandenburg-Mitte) in 1929.[44] In their reports the new authorities seriously sought a comprehensive plan for water provision and waste-water discharge which, until that time, had been technologically and institutionally separated.[45] Regional water resources were increasingly regarded as a

40. *Die private Stadtrandsiedlung. Untersucht am Berliner Beispiel, bearbeitet im Deutschen Archiv für Siedlungswesen*, Berlin, 1933, 12.

41. *Protokolle der Stadtverordnenten-Versammlung* Berlin 1931, 878.

42. *Protokolle*, 878

43. *Protokolle*, 23.

44. Engeli, *Landesplanung.*

45. *7 Jahre Landesplanungsverband*, 10.

precious common good, as Langbein, the director of the Berlin sewage administration, described them in 1935.[46]

The Nazi regime established in 1933 showed an ambivalent character as far as water policies were concerned. The anti-urban ideologies of blood and soil and for national autarchy sustained concepts for small-scale discharge of waste-water on agricultural fields. Sewage farms around Berlin should be decentralised and waste-water spread in lower quantities over large areas. Legal standards for sanitary utilities were lowered or handled more flexibly. The key strategy was an intended shift from 'intensive' to 'extensive' irrigation.[47] At the same time, modern technology like waste-water treatment plants expanded. Eventually some Nazi projects could have had some positive effects on the environmental status of water resources but most were not realised

The European Sanitary City: Egalitarian Concept or Driving Force of Environmental Inequalities?

Centralised water and sewage systems were the key element of the European sanitary city. Few contemporaries denied and few social scientists and historians have contested, until now, the fundamental improvements that they effected in terms of the environmental status and public health of urban citizens. Furthermore, the commodification of water and the rise of public management, with the help of new technologies, also established an egalitarian culture of water provision and discharge. The modern system can be considered, therefore, as a breakthrough for environmental justice. Traditional patterns of social differentiation in the field of water consumption, which, until that time, had been a key element of social inequality, lost their importance. In the case of Berlin, rising prices for piped water obviously did not meet with strong resistance, as was the case in other European cities.

But this convergence in the patterns of water consumption in the inner urban area introduced new environmental problems and social segregation at the periphery. Together with an egalitarian regime of urban hygiene, a certain colonial metabolic intervention in the hinterland occurred. Not only did the modern regime export waste-water as a negative external effect to the hinterland and create serious problems of pollution there; it also caused growing water scarcities in the entire region and a decline in agricultural revenues. Farmers and suburban villagers were victims of this new environmental segregation on the regional scale. Furthermore, in the course of the 1920s, the urban poor and especially the growing number of

46. F. Langbein, 'Der gegenwärtige Stand der Abwasserfrage' in *Der Bauingenieur*, H. 13/14, 1935, S. 147–55, 148.

47. *7 Jahre Landesplanungsverband*, 35; W. Engelmann, 'Die Wasserversorgung in ländlichen Gemeinden und im Einzelanwesen', in *Technisches Gemeindeblatt*, 1936, 251–53, here p. 251.

unemployed workers were excluded from the benefits of the European sanitary city during the world economic depression.

Despite these negative effects of social exclusion, environmental decline and segregation the model developed a strong logic of expansion. Together with the technological system, an entire legal, institutional and social complex was established. Municipal administrations and experts, together with public and private enterprises, developed a vision of connectivity between municipalities and rural areas. In fact, it was thought that the model should be imposed on many different types of settlement structures and serve as an authoritarian role model, even in places where it was not appropriate.

In the case of Berlin–Brandenburg, and of Germany in general, this strategy was continued throughout the twentieth century, across major political shifts and the socialist and capitalist regimes. Nevertheless small-scale local systems remained alive, especially in Brandenburg, until the end of the socialist period. It was only after the German unification of 1990 that mainly western companies and the Brandenburg government imposed centralised sewage systems even in rural areas, on a scale that was unknown before. These systems failed in economic terms because of an inefficient ratio of capital investment and low numbers of users, so that prices became extremely high and massive user protests arose.[48] What had become obvious for the first time in the 1920s seems to have been proved in the long run: despite its positive effects on public hygiene and environmental equality, the regime of the European sanitary city appears to be a model of limited value, in environmental, technological and social terms.

Select Bibliography

Ambrosius, G. 'Die wirtschaftliche Entwicklung von Gas-, Wasser- und Elektrizitätswerken (ab ca. 1850 bis zur Gegenwart), in H. Pohl, ed. *Kommunale Unternehmen. Geschichte und Gegenwart*, Stuttgart, 1987, 125–153.

Bärthel, H. *Wasser für Berlin*, Berlin, 1997.

Bärthel, H. *Geklärt! 125 Jahre Berliner Stadtentwässerung*, Berlin, 2003.

Baumeister, R. *Gemeinwohl und Sondernutzen im Städtebau*, Berlin, 1918.

Bayerl, G. 'Konsum, Komfort und Netzwerke. Die Versorgung mit Wasser', in R. Reith and T. Meyer, eds. *Luxus und Konsum – eine historische Annäherung*, Münster, 2003, 130–58.

Bernhardt, C. 'Regionaler Institutionenwandel im Wassermanagement in historischer Perspektive. *Das Beispiel des Oderraums in der DDR-Zeit', in T. Moss, ed. Das Flussgebiet als Handlungsraum. Institutionenwandel durch die EU-Wasserrahmenrichtlinie aus raumwissenschaftlichen Perspektiven*, Münster, 2003, 89–126.

48. Around 2008 Berlin newspapers reported on these conflicts continually. See for example 'Neue Kosten für alte Leitungen', *Berliner Zeitung*, 23 April 2008.

Bernhardt, C. 'Umweltprobleme in der neueren europäischen Stadtgeschichte', in C. Bernhardt ed. *Environmental Problems in European Cities in the 19th and 20th Century/ Umweltprobleme in europäischen Städten des 19. und 20. Jahrhunderts*, 2nd edition, Münster, 2004, 5–23.

Bernhardt, C. 'Die Vertreibung des Wassers aus der Stadt und aus der Planung. Zur Hygienisierung der öffentlichen Räume im 19. Jahrhundert am Beispiel Berlins', in C. Bernhardt ed., *Zur Geschichte der Planung des öffentlichen Raums*, Dortmund, 2005, 71–84.

Brüggemeier, F-J. *Tschernobyl, 26. April 1986. Die ökologische Herausforderung*, München, 1998.

Büschenfeld, J. *Flüsse und Kloaken. Umweltfragen im Zeitalter der Industrialisierung (1870–1918)*, Stuttgart, 1997.

Claude, V. 'Une coopération politique dans une mosaïque urbaine, le cas du service de l'eau en banlieue parisienne (1880–1923)', *Genèses*, 4, 2006, 92–111.

Eiden, C. *Versorgungswirtschaft als regionale Organisation. Die Wasserversorgung Berlins und des Ruhrgebietes zwischen 1850 und 1930*, Essen, 2006.

Hahn, H. and F. Langbein, *Fünfzig Jahre Berliner Stadtentwässerung 1878–1928*, Berlin, 1928.

Hoffmann-Axthelm, D. *Die dritte Stadt. Bausteine eines neuen Gründungsvertrages*, Frankfurt, 1993.

Krabbe, W. *Kommunalpolitik und Industrialisierung. Die Entfaltung der städtischen Leistungsverwaltung im 19. und frühen 20. Jahrhundert*, Stuttgart, 1985.

Mohajeri, S. *100 Jahre Berliner Wasserversorgung und Abwasserentsorgung 1840–1940*, Stuttgart, 2005.

Münch, P. *Stadthygiene im 19. und 20. Jahrhundert. Die Wasserversorgung, Abwasser- und Abfallbeseitigung unter besonderer Berücksichtigung Münchens*, Göttingen, 1993.

Tepasse, H. *Stadttechnik im Städtebau Berlins*, Berlin, 2006.

Vögele, J., Jörg and U. Koppitz, 'Sanitäre Reformen und der epidemiologische Übergang in Deutschland (1850–1920)' in S. Frank and M. Gandy, eds. *Hydropolis. Wasser und die Stadt der Moderne*, Frankfurt, 2006, 75–93.

Chapter Eight

German Cities and their Sewage Systems: Darmstadt and Dessau in the Nineteenth and Twentieth Centuries

Marcus Stippak

This chapter provides a condensed survey of two German urban sewage systems and the inequalities they caused between the late nineteenth and the late twentieth centuries.[1] The focus is on developments that took place within the cities of Darmstadt and Dessau and under consideration are the endeavours both city councils made in the 1870s to implement a modern, that is to say centralised, networked and comprehensive, sewage system in their effort to reduce environmental nuisance. The chapter also addresses those environmental conditions identified by contemporaries and how these were perceived before and after the new technical systems were implemented. There then follows an assessment of how German politicians, civil servants and councillors contextualised the technical changes in urban water management systems and modified their perceptions of the hygienic city. Finally, the political, social, urban and technical developments that both cities experienced from the early twentieth century will be sketched out. Although the environmental conditions within a particular urban area were generally improved, by no means all city dwellers enjoyed the benefit of their own bathroom or water closet. In short, in numerous cases the new technical facilities to discharge waste-water and excrement were not available to all and so environmental inequalities remained characteristic of many urban households for much of the twentieth century.

Once the municipal sewage systems were operational, environmental inequalities were shifted almost exclusively to areas beyond the city limits. In Darmstadt, peasants and farmers in the rural hinterland volunteered to treat waste-water on their sewage fields, whilst in Dessau almost all waste-water was discharged into two nearby rivers. After an examination of the new sewage systems of Darmstadt and Dessau and the relationship of both with their hinterlands, the deficiencies in these arrangements over time will be examined. More specifically, the issue of

1. This chapter is based upon M. Stippak, *Beharrliche Provisorien, Städtische Wasserversorgung und Abwasserentsorgung in Darmstadt und Dessau 1869–1989*, Münster *et al.* 2010. Originally PhD thesis, Darmstadt University of Technology, Germany, 2008.

environmental inequality returned to the urban area by the late 1940s, and more obviously in the early 1960s, and although both municipal authorities were eager to find practical solutions, their room for manoeuvre was restricted by political, financial and mental factors. Consequently, both municipalities tended maintain the existing arrangements and thus the inequalities were more or less perpetuated. It took many years before both cities were able to elevate their waste-water problems to a higher position on the political agenda. In this respect, and perhaps surprisingly, the fact that from 1945 Darmstadt was a West German capitalistic city and from 1949 Dessau was an East German socialistic city did not make a significant difference.

Background: Dessau and Darmstadt before the 1870s

Dessau was the capital of the rather small duchy of Dessau–Anhalt and Darmstadt was the capital of the slightly larger grand duchy Hesse–Darmstadt. Both territories were parts of the German Federation (*Deutscher Bund*); later, they belonged to the German Empire, founded in 1871. Until the 1850s, both cities were arguably backwaters. Their primary function was to serve as residences for the royal family and the royal court and as locations for the government, almost all ministries and a garrison. Many of the inhabitants were either civil servants or soldiers. In contrast to a relatively active cultural life, the performance of the local economy remained rather sluggish for many years. Local firms were primarily concerned with the consumption patterns of the court, the military and the bureaucracy. The pace of economic activity slowly increased after both cities were connected to the expanding German railway network during the 1840s, though even before this both cities had experienced a temporarily strong demographic growth, mostly due to immigration.

Table 1. Population and Average Annual Growth Rate in Darmstadt and Dessau 1816/18–1871 [2]

	1816/18	1834	1861	1871
Darmstadt	15,391	22,436	28,523	33,799
Dessau	9,136	11,251	15,613	17,459
Darmstadt		2.69%	1.00%	1.85%
Dessau		1.45%	1.44%	1.18%

Predictably, with these developments, new areas were established, as more space was needed both for tenements to house the expanding population and for new industrial areas. As elsewhere in Germany, and indeed in other European cities

2. Source: M. Stippak, *Beharrliche Provisorien*, 60, 90.

experiencing sustained population growth, there were concerns about the quality and quantity of drinking water and, in particular, the existing methods of disposal, based on old sewers, cesspools and carting, came under increasing pressure.

Piping Darmstadt: Implementation of a Combined Sewage System

How were these concerns addressed? As elsewhere,[3] after an animated local discussion, Darmstadt's new waterworks became operational in 1880. With no serious problems affecting its daily operations, public debate on the issue came to an abrupt end. However, at the same time as the controversy over the question of which water supply system should be chosen divided the local public, a prolonged dispute developed as to whether the city needed a new sewage system at all.

An early warning came from Karl Eigenbrodt, a physician and member of the city council.[4] Eigenbrodt published an article in 1869 in which he complained that the local sewers and cesspools were already in the same deplorable state as in many other German cities, principally because of their design. Their meandering course and square shape caused sedimentation as human waste passed through the sewers; and their physical condition deteriorated further because, even though it was forbidden, homeowners and landlords both connected their cesspools to the already overloaded sewers. Even law-abiding landlords intensified the problem. When their cesspools were brimful, they engaged farmers and rural labourers to remove the effluent and this procedure, according to Eigenbrodt, meant farmers often lost parts of their cart load, thus scattering human waste on the streets of the city. Their careless actions left behind dirty, stinking, unhygienic streets, causing serious environmental inequalities since access routes were worst affected. This, combined with the obvious lack of interest of the city council, meant that even minor improvements could not be effected. Another unforeseen aspect of defective sewering arose during the discussion of the water supply system. It was assumed, rightly, that leakage of waste-water and excrement would contaminate the local groundwater, an outcome that chemical analyses soon confirmed.[5]

3. See for example, J. von Simson, *Kanalisation und Städtehygiene im 19. Jahrhundert*, Düsseldorf, 1983; R. J. Evans, *Tod in Hamburg. Stadt, Politik und Gesellschaft in den Cholera-Jahren 1830–1910*, Reinbek bei Hamburg, 1990; J.A. Tarr, *The Search for the Ultimate Sink. Urban Pollution in Historical Perspective*, Akron, 1996; M.V. Melosi, *The Sanitary City. Urban Infrastructure from Colonial Times to the Present*, Baltimore, 2000.

4. K. Eigenbrodt, 'Kanalbauten in Darmstadt', in *Deutsche Vierteljahrsschrift für öffentliche Gesundheitspflege*, 1869, 137–40.

5. For doubts concerning the limited validity of quality criteria for water in the 1870s see M. Stippak, 'The Mental and Practical Impact of Pre-bacteriological Quality Criteria for Water in the 1870s', in R. Heil, A. Kaminski and M. Stippak *et al.* eds. *Tensions and Convergences. Technological and Aesthetic Transformations of Society*, Bielefeld, 2007, 295–303.

Arising from the public debate was the conclusion that it would be impossible to handle the expected amount of waste-water if the city continued to rely on the old sewers and cesspools. Eventually, Darmstadt's new sewage system was introduced, figuratively speaking, through the back door. The decision in favour of an expensive combined sewage system was facilitated by the prospect that polluted streets, as well as marshy soil within the urban area, could be prevented once and for all.

Contemporaries might have appreciated the new system because it seemed to promise a degree of social equality. As a local journalist wrote in 1874, a foul-smelling ditch in one of the newer city quarters was covered to please the eye and nose of its inhabitants. However, nothing comparable was undertaken in an older city district where residents lived along the upper stretches of the same ditch and suffered from its penetrating stink. Between the lines, the reporter blamed the city council for dealing with the older (poorer) and newer (wealthier) residential quarters unequally. From this point of view, the new system offered an opportunity to create political, social and environmental equality. Soon after, local manufacturers and craftsmen articulated the same view. In the journal of their association, they demanded a properly constructed sewage system for the entire city, which could effectively banish the omnipresent stench. Darmstadt's city councillors were confident of having enough water to flush Darmstadt's future sewers and so they voted for the combined sewage system.

Piping Dessau: Implementation of Two Separate Systems

Encouraged by the government of Dessau–Anhalt, Dessau's councillors decided in 1872 that a new sewage system should replace the existing sewers and cesspools. Moreover, this would enable the administration to keep all piped streets clean, regardless of whether these were in the old town or in one of the newly constructed quarters. At that time, numerous municipal directives, decrees and proscriptions were enacted, requesting the inhabitants to keep the streets tidy and their cesspools in good repair. However, contemporaries complained that the whole city stank.

Unlike the reform of the sewage facilities, a rearrangement of Dessau's water supply did not appear urgent. Since Dessau is situated between the Elbe River to the north and the Mulde River to the east, there was a sufficient volume of river water as well as local groundwater resources. Thus, this reform was postponed.[6] Two years later, however, in 1874, Dessau's city fathers argued that the priority

6. Another reason for this might have been the wish to stress the city's importance for the 1863 unified duchy. Both city council and government might have aimed for what Schott called a 'symbolic modernity' based on a new sewage system. See D. Schott, 'Symbolische Modernität und die elektrische Vernetzung der Stadt, oder: Stadtentwicklung als "Drahtseilakt"', in H. Uhl, ed. *Kultur–Urbanität–Moderne. Differenzierungen der Moderne in Zentraleuropa um 1900*, Vienna, 1999, 287–320. The same strategy was pursued by the Prussian king in his royal residence of

should be to modernise the local water supply facilities. They had become worried by chemical analyses that proved the unhealthy nature of local well water.[7] Councillors convinced the government that it was absolutely essential to make progress with new proposals to provide sufficient water supply in order to overcome the deficient removal of faeces and effluents and, eighteen months later, Dessau's first waterworks went into service.

In spite of the self-evident conclusion that the new waterworks would increase the consumption of water, no efforts were made to ensure sufficient sewage disposal. In addition, it turned out that the tap water provided was often unsuitable for drinking and cooking, and indeed for washing, due to its iron content. Dessau's second waterworks, operational from 1886, was also the source of grave problems since its water released lead from domestic water pipes, causing more than ninety persons to suffer serious lead poisoning within a few months. Consequently, there arose a new inequality that persisted as long as both waterworks provided harmful tap water. Paradoxically, then, the earlier, officially discredited wells provided qualitatively better water than the new municipal waterworks and it was not until appropriate counter-measures were introduced that this inequality ceased.

After the failure of Dessau's first waterworks, it became evident that the issue of sewage disposal had been managed in a rather short-sighted manner. In 1876 and 1886, consumers were asked to drain off several litres of water. Officials hoped this would reduce the contamination by removing stagnant water in the system. However, that request turned out to be impractical since several citizens pointed out that the existing tanks would overflow rapidly given that the capacity of many old cesspools was too small. This problem was intensified during the winter, as it was not permitted to pour water onto the streets.

Only after the disaster with the second waterworks and a devastating critique did Dessau's councillors move to implement a modern sewage system.[8] As elsewhere, public expenditure prompted many to be concerned about the cost of the new system, while others referred to the popular assumption that sewer gases were unhealthy and, under certain circumstances, even deadly. A strong group of opponents argued that a small city like Dessau would not need such a large or elaborate system. This resistance formed part of a widespread demand to find a balance between the city's present and future interests. Not surprisingly, perhaps, Dessau's councillors voted for a compromise. In September 1888, they decided that the entire city should be piped. While rain, industrial effluent and domestic

Berlin. Ignoring the authority and reluctance of Berlin's city councillors, he decreed in the 1850s that the city should be provided with a new sewage system.

7. M. Stippak, 'The Mental and Practical Impact', 295–303.

8. G. Wolffhügel, 'Wasserversorgung und Bleivergiftung. Gutachten über die zu Dessau im Jahre 1886 vorgekommenen Vergiftungsfälle', in *Arbeiten aus dem Kaiserlichen Gesundheitsamt*, 1887, 484–542.

waste-water should be discharged into the Elbe and Mulde Rivers, faeces should be collected on every single property. As before, farmers should be given the opportunity to manure their acreages with human excrement. This, however, required a regular and reliable method of transporting the faeces from the urban area. The primary sources and published literature suggest that Dessau's transport system did not meet these requirements.

Piped Cities around 1900

Soon after the fundamental decisions were made and the initial problems in Dessau were resolved, both cities quite quickly built up and extended their sewage systems. In Darmstadt, fourteen years after the waterworks went into service, the city's sewage system was almost completed. In 1894, it was about 74 kilometres long. By 1913, fifty more kilometres had been added and most rain, industrial discharges and domestic water were removed efficiently. The same applied in general to the removal of faeces, once the local authorities changed the regulations and permitted the discharge of excrement into one of the new sewer pipes. Therefore, the number of water closets quickly rose from 338 in 1885 to 4,139 in 1910. In other words, in 1885, about fourteen per cent of all inhabited buildings were equipped with a water closet; 25 years later, this had increased to almost seventy per cent. An even more dramatic increase occurred in the proportion of buildings with a sewer connection, since approximately 55 per cent of all inhabited buildings had such a connection in 1885 and virtually all had one in 1895.

Dessau's new sewage system was completed in 1893. At that time, it was around 41 kilometres long; by 1913, it measured about 84 kilometres. Like Darmstadt, Dessau possessed the technical facility to dispose all rain, industrial and domestic water. However, the ban on discharging liquid and solid excrement into the sewer pipes remained. Instead, a system was implemented which, for example, the cities of Leipzig and Dresden had already established: a domestic sewage facility had to be built for every property.[9] At least once a year, every tank had to be emptied by a municipally managed transport company. Increasingly, tenements were provided with a water closet and, in 1901, Dessau's mayor stated that water closets had been installed almost universally. Contemporary sources suggest this. The number of buildings containing a WC increased sixfold from about nine per cent in 1883 to over fifty per cent in 1895. The percentage of tenements connected to a sewer increased even faster: two years after the system was completed 91 per cent of these buildings had a sewer connection and, by 1906, all buildings were connected.

9. Dessau's chief archivist confirmed that these mini plants in principle functioned like traditional cesspools. See also, K. Müller, 'Die Wohlfahrtspflege und das Armenwesen in Dessau (1907)', in *Die Haupt- und Residenzstadt Dessau*, 2nd edition, Dessau, 1911, 95–164.

In the long run, the rate of water consumption in both cities increased almost continuously. As a result, a steadily growing volume of waste-water and excrement reached the land around the city. In 1880–81, Darmstadt's waterworks delivered approximately 96,000 cubic metres of water; around 1910 it delivered about 4.2 million cubic metres, an increase of over 4,275 per cent or a more than 142 per cent increase each year for thirty years. Dessau started with more than 444,000 cubic metres in 1881–82; by 1913–14, it delivered more than two million cubic metres, an increase of over 350 per cent or eleven per cent per annum. The average daily consumption per capita in Darmstadt rose from about six litres in 1880–81 to around 124 litres in 1913–14, a sustained average annual increase of sixty per cent; in Dessau the increase was a much more modest average 3.7 per cent per annum over the thirty year period. In 1904 each of the 80,000 inhabitants in Darmstadt produced between 150 and 175 litres of waste-water per day, annually equivalent to 54–63 cubic metres; in Dessau, on average, each of the 54,000 residents produced around 45 litres of waste-water daily in 1903–04 or about 45 cubic metres per year.

Of course, this expansion in the demand for water and the volume of waste disposal was in part a function of rising population levels in both cities (see Table 2). Both cities grew spatially after new residential and industrial areas were opened up between 1870 and 1910. In the same time period, further public facilities and those provided by private enterprises and backed by the local authorities had been introduced and enabled the administration to retain control of local development. In its 1901 published report, Dessau City Council informed its citizens that they would experience significant development in all services.[10] The same message was announced repeatedly, while Darmstadt's municipality displayed almost identical activities and publications. The message was clear: a competent, efficient and forward-looking city council had proved its value and was ready for the challenges that the future would bring.

Table 2. Population Growth in Darmstadt and Dessau 1871–1910 [11]

	1871	1890	1910
Darmstadt	33,799	56,931 (68.4 %)	87,089 (53.0 %)
Dessau	17,459	34,658 (98.5 %)	56,606 (63.3 %)

10. [Municipal Authorities of Dessau], *Bericht über die Verwaltung und den Stand der Gemeindeangelegenheiten in der Residenzstadt Dessau 1900/01*, Dessau, 1901 (henceforth abbreviated as VB DE; for Darmstadt this is VB DA).

11. Source: M. Stippak, *Beharrliche Provisorien*, 60, 90, 133.

Piped Cities around 1900: Old and New Inequalities

In 1913, the mayor of Darmstadt and his colleagues praised the state of the local sewage system.[12] The councillor for urban development claimed that the city's hygienic conditions had improved significantly because almost the entire urban area had been piped.[13] Similarly, the mayor of Dessau declared in 1914 that, due to the local waste-water management, the city had become a beautiful, healthy and convenient place to live.[14] More statements of this sort demonstrated confidence, pride and optimism. To rely only on these representations would lead to a wrong impression that all earlier inequalities had been resolved and none had arisen in the meantime. For example, the councillor who had praised Darmstadt's hygienic progress conceded that not all traditional cesspools had been replaced by water closets.[15] There were still buildings whose inhabitants did not have access to this kind of fixture. In short, even in a piped city, inequality persisted.

Reading the contemporary sources, data and reports carefully, some shadows are cast on the brightly shining image presented by the mayors. By relating, for example, the number of water closets installed to the number of inhabitants, a more variable picture emerges. As mentioned previously, about seventy per cent of tenements in Darmstadt were fitted out with a WC in 1910. This percentage alone marks a noteworthy achievement. The same applied to Dessau, where by 1895 around 54 per cent of all tenements already possessed a WC. But if we relate the number of water closets to the number of residents, lower ratios apply. For instance, the ratio of toilets to people in Darmstadt was 1:21 and in Dessau 1:30 in 1910. Thus, for many people it remained exceptional to have sole use of a toilet inside their own apartment. Instead, many still had to use toilets on the stairs or in the basement, or a traditional cesspool. Most likely, it depended on the decision of each landlord whether or not an apartment had its own toilet.

Moreover, Darmstadt had to make intensive efforts to purify and dispose of its sewage, which was eventually discharged into sewage fields to the north and north-west of Darmstadt. Due to the increasing amount of waste-water and excrement, these fields were enlarged over the years and, in 1896, the city councillors decided to establish and manage a public sewage farm. But even this extension did not suffice and it became necessary to cooperate with farmers who were interested

12. W. Glässing, R. Mueller and E. Stein, 'Darmstadt', in E. Stein, ed. *Monographien deutscher Städte. Darstellung deutscher Städte und ihrer Arbeit in Wirtschaft, Finanzwesen, Hygiene, Sozialpolitik und Technik*, vol. III, Oldenburg, 1913, VII–VIII.

13. Keller (initial unknown), 'Die Entwässerung der Stadt Darmstadt', in Stein, ed. *Monographien*, vol. III, 159–63.

14. E. Ebeling, 'Dessau – eine aufstrebende deutsche Mittelstadt', in Stein, ed. *Monographien*, vol. IX, Dessau, 1914, 81–95.

15. Keller, 'Entwässerung'.

in fertilising their acreages with the effluent. Consequently, the area of the sewage fields grew from 32 hectares in 1877 to 920 hectares in 1950.

Another problem was the amount of sewage sludge. This mud should have been removed from the sewage water before it got onto the sewage fields. But it was not. From 1902, the problem of the muddy and foul sewage water was repeatedly mentioned in the city's annual reports and even though receptor cesspits were planned and their construction was announced annually, these plans were not implemented before World War I.[16] Nonetheless, Darmstadt's municipality initiated the regulation, extension and consolidation of the sewage reception areas and so rural labourers and farmers had to cope with ever-greater volumes of muddy and foul sewage water. While the problem escalated, the councillors' interest in this complex of problems decreased significantly. Dessau's city fathers also displayed no interest in the areas surrounding their immediate jurisdiction. As soon as the sewage water was discharged into one of the nearby rivers, it was literally 'out of sight, out of mind.' The same applied to the excrement that should have been removed by a municipal transport company.[17] Neither a sewage treatment plant nor sewage farms had been established and, while the river water diluted the waste-water, farmers and rural labourers continued to use the city's excrement as fertiliser.

Exceptions or Normal Cases? An Incidental Remark

How did these developments fit into the overall picture presented by German cities around 1900? Shortly after the first German Cities Exhibition, the book *The German Cities* was published.[18] Most of its articles praised the achievements German city councils had made in the previous three decades. But there were critical voices, too. One critic declared that waste-water treatment would remain one of the most difficult and most important tasks for all levels of administration. Even communes that were considered to possess sufficient sewage water installations were to keep the situation under review.[19] The question is whether Darmstadt and Dessau had established satisfactory systems of waste-water management and excrement treatment.

According to an overall view published in 1913, a 'totally new era' had begun, as the possibilities arose to supplement the primary or mechanical treatment of sewage with chemical and biological treatment stages. Although many

16. [Municipal Authorities of Darmstadt], VB DA 1901/02; VB DA 1904/05; VB DA 1905/06; VB DA 1906/07; VB DA 1907/08; VB DA 1908/09; VB DA 1909/10.

17. E. Ebeling, 'Die Stadtverwaltung', in *Die Haupt- und Residenzstadt Dessau*, 2nd edition, Dessau, 1911, 176–230; [A./W.] Schmetzer, 'Städtische Betriebe', in Stein, ed. *Monographien*, vol. IX, 120–38.

18. R. Wuttke, ed. *Die deutschen Städte. Geschildert nach den Ergebnissen der ersten deutschen Städteausstellung zu Dresden 1903*, 2 vols, Leipzig, 1904.

19. H. Klette, 'Tiefbau', in R. Wuttke, ed. *Städte*, vol. I, 370–419.

different treatment methods existed at that time, Darmstadt and Dessau lagged behind, as neither had built up any of these options.[20] In this respect both communes displayed a 'conservative trend'.[21] But they were not the only ones who clung to older purification and disposal methods: like Dessau, 266 other German cities maintained a transportation system until the early 1930s and among them were major cities like Leipzig and Weimar.[22] With reference to technical aspects, Darmstadt's sewage fields were in fact an old-fashioned, if still popular, method of waste management and purification.[23]

Darmstadt and Dessau from the 1920s to the 1980s – Unequal Development

After World War I and the German 'November Revolution', both cities suffered a political reversal: they lost their historic functions when the grand duchy of Hesse–Darmstadt and the duchy of Dessau–Anhalt were dissolved. Furthermore, several institutions had been dissolved or transferred to other cities. Although, Darmstadt and Dessau remained the capitals of their particular territories until 1945, their significance as military and administrative centres dwindled. After Hitler's takeover (*Machtergreifung)* and the consolidation of the Nazi regime, Darmstadt's importance declined markedly. While Dessau became an important centre for the German aviation industry from the 1920s and a pillar of the Nazi re-armament policy after 1933, Darmstadt had to struggle to conserve its precarious status. The different trajectories were reflected by the population levels: up to 1944, the number of inhabitants in Dessau rose to 132,000, while the population of Darmstadt increased to only some 115,000 inhabitants. Even more dramatic than the political changes was the destruction both cities suffered at the end of World War II. In September 1944, Darmstadt lay substantially destroyed by bombing raids; the same happened to Dessau in March 1945. The city centres were almost totally in ruins. Due to war damage, and because many people left the cities, both Dessau and Darmstadt could estimate that only some 50,000 residents remained at the end of the war, a reduction of over fifty per cent.

Whilst the two cities focused their efforts on reconstruction, both lost their status as capitals. Darmstadt became just a normal city in the newly founded West German federal state of Hesse; Dessau, in turn, belonged to the Soviet zone of occupation after the withdrawal of the US military. When the East German federal

20. M. Strell, *Die Abwasserfrage in ihrer geschichtlichen Entwicklung von den ältesten Zeiten bis zur Gegenwart*, Leipzig, 1913.

21. D. Schott, *Die Vernetzung der Stadt. Kommunale Energiepolitik, öffentlicher Nahverkehr und die 'Produktion' der modernen Stadt: Darmstadt - Mannheim - Mainz 1880–1918*, Darmstadt, 1999.

22. J. Brix, K. Imhoff, and R. Weldert, eds. *Die Stadtentwässerung in Deutschland*, 2 vols, Jena, 1934, vol. II, 32–3.

23. Strell, *Abwasserfrage*.

states were replaced by districts in 1952, Dessau became a politically unimportant city within the Halle district. Regardless of the political setbacks, both cities then experienced a modest economic and demographic revival, though Dessau never again regained its earlier strength (see table 3). While the population of Darmstadt grew rapidly from its post-war levels up to 1970, thereafter it has virtually stagnated; Dessau, by contrast, has remained smaller; up to 1988 its number of inhabitants did not rise above 103,000.

Table 3. Population Growth in Darmstadt and Dessau 1944–1970 [24]

	1944	1950	1970
Darmstadt	115,196	94,788 (-17.7%)	141,224 (49.0%)
Dessau	c.132,000	91,973 (-30.3%)	98,261 (6.8%)

Another difference is displayed by the speed at which apartments were equipped with an integrated bath/shower and WC. In 1961, almost sixty per cent of all apartments in Darmstadt possessed an integrated bath or shower; in Dessau, this percentage was only reached after 1980. The time lag was much greater for the introduction of internal toilets to apartments: in Darmstadt, in 1961, about 82 per cent of apartments possessed an integrated WC, whereas in Dessau in 1980 only one in every six apartments had such an amenity.

Revolting Hinterland and the Collapse of Darmstadt's Sewage Fields

Due to the changes outlined previously, Darmstadt's sewage system did not attract much political attention in the first half of the twentieth century. It is said that, in the 1930s, the city council had intended to build a central sewage treatment plant, but the war delayed its implementation.[25] After 1945, efforts were focused on the need to repair extensive war damage. However, when confronted with the roundworm, an intestinal parasite, which was detected soon after the war not only in the city's effluents but also in the fruit and vegetables farmers cultivated on their acreages, Darmstadt City Council vigorously demanded the construction of a central sewage treatment plant to filter the parasite out of the waste-water and eliminate it from the fruit and vegetables sold and consumed in the city.[26] Darmstadt's councillors appealed to cut off the 'cycle of infection'.[27] However, the city could not overcome

24. Source: M. Stippak, *Beharrliche Provisorien*, 225, 326, 347.

25. H. Möll, 'Abwassermessungen und -untersuchungen in Darmstadt', in *Gesundheits-Ingenieur*, 1950, 387–90 and 1951, 39–42.

26. W. Baumhögger, 'Spulwurmplage und Abwasserbeseitigung in Darmstadt', in *Gesundheits-Ingenieur*, 1948, 40–3.

27. H.P. Waldmann, 'Die Ascaridenplage, insbesondere im Raume Darmstadt', in *Gesundheits-Ingenieur*, 1952, 50–1.

the resistance of the farmers, who insisted on the fulfilment of their contract to receive the sewage of the city so that they could fertilise their fields with untreated effluents. Given the reduced population of the city, the decline in the city's economy and the disagreement of scientists about the ways the parasite spread, the city then had little authority and less justification to argue for a new sewage treatment plant. Lacking financial resources on the one hand and political support on the other, the situation remained unchanged: the city's waste-water continued to arrive in the nearby sewage fields, largely untreated.

It was not until the Hessian government provided extensive financial aid in 1954 that plans were made for four new treatment plants. But in contrast to the three smaller facilities that included mechanical and biological treatment systems, the central treatment plant contained only a mechanical works which the city justified by referring to the continuing sewage fields, which they argued provided adequate purification. Newly erected slurry tanks, digestion towers, and a newly constructed facility, they argued, should break the 'cycle of infection' by killing the roundworm eggs. However, the optimism of the Darmstadt building officer that the new arrangements would suffice for many years faded quickly.[28] At first, even the farmers were satisfied because the merely treated sewage water still met their requirements as a fertiliser. But only eight years later this arrangement collapsed.

Several problems developed. Firstly, the capacity of the central treatment plant was too small. The engineers and politicians involved had not anticipated the considerable increase in the amount of sewage generated by the city. Secondly, the efficiency of the sewage fields had been over-estimated and, by 1960, it was noted that their capacity was already almost reached. So Darmstadt City Council asked the sewage farmers if they would be willing to extend their acreages. Yet the situation became more serious, as an increasing number of the population were affected by the city's effluent policy. In particular, communes along the lower reaches of Darmstadt's sewage ditch pressed for relief as they suffered most from the effluents that stank and repeatedly killed fish stocks in the watercourses.

In contrast to the late 1940s, by 1960, it was the surrounding areas rather than the city residents that demanded a new technical solution, while the city sought to buy time. Problems stemmed not only from the volume of effluents to be dealt with but also from their consistency, because increasing levels of fuel and engine oil were present. Sewage farmers also experienced a shortage of recruits for their unpleasant work, as younger people were looking for better work opportunities – that is to say, more reputable and profitable jobs. As a result of these structural changes, Darmstadt's urbanising hinterland felt particularly disadvantaged. By pointing to the environmental problems that Darmstadt's effluents caused, rural labourers,

28. M. Haury, 'Darmstadts Abwassersorgen und durchgeführte Massnahmen. Sonderdruck 'Städtehygiene' 11/1958', in Municipal Archive of Darmstadt, 12h Tiefbau und Energie – Stadtentwässerung und Kläranlagen, file 'Historisches zur Stadtentwässerung'.

farmers and communes campaigned for improvements in sewage treatment and, in the light of political and financial sanctions imposed by superior administrative authorities, Darmstadt finally conceded and agreed to upgrade its central sewage treatment plant. Thus, after almost seventy years, the era of Darmstadt's sewage farms ended.

Revolting Residents in a Powerless City: Environmental Inequality in East Germany

Dessau's first sewage system outlasted its counterpart in Darmstadt. Since all East German communes were to a great extent politically and financially disempowered in the early 1950s, their abilities to plan and to operate independently were severely restricted. In 1949, for example, Dessau lost responsibility for the management of its waterworks and sewage system. Both were transferred to what was later known as a nationally owned enterprise (*Volkseigener Betrieb*). Accordingly, Dessau had to cooperate with superior bodies and so depended on decisions made elsewhere. The potential of Dessau to influence certain issues was correlated with access to influential civil servants and politicians.

The first central sewage treatment plant in Dessau only became operational in 1976. Evidently the city's political influence was limited, as its economic significance was modest. Had it been otherwise, more serious efforts would have been made to improve environmental conditions in Dessau and its surrounding area. Very probably, Dessau shared this fate with most other East German communes. As an expert had worked out in 1955, almost all rivers in the middle German industrial region (*Mitteldeutsches Industriegebiet)* were polluted to such an extent that their water could not be used for drinking. The Elbe and Mulde Rivers were heavily polluted and in a poor condition when they reached Dessau.[29] Even though the Construction Law (*Aufbaugesetz)* that was passed in 1950 described Dessau as one of the country's most important industrial centres, the city had to wait almost three decades before a central sewage treatment plant was built. Even this did not result in an improvement, as the stench that originated from the city's effluents could not be reduced. It was both an environmental problem and a political burden.

In the early 1950s, plans were discussed based on the factors applying in the late 1930s and early 1940s. These proposed that at least a part of the waste-water should be used as fertiliser on acreages in the hinterland. Quite soon, the central sewage treatment plant project was revived. Furthermore, it was proposed to connect three suburbs that had been incorporated in the 1920s to the network of pipes. Optimism quickly evaporated, however, when the superior bodies learned that the inhabitants of Alten, a western suburb of Dessau, were very keen on comprehen-

29. H. Hübner, 'Die Wasserversorgung Mitteldeutschlands – ein Verbundplan der Wasserwirtschaft', in *Wasserwirtschaft-Wassertechnik*, 1955, 138–45.

sive plans to develop the network of pipes in the suburb. The same demand was articulated in other quarters too. Additionally, a very large number of complaints was sent to members of both the local and the district council; local politicians and civil servants could do little more than put off the complaining residents.

So long as no financial, technical, and logistical resources were assigned, a large number of people had to live in close proximity to the domestic sewage treatment plants, which were already antiquated. Even though local decision makers expressed their concerns to political superiors, action was not noticeably accelerated, despite the fact that district councillors knew from an early stage about the problems. The priorities of the East German central government were notoriously difficult to influence.

Even after the prestigious central sewage treatment plant, situated close to the Elbe River, was operational, the situation did not improve. In a sense it got even worse because the capacity of the plant could not cope with the consistency of the industrial effluents. Therefore, it could not reduce the malodorous nuisance that developed particularly at the places where the city's main sewers discharged their contents into the Elbe and Mulde Rivers. Further on, only the effluents of one sewer sub-system were directed to the new plant while the other still discharged its untreated waste-water into the Mulde River. Thus, a stink permeated both and annoyed more than 10,000 residents. Accordingly, the central sewage treatment plant became a political nightmare. Exposed to a persistent and intolerable stench, the people most affected lost confidence in the ability of the political system and its expressed intent to abandon inequalities of an environmental and social character. It was not until the decline of the German Democratic Republic that this situation changed.

Concluding Remarks

It has been shown that Darmstadt and Dessau operated initially under similar general conditions. Against the background of demographic, economic and structural changes both city councils decided to make use of the expertise that contemporary hygienists and engineers offered. As noted, this knowledge was not the only factor in the rearrangement of the urban environment because, in addition, there was a corresponding predisposition to identify environmental problems and a readiness to resolve them. Thus, a combination of several factors led to these changes.

Different harmful environmental conditions existed years after the systems in question had been put into service. Although many city councils enhanced their political status internally and externally, their effective real power remained limited. As long as a landlord did not approve, his tenants had to live without an integrated bathroom, shower or WC. Correspondingly, the environmental conditions that affected people varied appreciably. It is remarkable that this finding applies for East Germany too, where the greater proportion of the tenements was owned by the state.

This leads to the more fundamental conclusion that the cities' interests and room for manoeuvre changed dramatically over the years. After the sewage systems had been implemented, other technologies, as well as political and economic developments, attracted much of the cities' attention. Consequently, the communes' political, technical, economic and mental resources did not suffice to solve the issue satisfactorily. This situation did not change until the sewage system caused nuisances that other people were no longer willing to accept. According to the different political systems established in West and East Germany, the potential for the cities to adapt to these changes diverged considerably: Darmstadt could be forced to modernise its technological infrastructure; Dessau, by contrast, could do no more than manage the technical and environmental *status quo*.

Select Bibliography

Bernhardt, C. 'Towards the Socialist Sanitary City: Urban Water Problems in East German New Towns 1945–1970', in D. Schott, B. Luckin, and G. Massard-Guilbaud, eds. *Resources of the City. Contributions to an Environmental History of Modern Europe.* Aldershot, 2005, 185–202.

Bernhardt, C. 'Zwischen Industrialismus und sanitärer Wohlfahrt: Umweltprobleme im Sozialismus am Beispiel der Wasserfrage in der DDR', in T. Meyer and M. Popplow, eds. *Technik, Arbeit und Umwelt in der Geschichte*, Münster, 2006, 367–80.

Frank, S. and M. Gandy, eds. *Hydropolis. Wasser und die Stadt der Moderne*, Frankfurt, 2006.

Evans, R.J. *Tod in Hamburg. Stadt, Politik und Gesellschaft in den Cholera-Jahren 1830–1910*, Reinbek bei Hamburg, 1990.

Hardy, A.I. *Ärzte, Ingenieure und städtische Gesundheit. Medizinische Theorien in der Hygienebewegung des 19. Jahrhunderts*, Frankfurt, 2005.

Melosi, M.V. *The Sanitary City. Urban Infrastructure from Colonial Times to the Present*, Baltimore, 2000.

Münch, P. *Stadthygiene im 19. und 20. Jahrhundert. Die Wasserversorgung, Abwasser- und Abfallbeseitigung unter besonderer Berücksichtigung Münchens*, Göttingen, 1993.

Rodenstein, M. *'Mehr Licht, mehr Luft' – Gesundheitskonzepte im Städtebau seit 1750*, Frankfurt, 1988.

Simson, J. von, *Kanalisation und Städtehygiene im 19. Jahrhundert*, Düsseldorf, 1983.

Schott, D. 'Symbolische Modernität und die elektrische Vernetzung der Stadt, oder: Stadtentwicklung als "Drahtseilakt"', in H. Uhl, ed. *Kultur–Urbanität–Moderne. Differenzierungen der Moderne in Zentraleuropa um 1900*, Vienna, 1999, 287–320.

Schott, D. *Die Vernetzung der Stadt. Kommunale Energiepolitik, öffentlicher Nahverkehr und die 'Produktion' der modernen Stadt: Darmstadt - Mannheim - Mainz 1880–1918*, Darmstadt, 1999.

Stippak, M. *Beharrliche Provisorien, Städtische Wasserversorgung und Abwasserentsorgung in Darmstadt und Dessau 1869–1989*, Münster *et al.* 2010 (originally PhD thesis, Darmstadt University of Technology, Germany, 2008).

Stippak, M. 'The Mental and Practical Impact of Pre-bacteriological Quality Criteria for Water in the 1870s', in R. Heil, A. Kaminski and M. Stippak *et al.* eds. *Tensions and Convergences. Technological and Aesthetic Transformations of Society*, Bielefeld, 2007, 295–303.

Tarr, J.A. *The Search for the Ultimate Sink. Urban Pollution in Historical Perspective*, Akron, 1996.

Wuttke, R. ed. *Die deutschen Städte. Geschildert nach den Ergebnissen der ersten deutschen Städteausstellung zu Dresden 1903*, 2 vols, Leipzig, 1904.

Würth, G. *Umweltschutz und Umweltzerstörung in der DDR*, Frankfurt, 1985.

PART IV

Waste and Inequalities

Chapter Nine

Settling Urban Waste Disposal Facilities in France c. 1900–40: A New Source of Inequality?

Stéphane Frioux

In 2008, garbage from Naples was sent to Switzerland to be treated. It was a modern day option to solve an age-old problem: how to dispose of urban waste and where to locate waste facilities? These highly sensitive issues gained in significance as a direct consequence of industrialisation and urbanisation. In France, as elsewhere, what to do with the city's refuse has become a serious practical and political problem for urban authorities. Whether waste disposal is in municipal landfill sites, which are usually full after a few years and produce complaints from those who live nearby; or is composted in sealed containers; or industrially treated by incineration remains contentious and attracts criticism. Incineration, for example, often thought indispensable as an urban disposal practice, came to be suspected some time ago of disseminating dioxins into the urban and suburban atmosphere and has increasingly become the object of criticism from environmental activists, journalists and local residents.

It was at the beginning of the twentieth century, when incineration was still in its infancy, that the choice of site emerged as a problem for urban managers. Before building new facilities, most French local authorities carefully assessed the costs and benefits, particularly the costs of construction, which raised major objections. Alongside these financial considerations were additional concerns about potential nuisances for the neighbours of the projected plant or dumping site. 'NIMBY' ('Not in my backyard') arguments and political strategies conspired to delay local implementation of technical solutions to an enduring problem.[1] At the turn of the nineteenth century, industrially based options – 'biological' sewage treatment plants, garbage 'destructors' and composting treatment plants – offered municipalities new ways to solve the waste disposal problem.[2] This chapter exam-

1. At present, a right-wing majority administration in Marseilles wants to create an incinerator in another *département*, controlled by the left-wing *Parti socialiste*, which rejects the Marseilles strategy and prefers a controlled dumping alternative.

2. C. Hamlin, 'William Dibdin and the Idea of Biological Sewage Treatment', *Technology and Culture*, 29, 1998, 189–218; M.V. Melosi, 'Technology Diffusion and Refuse Disposal: The Case of the British Destructor', in J.A. Tarr and G. Dupuy, eds. *Technology and the Rise of the*

ines how the choice of location for waste treatment plants was addressed by French municipal authorities, and this is considered in relation to comparable studies in Great Britain and the United States.[3] Across the Atlantic and until the 1980s, about 75 per cent of waste disposal facilities were located in or near African American neighbourhoods: 'NIMBY issues were too often ignored, and there was little or no public discussion of siting decisions that raised questions about placement in poor or minority communities.'[4] So the question is, did similar forms of social discrimination operate in the French choice of sites for waste disposal facilities? Was the implementation of waste treatment facilities a source of environmental inequality in France?

By comparison with British and American cities, the development of garbage and sewage treatment plants in France was very limited before the second half of the twentieth century. This limited diffusion of such facilities in the French urban network was not linked, however, with a negative discourse about them. On the contrary, they were seen as 'municipal modernity' and were rarely suspected of being environmentally harmful by those charged with how to manage waste. However, by scrutinising the local scale and analysing the issues at stake in the choice of location for a new garbage treatment plant, it is possible to illuminate the factors behind the municipal decisions and, where they existed, to examine the nature and intensity of protests associated with the decisions.

From Dumps to Incinerators: The Coming of Modernity

Urban environmental history studies have already focused on the deterioration of living conditions of urban dwellers during the decades that followed the industrial revolution and the resultant economic and demographic dynamics of western cities. The 'search for the ultimate sink'[5] in which to dispose of the increasing amounts of waste produced by the dramatically increased population led urban managers to seek new solutions and locations.

Networked City in Europe and America, Philadelphia, 1988, 207–26; S. Barles, *L'invention des déchets urbains: France 1790–1970*, Seyssel, 2005.

3. J. Clark, '"The Incineration of Refuse is Beautiful": Torquay and the Introduction of Municipal Refuse Destructors', *Urban History*, 34:2, 2007, 255–77; M.V. Melosi, 'Sanitary Services and Decision Making in Houston, 1876–1945', *Journal of Urban History*, 20:3, 1994, 365–406, notes racial segregation produced another dimension.

4. Melosi, 'Sanitary Services and Decision Making'; M.V. Melosi, *Garbage in the Cities: Refuse, Reform and the Environment, 1880–1980*, 2nd edn, Pittsburgh, 2004, 187. See also the case study on Houston in M. V. Melosi and J.A. Pratt, eds. *Energy Metropolis: an Environmental History of Houston and the Gulf Coast*, Pittsburgh, 2007.

5. J.A. Tarr, *The Search for the Ultimate Sink: Urban Pollution in a Historical Perspective*, Akron, 1996.

The decades preceding the First World War saw the development of what, in Britain from the 1870s, was called the 'sanitary engineering' industry, in which new branch of engineering the most important fields were water purification and waste-water treatment. At approximately the same time, a 'refuse revolution' occurred. In Britain and in the United States, private firms built incineration plants and sewage treatment facilities.[6] In Europe, the German city of Hamburg became the first continental city (in 1895) to adopt incineration, because of the failure of a liberal city management to combat the conditions that produced a serious cholera epidemic affecting the inhabitants of the Elbe river bank harbour area in 1892, which discredited their non-interventionist policies.[7] In the case of French waste disposal methods, a technical turning point in refuse management is difficult to detect before 1900 and developments remained very limited before 1914. Even among high-ranking public health specialists, the preference remained for the conventional use of agricultural sites for city refuse; this meant allowing organic waste to putrefy in open fields and rural sites.[8] Medical elites, backed by commercial interests, agreed on the importance of maintaining a certain complementarity between the urban world and the countryside. In 1896, Professor Paul Brouardel, chairman of the *Comité consultatif d'hygiène publique* (National Board of Public Health), wrote a pamphlet along with the physician Octave du Mesnil, in which the two famous Parisian hygienists claimed that using the railway to carry garbage from the capital to open fields forty miles away was preferable to its mere destruction by fire.[9]

Why did a change occur? In the mid-1890s, the authorities of Paris considered the adoption of a new method of garbage disposal. The city faced serious problems. No longer was there a site within the municipal area with the capacity to dispose of the volume of waste, and mayors in suburban townships initiated by-laws prohibiting Parisian garbage carts from using their streets. Almost all French cities were in a similar situation, surrounded as they, too, were by dumps perceived by hygienists as a 'necessary evil' and the subject of numerous complaints from suburban residents.[10] From 1895 and 1896, after an unsuccessful attempt to burn limited amounts of garbage in the municipal plant of the quai de Javel near the river Seine, some entrepreneurs tried to make a profit from garbage by using

6. M.V. Melosi, *The Sanitary City: Urban Infrastructure in America from Colonial Times to the Present*, Baltimore, 2000.
7. R.J. Evans, *Death in Hamburg. Society and Politics in the Cholera Years 1830–1910*, Oxford, 1987.
8. See the chapter by Marcus Stippak in this volume for an account of similar practices in Dessau and Darmstadt.
9. *Immondices – transport par chemin de fer – destruction par le feu. Rapport présenté par les Dr P. Brouardel et Du Mesnil, extrait du tome XXVI (année 1896) du recueil des travaux du Comité consultatif d'hygiène publique de France et des actes officiels de l'administration sanitaire*, Melun, 1896.
10. The dumps had to get an administrative authorisation, according to the decree of 1810 concerning insalubrious industries.

industrial grinding plants which produced *poudro*, a thin organic matter able to fertilise fields and pastures.[11] They established their facilities in neighbouring towns, in spite of some protests from municipal councils and local citizens and, because within a few years the agricultural treatment of refuse was no longer capable of absorbing all waste products, incineration became a necessity. Four grinding and incinerating plants were operational in Parisian suburbs by 1914, with about six others elsewhere in France at the start of the First World War.[12] In most other cases, city councils were content to operate open dumps or contract out refuse collection to local entrepreneurs – often these contractors were farmers in the small towns or businessmen exploiting dumps and paying people to sort refuse in a rather casual way and letting organic matter putrefy until it became usable as a fertiliser.

For all these reasons, including some financial factors that delayed the decision to construct a municipal garbage treatment plant, France appeared, before 1914, to lag behind other industrial countries, especially Anglo-Saxon ones. Though an accurate national survey of French facilities is not available, research on some sixty French cities and evidence drawn from technical journals suggests that about forty garbage treatment plants were operating in 1939.[13] These are very modest figures compared to those of Britain where, in 1912, there were more than 338 incinerators, eighty of which (24 per cent) produced electricity.[14] In the United States, the New Deal programme boosted the adoption of such facilities, with about 700 incinerators operating in the 1930s, despite complaints about the smoke and odours they produced.[15] The publicity regarding the potential nuisance of such innovatory disposal facilities is thus very different to that where they remained scattered thinly across the country.

Incinerators: The Modern Way of Managing Waste

From the mid-nineteenth century, numerous documents attest to the fact that cities were covered with industrial smoke and fog (smog) that degraded nature and urban buildings alike.[16] Yet it is very difficult to find objective criticism regarding smoke produced by incinerators in the first half of the twentieth century. In all cases, the French reluctance to adopt innovative methods of waste disposal was not

11. Barles, *L'invention des déchets urbains.*

12. Saint-Ouen, Romainville, Ivry, Issy-les-Moulineaux. There were also failures in Boulogne and in Créteil to establish incineration plants.

13. S. Frioux, *Les réseaux de la modernité*, PhD Thesis, Université Lyon 2, 2009.

14. Clark, '"The incineration of refuse is beautiful."'.

15. Melosi, *The Sanitary City*, 240, 277.

16. S. Mosley, *The Chimney of the World. A History of Smoke Pollution in Victorian and Edwardian Manchester*, London, 2008. F. Uekotter, *The Age of Smoke. Environmental Policy in Germany and the United States, 1880–1970*, Pittsburgh, 2009.

due to a specific local or national sensitivity to any inconveniences produced by, or attributed to, waste disposal facilities. The tone of the numerous publications and articles about incineration is more that of a gospel *credo* than of an objective critic, as the emphasis is firmly on the benefits of incinerating plants as an element in urban modernisation policies. Around 1910, a famous French hygienist, Dr Albert Calmette, publicly supported the incineration method, as he did biological sewage treatment.[17] The idea of progress associated with the cremation of garbage seeped into society; in 1929 in Toulouse, for example, the pupils were invited to visit the municipal plant located on an island of the Garonne River, near a *promenade* and a stadium; once back in their classrooms, they praised the cleanliness of the plant.[18]

The faith placed by hygienists in the sanitary efficiency of incinerators influenced many engineers and even some mayors. Consequently cities, both small and large, increasingly decided to choose the path of incineration, especially once French patents and companies appeared on the market after 1918. The inter-war period saw the first expansion of industrial treatment of garbage, with three different methods adopted unequally. Firstly, incineration was the technology preferred by French cities – at least 35 cities operated a plant by 1938; secondly, composting convinced only five southern cities, with the exception of a few experimental plants near Paris;[19] thirdly, in the 1930s, Parisian engineers tested the 'controlled tipping' or 'sanitary landfill' process that addressed the inconveniences associated with former dumps, such as rats, flies and spontaneous fires. This method was very popular in Britain, where the Ministry of Health itself gave official recommendations to municipal engineers and surveyors.[20] Documents produced by companies, such as brochures and leaflets with lists of references, and the exchange of information between cities contributed to the dissemination of incineration technology (see map 1). The size of French cities was instrumental in the adoption of incinerators – almost every city with more than 100,000 citizens adopted an industrial form of waste treatment[21] – as was the desire of elected officials to show that they were concerned with the notion of 'modernity' in every aspect of urban policy. In ad-

17. AM (Archives Municipales) Lyon, 923 WP 273, report by the engineer in chief of Lyon, 1 Feb 1930.

18. AM Toulouse, 1R 291.

19. At that time, the method, invented by an Italian physician, Dr Beccari, consisted in letting garbage ferment in concrete cells during several weeks, and then produce a fertiliser for agricultural use. Plants were settled between 1928 and 1936 in Cannes, Aix-en-Provence, Avignon, Valence and Narbonne. Experiments were conducted not far from Paris, in Versailles and Villeneuve-Saint-Georges.

20. J.C. Dawes, 'Public Cleansing Service in England and Wales', *Conférence internationale de l'Union internationale des villes*, 2e partie, 'La Collecte et la Destruction des Ordures Ménagères', Lyon, 1934, 8–81. On Paris, see J. Partridge, *Les décharges contrôlées d'ordures ménagères*, Paris, 1937 (excerpt from *Annales d'hygiène publique*, 1937).

21. Exceptions were Nantes and Strasbourg.

dition, the imperative of preserving the 'social tone' of the city in order to attract tourists to spa and seaside resorts was also a powerful public relations influence.

Harmless Facilities?

An array of accounts described pioneering plants working without inconvenience in densely populated areas and near public facilities. The municipal commission of Rouen, which in 1908 visited several incinerators in England and on the continent, noted that some facilities were located in wealthy neighbourhoods, as in Liverpool.[22] A decade earlier, Geneva's councillors were also optimistic, stating that in Edinburgh the plant was built right beside the biggest flour mill in the world.[23] In the overwhelming majority of experts' reports, the presumed harmlessness of incinerators made it possible to set them close to dwellings, in contrast to former dumps which had to be removed as far as possible from houses because of the written complaints they elicited.[24] Incinerators were seen as hygienic and economical, because they saved money and time in garbage collection operations[25] and because their emanations were less noticeable than those from other factories. As one observer remarked: 'It is always an interesting experience to compare the thin plume of light colour that emerges from the stacks of incinerators to the heavy black clouds coming out from the surrounding plants' stacks.'[26]

Often quoted in contemporary publications was the case of Monaco. In 1909, according to Professor Chérioux, member of the Public Health Council of the Seine:

> These plants, whatever the type of furnace they have, can be laid-out and operated in such a way that they do not represent a cause of annoyance or unhealthiness for the neighbours or for the workers … In London, in one of the smartest areas, near the Cecil Hotel, one finds an incinerating plant … Nobody complains and people have built elegant mansions all around. In Monaco, the garbage treatment plant is situated right by the Prince palace, almost under its windows, and the unhealthiness it could entail was certainly not very high, since, after the plant's construction, the new Hotel-Dieu was built right next to it.[27]

22. AM Rouen, 5M 1, deliberation of municipal council, 29 Jan 1909.
23. Archives of Paris, VONC 1484, extract from the minutes of Geneva city council, 21 Mar 1899.
24. AM Saint-Etienne, 4O 1, statement from a councilor to the city council, 9 Jun 1906.
25. Garbage dumps required a lengthier transport journey.
26. AM Lyon, 923 WP 269, memo from CAMIA to the mayor of Lyon, 13 Aug 1929. This statement may have been true when the incinerator was working without any addition of coal, because coal was the main producer of heavy smoke, or when the combustion was complete.
27. Quoted in Calmette, Imbeaux, Pottevin, *Egouts et vidanges, ordures ménagères, cimetières*, Paris, 1911, 497. Reproduced by the SEPIA in a brochure (AM Lyon, 923 WP 273).

The example of Monaco was quoted several times by private companies promoting incineration throughout the period, up to the 1950s.[28] During the inter-war years the firm CAMIA used other examples showing that garbage incinerators did not hinder urban development to convince French mayors to embark on a contract. Its brochures mentioned that a furnace was operating in the Grand Hotel of the ski resort of Font-Romeu:

> Without the slightest hesitation, the engineer-constructors settled the furnace in the hotel itself, near the bakery, under the great dining hall ... The fumes, evacuated by the kitchen's chimneys, are odourless and there is not a single grain of dust to soil the purity of the air, nor the whiteness of the snow.[29]

CAMIA stated that other incinerators were located near a nursery school and a hospital (in Elbeuf), and the laundry of the Grand Hotel in Cabourg[30]. But, according to Clark who has studied the British case, 'the very fact that these engineers took explicit pains to link destructors to "high-class property" raises a suspicion that Gloucester and Monte-Carlo were great exceptions to the general rule.'[31] A shadow of doubt existed since incinerators did not work efficiently every time or everywhere, especially in the United States.[32] Some of them produced smells and smoke when the combustion was incomplete, though it is important to bear in mind that their chimneys were often scattered among dozens of others in the urban landscape and that other techniques of garbage disposal also had their inconveniences: open dumps were associated with rats, flies, spontaneous fires, and reduction plants were accused of producing noxious smells.

Another promotional idea found in some documents is that the incinerators were so hygienic, so clean, that it would be possible to have lunch on the floor of the plant.[33] This propaganda did not prevent some people from expressing their concerns about incineration, though they were rare. After the Second World War, a private contractor wrote: 'If the natural psychological reaction due to the bad reputation of old incinerators could be set aside, we contend that such a plant could be constructed in the very centre of the agglomeration.'[34] Despite the relative absence of negative writings from engineers about these waste disposal facili-

28. AM Chambéry, 1O 93, letter from E.A.T. to mayor of Chambery, 2 Jan 1951.
29. *La Technique Sanitaire et Municipale*, May 1926, 115.
30. AM Lyon, 923 WP 269, memo quoted.
31. Clark, '"The Incineration of Refuse is Beautiful"', 261.
32. Melosi, *Garbage in the Cities.*
33. AM Chambéry, 1O 93, newspaper clipping about incinerator of Berne, *Journal de Genève*, 1 July 1955. Same image used about Nancy: AM Dijon, SG 58G, letter from Georges Connes to the mayor, 1 Dec 1952.
34. AM Chambéry, 1O 93, anonymous note from the EAT: 'principe et avantages du procédé d'incinération des ordures ménagères USP', 1951 or 1952.

ties, this sentence reveals that there could have been a problem, hidden behind the 'gospel of modernity' preached by the hygienists and the builders. As noted earlier, incineration started late in France, compared to Britain, and so engineers were able to benefit from the technical improvements achieved since the 1880s. The French firm SEPIA claimed in one of its brochures that 'the first furnaces, with natural draughts, did not allow the necessary conditions to have complete combustion and [that] the chimneys emitted smelly smoke which contributed to the bad reputation that affected incinerating plants for a while.'[35] The municipal commission of Rouen had already noted that 'the smoke produced by incinerating plants presents itself under a whitish aspect and absolutely does not smell' and had concluded that 'the functioning of an incinerating plant did not constitute any nuisance for the neighbouring areas.'[36] One of the few municipal statements against incineration can be found in a letter from the mayor of Nancy, in answer to an enquiry from his colleague in Rouen:[37]

> Given the bad work of the plant, the inconveniences of collection and in order to suppress the dust which prompts complaints from a whole neighbourhood, the Municipality is studying the construction of a new plant, to be erected far away from densely populated areas.[38]

In Southern France, where garbage burned less easily since it contained a greater proportion of organic matter and smaller quantities of ashes and combustible matter, difficulties were found in Toulon, a fact that persuaded Aix-en-Provence to reject the incineration alternative as 'a very flawed operation because it emits unbearable odours for the neighbourhood'.[39]

The general impression obtained from public records is that incinerators were not like other plants because they were municipally owned and managed and were perceived as a sanitary necessity for the urban environment. Of course, that fact did not prevent them from being the subject of complaints,[40] but it is likely that most citizens' petitions against them ended up in the dustbin of a municipal office. The situation may have been different elsewhere, for instance in the United States, where some garbage disposal plants, like reduction plants, were mainly operated by private contractors. More generally, in France as in Germany[41] until the 1960s

35. AM Lyon, 923 WP 273, brochure *La destruction des ordures ménagères et de tous autres déchets* (1924).

36. AM Rouen, 5M 1, city council minutes, 29 Jan 1909.

37. Both cities had plants equipped with British furnaces patented by the firm Heenan and Froude.

38. AM Rouen, 1I 17, letter from Nancy's mayor, 11 May 1931.

39. AM Aix-en-Provence, I6 70, note to the *sous-préfet*, n.d.

40. In Germany, municipally operated electricity works often provoked complaints (Uekoetter, 64–65).

41. Uekoetter, 185.

at least, incineration was not considered to be a major source of air pollution. At that time, only the level of carbon monoxide and the weight of dust in the fumes were measured.[42] In the 1970s, several people complained about the soot emitted by the incinerator in Biarritz, an ageing plant erected more than 45 years earlier. Some of the complainants belonged to the upper middle class and owned summer residences in the seaside resort:

> As a houseowner living for several summer months in your town, I had invited some friends in September and October. We had to leave the place, despite the wonderful Autumn. The smoke from the garbage, and the partially consumed rubbish came on to my terrace and forced us to keep doors and windows shut.[43]

The Biarritz municipality was not in a hurry to make improvements: the opening of a metropolitan garbage collection and disposal service occurred only in 1978. Generally speaking, local authorities were favourably disposed to incinerators during the decision-making process and the implementation stage and only once the plants had been operational for years did they explicitly admit to some of their nuisances. For instance, the administration of Cherbourg admitted, almost two decades after the creation of the city's incinerator near the slaughterhouses, that 'the dreadful odour about which the vicinity in our agglomeration is complaining with good reason' comes from a plant whose 'patents [were] outdated'.[44]

Locating the Incinerators

Before 1914, during the first era of industrial garbage treatment, plants were located in zones with scattered population. Consequently, complaints existed but were sporadic. They appeared in Toulon, one of the first cities to authorise a grinding plant, and later an incinerator, in 1908–09. A petition filed by a Mr Bonelli stated that after less than one month of full operation of the facility, one of his two tenants had left the house. The other tenant stayed, but he was working at the plant. Bonelli wrote: 'The house is invaded by the smoke and the dust, the furniture and clothes are soiled … if the slightest wind rises, living in the house, even shut, is unbearable.'[45] In addition to the house, the property contained an orchard, a vine and a vegetable garden. At least one of the two landowners whose properties were next to the plant and Bonelli's house also filed a lawsuit against the municipality. The case of Toulon is typical of a situation where garbage disposal facilities affected an area of rural and rented housing where people sought clean and pure air, and of

42. AM Villeurbanne, 1J 21, circular of 14 Apr 1962. On the proportion of dust in the smoke, see also AM Clermont-Ferrand, 1I 70, letter from USP, 1 Jun 1938.
43. AM Biarritz, 1M 45, letter from M. Maltot, Oct 1973.
44. AM Limoges, 3D 112, letter from mayor of Cherbourg, 2 Jan 1952.
45. AM Toulon, 1O 9, memo 11 Oct 1911.

the social characteristics of the plaintiffs, often landlords or middle class people. This makes it too difficult to assess the actual injustices inflicted on the workers renting houses near the waste disposal facilities.

The inter-war period saw the dawn of French city planning. In 1919 and 1924 two acts were passed to create improvement plans in the context of a rapid urban sprawl (*Plan d'aménagement, d'embellissement et d'extension*) in cities of over 10,000 inhabitants and in the spas. The general idea was to allow a rational spatial expansion of the city with a separation or zoning of residential, leisure, and industrial functions.[46] This planning development could have led to a radical separation between garbage treatment plants and residential areas. Indeed, most projects put the waste treatment plants in the suburbs, three to four kilometres from the downtown area, even though many of them were not carried out by 1940.

Biarritz is representative of the situation that confronted various cities by virtue of their spatial growth. The municipal engineer of this famous seaside resort explained that the expansion of the city forced the administration to give up the previous way of dealing with garbage disposal, in dumps producing fertiliser for surrounding farmers. From 1927 onwards, the city developed an ambitious project for a central industrial district (*Le nouveau centre industriel de la ville de Biarritz*); the planned incineration plant was conceived as the keystone of the complex.[47] The site was seen as well-suited for its purpose. Located at the geographical centre of the municipal territory, in an area where land was unsuitable for residential buildings, it was 'sufficiently removed from leisure areas and human concentrations and should entail no trouble for neighbours'. No dust was anticipated from the incinerator chimney and the project included the construction of a laundry, public baths and a housing programme for municipal employees just next to the garbage treatment plant.[48]

Biarritz, like Tours, Belfort, Lyon, and Nancy, sought to make garbage incineration profitable by utilising the vapour produced during combustion, a practice that had been common in British cities two decades earlier but which was no longer seen as efficient there.[49] It often led municipal officials to locate their plants on the outskirts of the city, more precisely near the slaughterhouses (Belfort, Tours, Marseille) and projected sewage treatment plants (Belfort, Lyon). In Elbeuf,

---

46. J-P. Gaudin, *L'Avenir en plan. Technique et politique dans la prévision urbaine, 1900-1930*, Seyssel, 1985. For case studies of the ways industrial interests interfered in urban planning in Interwar France, see G. Massard-Guilbaud, 'The Struggle for Urban Space: Nantes and Clermont-Ferrand, 1830–1930', in D. Schott, *et al. Resources of the City. Contributions to an Environmental History of Modern Europe*, Aldershot, 2005, 113–131.

47. AM Biarritz, 1M 41, Le nouveau centre industriel de la ville de Biarritz, dactylog. report, ca. 1927–1928.

48. AM Biarritz, 1M 41, Le nouveau centre industriel de la ville de Biarritz.

49. Dawes, 'Public Cleansing Service in England and Wales.'

one of the few French cities that were pioneers in this domain before 1914, the location chosen was next to a public school and the municipal waterworks, evidence that incinerators were not seen as polluting facilities.[50]

Were less affluent neighbourhoods more frequently victims of choices made as to where to set up incineration plants or sewage treatment stations? Some studies have already highlighted this aspect of urban inequalities. Studying the United States, Melosi noted that the concern of city leaders over neighbourhood complaints was minimal in the first half of twentieth century, as incinerators were mainly situated in non-white districts. Most of the Houston incinerators in the 1930s were located in black or Hispanic neighbourhoods, populated by people with no political rights or, at best, the weakest political voice. Complaints were not taken into account by engineers, who assessed the efficiency of waste disposal facilities only on operational costs and not in terms of air pollution.[51] In the case of French cities, when incinerators were distanced from residential areas, it was because of the location of existing municipal facilities rather than because of the smoke or odour city officials imagined would result from the garbage incinerator itself.[52] They were commonly built on the eastern side of towns and cities, as in Rouen, because the prevailing westerly wind direction in many French regions meant pollutants would be blown away from the urban area.[53]

At least one important exception to this policy of settling incinerators in industrial zones can be noted in the case of Villeurbanne. This city had preserved its independence in spite of several attempts by Lyon to annex it. Villeurbanne experienced rapid demographic growth in the space of a few decades, increasing from about 30,000 inhabitants in 1900 to 60,000 in 1926 and 80,000 in 1931. A socialist majority ruled the city council from 1892 onwards.[54] As early as 1911, the administration of Jules Grandclément, a physician, decided to municipalise the service of garbage collection and to build an incinerator.[55] The modest plant started operating in 1912 and was located in the geographical centre of the town, a place that was not very densely populated at that time. Fifteen years later, the new mayor, also a socialist and a physician by training, Lazare Goujon, decided to improve

50. Archives of Seine-Maritime, 2 OP 716/ 35.

51. Melosi, ed. *Energy Metropolis*, 125–126. In the same book, see also the contributions by Robert Bullard and Elizabeth Blum on the environmental justice movement.

52. For statements about the innocuousness of incineration: AM Marseille, 478 W 62.

53. In French towns, as in British ones, wealthier suburbs were generally to be found in the west as a result.

54. B. Meuret, *Le socialisme municipal, Villeurbanne, 1880–1982: l'histoire d'une différenciation*, Lyon, 1982.

55. AM Villeurbanne, 1J 21. At that time, only the four Parisian suburban plants were working in France and a few others were planned (Le Havre, Rouen, Elbeuf, Nancy). The model was taken from the suburban Belgian city of St-Gilles, near Bruxelles.

the plant and to build a completely new incinerator just next to the old one. His municipality envisaged the use of the steam to provide heating for 222 social housing apartments located less than 300 metres from the plant.[56] Goujon contracted with the most important French firm in the field, SEPIA, and the technologically pioneering plant was ready for operation by 1929 and started providing heating to the Villeurbannais people in the following months. So far, so unsurprising. But *after* the renovation and extension of the incinerator, Lazare Goujon launched what was to be a very important initiative: the building of new civic centre, named 'Les Gratte-Ciel', the French equivalent for 'Skyscrapers'. The programme contained a very large city hall, about 1,700 apartments of social housing, a swimming pool, a theatre and a centre for public health services. The incinerator was just across the street from the City Hall and a stone's throw from the windows of the closest apartments (see Fig. 1).

Figure 1. Aerial View of Gratte-Ciel, Villeurbanne in the Mid-twentieth Century[57]

Note the two incinerator chimneys and the urban central heating plant in the foreground.

56. AM Villeurbanne, 1O 84, draft letter to specialised companies, 4 June 1927.

57. Source: Courtesy of Archives Municipales de Villeurbanne.

It was extended in 1952 by the addition of two furnaces and, surprisingly, few written complaints about it can be found in the municipal archives for the 1960s and1970s, although it is known that residents were very upset at the odours coming from the plant, despite its seventy metre tall chimney which remained in the civic centre until the 1980s.[58] Given the significance of building this new central administrative and residential district, it is difficult to think of a more convincing illustration that an incinerator was not, at that time, considered to be a nuisance or an element degrading a cityscape, but rather a modern facility provided for the civic community.

Building Inequalities? NIMBYism in the *Beaux Quartiers*

Documents of citizens' complaints against the waste disposal facilities are not always easy to find and often relate to public debate about a projected plant. Consequently, they refer to *potential* nuisances. Some cases relate to problems associated with older waste facilities and these provide some sense of the social characteristics of the plaintiffs through their addresses and occupations, although they are probably filtered by the selective views of municipal engineers and hygienists who disregarded complaints as tirades against modernity.

Despite the praise in engineers' writings, most people were not happy when they learned that a garbage treatment plant was planned for their neighbourhood. This was the case in the nineteenth century when the main mode of treatment was crude dumping, and even if early twentieth century experts were optimistic about incineration and not very concerned about potential hazards produced by chimneys, garbage incinerators still raised opposition. During the inter-war period, twin developments were noticeable: on the one hand, petitions against garbage treatment plants did not seem to increase in the cities studied; on the other hand, the spatial proximity between industrial facilities and housing increased. Though this raised no major doubts among experts and administrators, municipal administrations increasingly launched campaigns against smoke, as was the case in Villeurbanne.[59]

In inter-war French cities, environmental inequalities of a socio-economic character were evident. For instance, in Cannes, residents of the La Bocca suburb had to contend with garbage dumped in open spaces and asked sarcastically in local newspapers why the private firm contracted to collect garbage was not as negligent in its service for the Croisette promenade and the municipal Casino as it was for their neighbourhood.[60] Their petitions against locating the composting plant in their suburb were not taken into account by the sanitary inspector, who claimed

58. AM Villeurbanne, 5J 2. Interview with Mme Gros (April 2009).

59. AM Villeurbanne, 5J 6. The 20 Apr 1932 law, said *Loi Morizet*, took care of the smoke problem after three decades of ineffective municipal by-laws.

60. AM Cannes, 1J 74, article from *L'Eveil*, 11 Jul 1931.

that it was private dumping that was responsible for the odours about which La Bocca residents complained.[61] Other suburbanites, having consciously fled the city in order to breath 'fresh air', also protested against incinerator projects in their vicinity. This was the case in Aix-en-Provence around 1930 when protests came from middle class citizens such as landowners living in villas with large gardens, high school teachers and entrepreneurs who wrote as spokespeople for others of more modest means in their areas to protest at the to protest at the smells that would come from the future composting plant.[62]

Occasionally, people who lived in an area where the construction of an incinerator was scheduled were very optimistic. In 1935, a citizen from Marseille travelled to Lyon to see the incinerator already operating there. He returned and shared his reactions with the mayor: 'While the plant is located a little outside the city, were it in the heart it would not affect the public health at all.'[63] In 1912–13, an elite group living in Boulogne, near Paris, argued that the type of plant for which authorisation was sought in Boulogne had attracted no protest in another suburban community, though that was possibly because this town (Saint-Ouen) was already overwhelmed with noxious industries, such as the first grinding plant ever built in the outskirts of Paris and later transformed into an incinerator.[64]

Very few successful campaigns in opposition to new treatment plants can be found originating in a working class neighbourhood, with the exception of one in Avignon. In this southern city, where by 1928 the municipality had explored the option of incinerating garbage, the new socialist mayor decided, in 1929, to contract a Mr Verdier, an engineer and industrialist proposing a zymothermic plant modelled on the first facility set up by him, in Cannes. Throughout 1930 a complex process of enquiries and protests took place, ending in a rejection of the location chosen by the municipality, though the plant itself was authorised in another location in 1931.[65] Initially, protests were headed by commercial and taxpayer interests, with political support from the previous 'radical' city council. They held public meetings to contest the municipal decision and wrote to the *préfet* that the project was too expensive and lacked foresight in certain details. Among the various arguments used in the numerous letters sent by the taxpayers' association to the *préfet* was the obvious lack of publicity for the official public enquiry in July 1930: the mandatory public notices had not been posted in the neighbourhood of the

61. *Ibid.* letter from the *Inspecteur départemental d'hygiène* to the *Préfet*, 2 Oct 1933.

62. AM Aix-en-Provence, I6 69, several protestations, Feb–Mar 1930.

63. AM Marseille, 478W 62, letter from the *Groupe philanthropique et Comité de Défense d'intérêts de Quartier Abattoirs Bd Bernabo et voies environnantes*, to Mayor of Marseille, 1935.

64. AM Boulogne-Billancourt, 3I 75, poster published by the *Groupe d'étude des intérêts de Boulogne*, n.d. [1912 or 1913]. On Saint-Ouen, A. Guillerme and A-C. Lefort, *Dangereux, insalubres et incommodes. Paysages industriels en banlieue parisienne, XIXe–XXe siècle*, Seyssel, 2004, 232–41.

65. AD (Archives Départmentales) Vaucluse, 2O 7/47.

projected treatment plant and when some of them *were* stuck on walls, they were quickly covered by posters announcing the national feast. Officials consulted by the *préfet* confirmed these facts. Despite the rather optimistic view of the official conducting the enquiry, whereby explanations given by the industrialist were accepted uncritically, the inspector of factories stated that late petitions arriving after the end of the public enquiry had to be taken into consideration by the sanitary commission. The two local commissions conducted their own surveys: the first one sent a member to the plant operating in Cannes, the second conducted two hearings of the main protagonists in December 1930 with the mayor, industrialist, president of the taxpayers' association and members of the neighbourhood committee giving evidence. It was this latter group of people who partly convinced the sanitary expert, by showing that the choice of the projected location for the plant would render it impossible to build a much-needed school in the increasingly densely populated area. Petitions were signed mostly by artisans and shopkeepers, whose motives were probably as much tax-related as environmental, since the municipality would have had to borrow money and raise taxes to finance the plant. The argument about protecting the growth of a neighbourhood and its children seemed to have convinced the sanitary council.

Another interesting case occurred in the western Parisian suburbs, in a residential area enjoyed by the nobility and bourgeoisie since the eighteenth century. Among dozens of municipal administrations contacted by private companies trying to sell incinerators, Neuilly, one of the towns in the western suburbs of Paris, contemplated burning its garbage. In a report in 1933, mayor Bloud stated that 'garbage must be destroyed in a place as close as possible to the location where it is gathered and to the places where their by-products can be consumed. The municipal administrative framework corresponds perfectly to this necessity.'[66] At that time, the garbage of Neuilly was burned in the same large plants that served Paris and its neighbouring communities. Bloud continued:

> Public opinion is hostile to establishing an incineration plant next to housing. People fear smoke, emanations and odours. And it is true that, if we consider as normal the way garbage treatment is operated in the Paris area, this opinion is undoubtedly grounded. But no technician – as some of you are – will contradict me, if I argue that in a rational and modern operating mode, those disadvantages can and must be suppressed.

According to Bloud, the reform would improve the environmental conditions of the whole Seine *département*.

> It does not matter, they think, that the content of our garbage cans go to poison the poor and the working populations of the Northern or Eastern suburbs. They are mistaken. This situation does not leave us indifferent … By our actions, we are

66. AM Levallois-Perret, I 51.

> happy to contribute to the cleaning up of distant Paris suburbs, to free people who deserve all our sympathy from a heavy and unfair bondage, and we are happy to serve social progress and humanity especially since there will be no harm for us.[67]

Neuilly was the only town on the western border of Paris that lacked industry and, once more, the planned location for the plant was not in Neuilly itself, but just a few metres to the north, in the town of Levallois-Perret. Once this project became known, the Levallois citizens protested vigorously. The bourgeois and middle class population of the targeted area printed a flyer with a picture showing the anti-hygienic operation of refuse sorting, as done in existing plants in the Paris suburbs (see Fig. 2). The flyer proclaimed: 'Here is a photograph of a plant such as the one they want to build. And this plant will not burn Levallois' but Neuilly's garbage. Against this scandal, let's protest.'[68]

Aux Habitants de Levallois

Chers Concitoyens,

On a formé le projet d'édifier **dans le Quartier de Villiers,** une Usine pour *l'incinération des ordures ménagères*

Voici la photographie d'une *usine semblable* à celle que l'on veut construire

Et cette Usine ne brûlera pas les ordures de Levallois, **mais celles de Neuilly.**

Contre ce scandale protestez...

Figure 2. First Page of a Four-page Flyer Published by Inhabitants Oppposed to an Incinerator Planned in Levallois [69]

67. AM Levallois-Perret, I 51.

68. AM Levallois-Perret, I 51.

69. Source: *Ibid.*

On the fourth page of the flyer, readers could discover an illustration of another annoying consequence of garbage incinerators: the traffic of carts and trucks full of garbage (see Fig. 3). The leaflet stated:

> Dear fellow citizens, you will have, in addition, the advantage of seeing every morning, and even for a great part of the afternoon, a line of trucks like the one shown below, parked, which will spread into our neighbourhood the sweet smell of the neighbouring city's garbage.[70]

The Levallois leaflet concluded with the exhortation: 'For the Hygiene of the area. For the Health of everyone. For the future of Levallois–Neuilly. All sign the Petition.'[71]

Chers Concitoyens,

Vous aurez, en outre, l'avantage de voir stationner tous les matins, et même une grande partie de l'après-midi, une file de camions semblable à celle représentée ci-dessous, qui répandra dans notre quartier les douces odeurs des ordures de la ville voisine.

Nous attirons également l'attention des Habitants de Neuilly *(rue de Villiers, boulevard du Château, boul. Victor-Hugo, etc.)* qui auront à supporter les inconvénients de la proximité de l'usine et nous les invitons cordialement à *protester avec énergie.* Ils peuvent signer la pétition ouverte ou *déposer leur protestation à la Mairie de Levallois-Perret.*

Pour l'Hygiène du quartier.
Pour la Santé de Tous.
Pour l'Avenir de Levallois-Neuilly.

Signez tous la Pétition

Figure 3. Fourth Page of a Four-page Flyer Published by Inhabitants Oppposed to an Incinerator Planned in Levallois [72]

---

70. *Ibid.*
71. *Ibid.*
72. Source: *Ibid.*

This visual 'nuisance' was unavoidable and also existed when the earlier dumping system was in practice. A strong objection was forthcoming from residents opposing the project in their district and was even voiced by the rural population. This factor may have been as important in the minds of local administrators as the potential odours and dust from a treatment plant, because it would impede daily traffic in the streets. The aesthetic value was also considered seriously by upper middle class people and local administrators: the defence of real estate interests was thus assured because most plants were located in zones already devoted to large industrial facilities. In many cases, inequalities were not the result of a deliberate will to preserve rich areas from the proximity of noxious smoke and smells, since waste treatment plants were presented as modern and clean installations. Nevertheless, it is true that wealthy people generally avoided incinerators or composting plants, just as they avoided other municipal installations like power plants and slaughterhouses.

Conclusion

During the first decades of the twentieth century, urbanites were still living in an 'age of smoke' whose industrialists were not always sympathetic to developing smoke abatement strategies.[73] In France, garbage treatment facilities hardly contributed to this urban experience, at least in the minds of expert hygienists and sanitary engineers, whose unswerving faith in incineration ensured very little criticism against the potential inconveniences of these waste disposal facilities. This positivity contrasts with their critical attitude towards other types of sanitary disposal plants like sewage treatment works.

Garbage incinerators could be located in different sorts of places but, perhaps unsurprisingly, were often located in industrial zones. The inter-war period saw the development of planning and the birth of zoning in France and, because they could provide energy for other municipal works – electricity works, waterworks, slaughterhouses – incinerators tended to be located near them, thus reinforcing the effect of zoning in French cities. Some, however, though probably fewer than in Britain during its first age of incineration (1880–1910), were constructed in the very centre of the urban area, as in Villeurbanne or Monaco. Other waste treatment facilities were sited in peripheral areas attractive to middle class residents wishing to live near the countryside, as in Aix-en-Provence.

From the materials gathered for this study, it is difficult to determine if there was any sort of environmental discrimination in locating waste disposal facilities. The industrial structures – incinerators as well as composting plants – represented progress from the age of open dumps, and thus, for several decades, were not considered a hazardous source of air pollution. They were also constructed later than

73. Uekotter, *The Age of Smoke*; P. Thorsheim, *Inventing Pollution. Coal, Smoke and Culture in Britain since 1800*, Athens, Ohio, 2006.

other insalubrious plants, such as the gasworks or chemical factories that had provoked complaints since the mid-nineteenth century, and their municipal status may have protected waste treatment plants from a potentially larger number of NIMBY campaigns. What appears, nevertheless, is an unequal ability to protect oneself for wealthy populations and the lower classes. The upper middle classes had money, time and the cultural background to develop strategies and arguments against the projected sites when they endangered their neighbourhood, their promenades and the value of their apartments. Workers were already living close to industrial zones in which the settling of a waste treatment plant was seen as normal; they were thus less able to make their voices heard.

Select Bibliography

Barles S. *L'invention des déchets urbains: France 1790–1970*, Champ Vallon, 2005.

Clark, J. '"The Incineration of Refuse is Beautiful": Torquay and the Introduction of Municipal Refuse Destructors', *Urban History*, 34:2, 2007, 255–77.

Frioux S. *Les réseaux de la modernité. Amélioration de l'environnement et diffusion de l'innovation dans la France urbaine (fin XIXe-années 1950)*, PhD dissertation, Université Lyon 2, 2009.

Guillerme A., G. Jigaudon, A-C. Lefort, *Dangereux, insalubres et incommodes. Paysages industriels en banlieue parisienne, XIXe–XXe siècle*, Champ Vallon, 2004.

Melosi, M. 'Sanitary Services and Decision Making in Houston, 1876–1945', *Journal of Urban History*, 20:3, 1994, 365–406.

Melosi, M. 'Environmental Justice, Political Agenda Setting, and the Myths of History', in M. Melosi, *Effluent America: Cities, Industry, Energy and the Environment*, Pittsburgh 2001.

Melosi, M. *Garbage in the Cities, Refuse, Reform and the Environment, 1880–1980*, Pittsburgh, 2004 (1981).

Mosley, S. *The Chimney of the World. A History of Smoke Pollution in Victorian and Edwardian Manchester*, New York, 2008 (2001).

Tarr J., ed. *Devastation and Renewal, an Environmental History of Pittsburgh and its Region*, Pittsburgh, 2003.

Uekoetter, F. *The Age of Smoke. Environmental Policy in Germany and the United States, 1880–1970*, Pittsburgh, 2009.

PART V

Energy and Industry

Chapter Ten

Social Inequality in the Supply and Use of Fuel in Scottish Towns c.1750–1850

Richard D. Oram

The dramatic increases in wholesale and retail fuel prices experienced globally in the months preceding the economic crash of 2008 brought the issue of 'fuel poverty' to the forefront of social policy agendas worldwide. Modern British definitions state that fuel poverty occurs

> Where a combination of poor housing conditions and low income mean that the household cannot afford sufficient warmth for health and comfort ... [defined as] ... where a household needs to spend 10% or more of its income to meet fuel costs.

In energy consumption terms, the definition of a 'fuel poor' household is:

> One needing to spend in excess of 10% of household income to achieve a satisfactory heating regime (21°C in the living room and 18°C in other occupied rooms).[1]

While the contemporary definition of a satisfactory heating regime is inappropriate for a discussion of conditions in pre- and early modern Scotland, the ten per cent expenditure threshold, which most late eighteenth and early nineteenth century Scottish accounts also offer as the average proportion of household income spent in this context, is a useful benchmark against which to measure past evidence for social inequality in access to and consumption of fuel. Given the accepted central importance of access to adequate fuel for economic development and the maintenance of basic living standards, it is striking that fuel poverty and wider questions of social inequality in fuel supply and use have, to date, figured little in socio-economic analyses of the period of rapid urbanisation in Scotland which accompanied the first phase of industrialisation in the late eighteenth and early nineteenth centuries. The nature of this reconfiguration had a profound impact on patterns of domestic as well as commercial or industrial fuel demand and consumption and triggered a major transition in the forms of fuel exploited by and available to the majority of town dwelling Scots.

1. http://www.worcester-bosch.co.uk/homeowner/environment/new-technologies-glossary Accessed 27 June 2008.

Most discussion of fuel resource exploitation in Scotland has focused on its use by industry rather than on domestic consumption patterns, where comment has been limited principally to transitions made from one form of fuel, usually peat or wood, to another, coal; and regional variations in the dating and extent of that transition.[2] The drivers behind such changes and the consequences of them, however, have received scant consideration from social historians, despite general recognition of the impact on already contracting incomes of new and rising fuel-related costs as the opportunities to obtain fuel supplies with family labour diminished. It is the aim of this chapter to outline in its first half the broad trends in the social patterns of urban domestic and industrial fuel use in the late eighteenth and early nineteenth centuries and, in the second half, to consider the influence upon those trends of crises of supply in traditional domestic fuel forms and to examine the social inequalities in fuel supply to and use of fuel by Scottish households which were created or reinforced in that period.

Table 1. Estimated Total Population in Western Europe Living in Urban Centres With over 10,000 inhabitants, 1700–1850[3]

	1700	1750	1800	1850
England and Wales	13.3	16.7	20.3	40.8
Scotland	5.3	9.2	17.3	32.0
Netherlands	33.6	30.5	28.8	29.5
Belgium	23.9	19.6	18.9	20.5
Central Italy	14.3	14.5	13.6	20.3
Spain	9.0	8.6	11.1	17.3
France	9.2	9.1	8.8	14.5
Portugal	11.5	9.1	8.7	13.2
Germany	4.8	5.6	5.5	10.8
Ireland	3.4	5.0	7.0	10.2
Poland	0.5	1.0	2.5	9.3
Switzerland	3.3	4.6	3.7	7.7
Austria-Bohemia	3.9	5.2	5.2	6.7
Scandinavia	4.0	4.6	4.6	5.8

Industrial and commercial development and their attendant enhanced employment prospects triggered a phase of urban expansion in Scotland between

2. I. Levitt and T.C. Smout, *The State of the Scottish Working Class in 1843*, Edinburgh, 1979, 56–7 summarises regional variation in fuel preference based on questions of price and opportunity value rather than general availability.

3. Source: data from J. de Vries, *European Urbanisation, 1500–1850*, London, 1984, 39–48.

1750 and 1850, second in Europe only to England and Wales in its rate and scale of growth. It is against that background that changes in fuel types used by urban populations must be considered. In 1750, fewer than one in ten Scots – around nine per cent – lived in towns but by 1850 this had risen to 32 per cent (see table 1), largely through migration from rural districts where agricultural 'Improvement' had seen the transformation of many small tenant farmers into landless, waged labourers. The seasonality and social instability of agricultural labouring encouraged a move into full time waged employment in urban-based industries. Urban expansion and industrialisation, however, were not spread equally across Scotland and not all towns saw the same influx of economic migrants. Those which did experience the greatest growth generally not only had access to plentiful sources of energy and raw materials but were also better connected to distribution networks.[4]

Much of this urban growth was stimulated by the new employment prospects presented by industrialisation. One of the first sectors to experience major expansion from the late 1700s was textiles, principally cotton in west central Scotland,[5] but also linen in parts of the east and woollens in the south.[6] Large mill complexes marked a decisive shift from rural, part time and largely domestic weaving to full time, factory-based employment and a move from manufactures that depended primarily on human muscle power to mechanised production. In the first phase of mechanisation after c.1775, water provided the main energy source for the looms and mill location was determined by access to a reliable year round supply.[7] The increasing application of steam power for weaving machinery, however, a transition largely completed by 1839, moved the dependence from water to one on adequate combustible material for heating the boilers.[8] In Scotland, that material was coal, which had been mined since at least the early thirteenth century in parts of the central Lowlands.[9]

---

4. For an overview of urbanisation in the eighteenth and nineteenth centuries, see T. Devine, *The Scottish Nation 1700–2000*, London, 1999, chapter 8.
5. http://canmore.rcahms.gov.uk/en/site/40417/details/bute+rothesay+cotton+mill/ Accessed 23 July 2009. J. Ferrier, 'Robert Thom's Water-cuts', in *Transactions of the Glasgow Archaeological Society*, new series 15:3, 1966, 129–38.
6. C.A. Whatley, 'The Making of Industrial Scotland 1700–1900: Transformation, Change and Continuity', in M. Mulhern, J. Beech and E. Thompson, eds. *A Compendium of Scottish Ethnology*, vol. 7, *The Working Life of the Scots*, Edinburgh, 2008 27–47, here 29–30.
7. See, for example, R. Buchanan, *Practical Essays on Mill Work and Other Machinery*, revised by T. Tredgold, 2nd edn, London, 1823.
8. R.J. Findlay, '1770–1850: The Age of Industrialisation', in M. Lynch, ed. *The Oxford Companion to Scottish History*, Oxford, 2001, 198–99.
9. *Liber Cartarum Sancte Crucis*, Bannatyne Club, 1840, no.41 for mines at Carriden on the Firth of Forth c.1210; J Hatcher, *The History of the British Coal Industry*, vol. 1, *Before 1700: Towards the Age of Coal*, Oxford, 1993, 96–111.

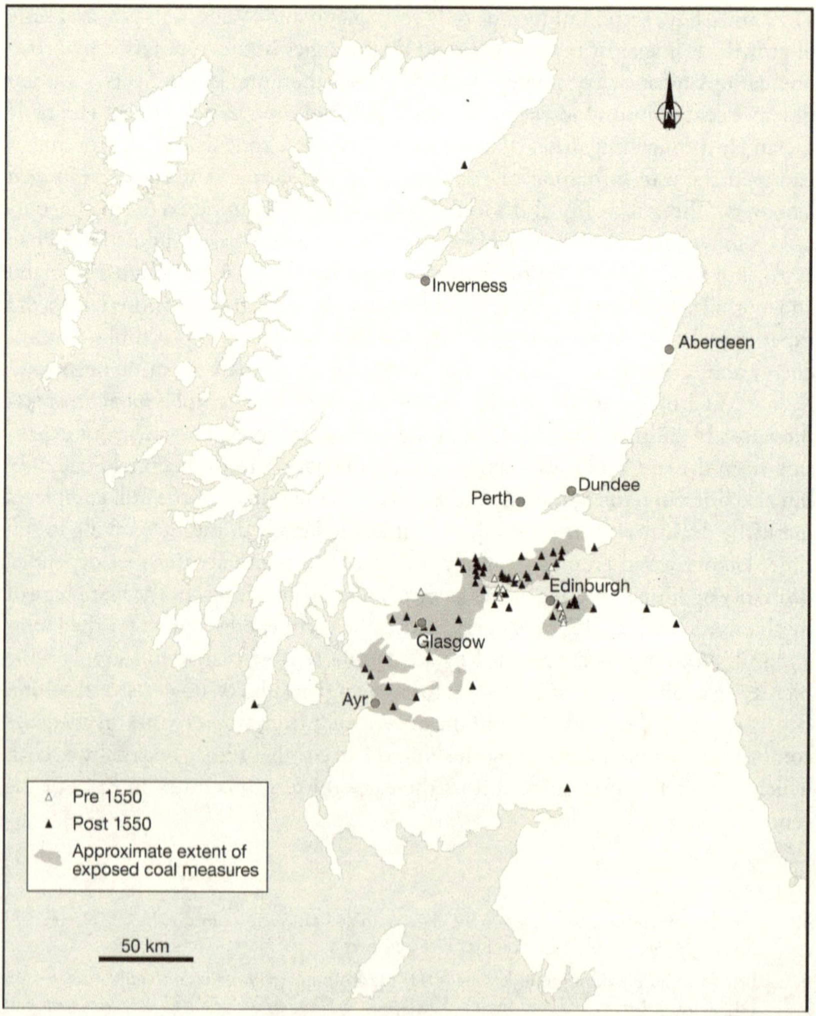

Figure 1. Scottish Coalmining

By the early 1700s, Scottish coal mining had begun the transformation from largely landowner controlled ventures run as adjuncts to other estate activities into professionally managed, cost-efficient private businesses capitalised by investors looking for profitable returns.[10] Traditionally, much of the coal mined was con-

10. B.F. Duckham, *History of the Scottish Coal Industry: 1700–1815*, vol. 1, Newton Abbot, 1970.

sumed by the sea salt industry[11] but technological innovation, rising output and keen pricing made coal more widely available for other industries; by 1800 coal had replaced charcoal as the favoured fuel in iron smelting. The domestic use of coal, already established during the Middle Ages in towns within coal producing districts and where pre-modern population expansion rendered reliance on other fuels unrealistic – for example in Edinburgh[12] – also increased steadily across the century but factors ranging from hearth design to the difficulties and cost of landward distribution at first limited the social spectrum of households burning coal in towns where viable alternatives existed.

Edinburgh, with a population in excess of 12,000 in the sixteenth century, was early modern Scotland's largest town and its inhabitants were already largely separated from direct agricultural production of their own food and finding of their own fuel by c.1500. For its inhabitants, whose incomes were wage-based, locally available coal was a cheap alternative which by that date had replaced dependence on wood and peat. Edinburgh, however, was a unique entity until the era of rapid urban growth and industrialisation in the second half of the eighteenth century. Until then, farming and fuel-gathering probably formed the main economic activities of most Scottish town dwellers after the specialist craft and trade niches that they occupied.[13] That traditional division of economic activity gradually disappeared as urban population levels climbed in the eighteenth century but a clear shift away from self-sufficiency in food and fuel production by townsmen was already evident in Scotland's urban centres in the 1500s.

In respect of this increased dependence on the purchase of supplies of food and fuel, two key factors emerge as important drivers behind significant changes in the exploitation of fuel resources and the social pattern of fuel use in Scotland. The first is fundamentally a material condition: in common with much of northern Europe in the early modern period, from before the steep rise in general population levels after c.1550, Scotland experienced a protracted perception of crisis in respect of fuel supply for its main population centres in Lowland areas.[14] This crisis affected

11. C.A. Whatley, *The Scottish Salt Industry 1570–1850: An Economic and Social History*, Aberdeen, 1987; Hatcher, *British Coal Industry*, 106.

12. Hatcher, *British Coal Industry*, 107.

13. P. Dennison, 'The Occupational Structure of Towns c.1100–1700', in Mulhern, Beech and Thompson, eds. *Working Life of the Scots*, 11–26, here 14–15.

14. R.D. Oram, 'Abondance inépuisable? Crise de l'approvisionnement en combustible et réactions en Écosse du Nord entre environ 1500 et environ 1800', in J-M. Derex and F. Grégoire, eds. 'Histoire économique et sociale de la tourbe et des tourbières', *Æstuaria: cultures et developpement durable*, Cordemais, 2009. For a comparative study of the Netherlands, see C. Cornelisse, 'The Economy of Peat and its Environmental Consequences in Holland During the Late Middle Ages', *Jaarboek voor Ecologische Geschiedenis 2005/2006: Water Management, Communities, and Environment. The Low Countries in Comparative Perspective, c.1000–c.1800*, Ghent, 2006, 95–121.

urban and rural communities alike and led to the adoption of a range of strategies to best manage the problem of a dwindling supply of traditional fuel resources and limited opportunities for exploitation of alternative fuel types. Where possible, this problem was resolved by a shift towards the use of coal; access to supplies for domestic use was becoming more widespread as output to serve industrial needs increased.

A second factor affecting fuel use was principally an intellectual shift founded in the eighteenth century economic philosophies of modernity and 'Improvement'. The use of one particular fuel type over another – in Scotland coal rather than peat – was a signifier of acceptance of that philosophy and a marker of advancing societal sophistication. The economic transformation that was central to Improvement made this transition logical and relatively straightforward in areas where rising demand for fuel and the availability of alternatives had seen commercial development of coal mines and increases in output had established an abundant fuel supply for local consumers. In towns close to coal producing areas and where the majority of the population was already separated from the primary production of fuel for personal domestic needs, this shift was probably inevitable. Paradoxically, however, the spreading influence of Improvement principles also led, in places remote from the coal fields, to a worsening of already problematic fuel supply issues and resulted in the consolidation of a social hierarchy of fuel use and the polarisation of social inequalities in physical and economic access to supplies.

By the later eighteenth century in Britain the notion of 'modernity' was a central theme of socio-economic theorists and commentators. One central strand in the modernity thesis was the idea of Improvement, principally with regard to land management and agricultural regimes but also in terms of industrial production and general structures of socio-economic organisation. Advocates of Improvement saw the introduction of what were considered to be the most economically efficient technologies and productive methods as a fundamental building block of a progressive, culturally advanced society. A broad range of traditional methods and materials was commonly presented in Improvement literature as impeding the path to modernity and these became veritable icons of primitivism, embodying all that was wrong with pre-Improvement society. Conversely, an alternative range of commodities was advanced as the *sine qua non* of modernity.

One of these new icons at the heart of the Improvement revolution was fuel. A distinction, however, needs to be made between domestic and industrial use, for the latter had significantly different thermal requirements and supply-volume demands from the latter. Three main forms of fuel were available for domestic and industrial use in pre-1750 Scotland: wood (including charcoal), peat and coal. Urban use of firewood declined rapidly in the thirteenth and fourteenth centuries as woodlands accessible to town dwellers were consumed. While wood remained abundant in some Highland areas into the eighteenth century, by the late twelfth century most of lowland and southern Scotland was already almost treeless. Good quality wood for fuel was largely the preserve of those who enjoyed legal possession of surviving areas of woodland, although their tenants often received rights to collect windfall and deadwood for their own fires. Charcoal produced in areas of surviving woodland was transported to the towns, mainly for specialist metalworking. Access to oak woods for the manufacture of good quality charcoal for iron working led to the development, between 1607 and the 1750s, of iron smelting concerns at points along the Atlantic seaboard where such woodland survived. Charcoal, however, was principally an industrial fuel and there is little evidence from Scotland of its domestic use.

With the decline of wood use, peat had been the main fuel available away from the zones of early coal exploitation around the Firth of Forth and in Ayrshire or remote from ports through which English coals were imported. While coal or charcoal were the favoured fuels for industrial purposes, peat was also used in industrial processes, especially evaporation of sea water for salt, but the best peats can also produce the temperatures necessary for metallurgy. Dried peat has an effective calorific value higher than wood but lower than coal,[15] but has a thermal efficiency capable of smelting iron. However, its significantly lower calorific value per unit volume (between one quarter and one sixth that of coal) rendered it impractical for large-scale industrial use. The much greater volume of peat over coal necessary for industrial processes, and the attendant problems of bulk drying and storage arrangements, had already seen its replacement by coal as the main fuel employed for industrial energy needs in much of east-central Scotland before the end of the Middle Ages; when Scottish industry expanded in the eighteenth century, coal was already established as its principal calorific source. Domestic consumers had significantly lower requirements in terms of fuel calorific values but the differential costs in terms of thermal efficiency and volume storage issues rendered coal as attractive to them as to industrial consumers. The chief restrictions on domestic coal use were ready access to dependable supplies and cost and availability relative to other fuel types. Where personal (that is free) or cheap waged labour was available for fuel-gathering and alternative fuel types were readily accessible, chiefly in those smaller urban communities away from the central Lowlands that lacked a substantial industrial sector, the purchase of coal was unnecessary and inflated levels of duty charged on the commodity plus high shipment costs rendered it uneconomic for most consumers.

By the 1790s and early 1800s, fuel resources were a central issue in Improvement literature directed at both urban and rural readerships. Some forms of that 'necessary article', fuel, were presented as almost an embodiment of progressive thinking and others as symbolic of a tendency wedded to ignorant cultural conservatism. Amongst traditional fuel types, the use of peat was considered by most

15. Peat, effective calorific value of dry substance MJ/kg = 20–23; wood, 17–20; coal, 28–33. One MJ/kg = 239 Kcal/kg.

Improving opinion formers to be especially primitive and economically wasteful, the very antithesis of modernity. The argument was founded principally on the best application of household labour resources, with the use of the supposedly 'free' labour of household members to gather fuel for personal domestic use rather than expending it in market-oriented production being presented as a real charge against notional labour credit – that is, a loss of potential externally generated earnings. When a cash value was assigned to the time and labour spent in obtaining it, peat was easily presented as more costly than the equivalent bulk of coal.[16] Not only could cash savings be made, but also the opportunity cost in time and labour expended on peat cutting could instead be devoted to more productive activities, thereby improving the economic wellbeing of the individual and bettering the socio-economic state of the nation. As a consequence, throughout Britain generally the use of coal came to be seen as a marker of social advance and modern thinking while, conversely, adherence to alternative fuel types was stigmatised as signifying socio-economic backwardness, indolence and under-employment of labour, vices of the Highland Scots and Gaelic Irish.

Issues of accessibility of supply and relative cost rarely figured in the proselytising literature that advocated the domestic as opposed to the industrial use of coal, unless to present the newly fashionable fuel type more favourably than traditional forms. The targeted audience in this coal advocacy was the expanding middle class, which was enjoying increasing levels of disposable income in the later eighteenth century, at a time when staples like basic foodstuffs and fuel were becoming more expensive and luxury commodities, including household servants and labourers, were becoming cheaper.[17] For members of this rising social group, the obtaining of fuel had long ceased to be a primary domestic economic concern as, despite price inflation for some basic commodities, the proportion of expenditure on fuel declined relative to their overall income and outgoings. For them, the one economic concern raised by the Improving literature was that of the opportunity cost of the waged labour they employed that was expended on obtaining fuel from traditional sources. The diversion of that labour to other tasks and the direct purchase of fuel on the market were presented as a more efficient use of capital and labour.

Along with arguments in favour of greater economic efficiency, social fashion was also encouraging a transition from traditional fuel types to coal amongst Scottish burgh elites. By the time of the production of the *Statistical Account of Scotland* in the 1790s, a significant number of its contributors reported that coal use was predominant amongst the 'better sort' in both urban and rural districts, even in areas where peat was still abundant. In major urban centres, like Dundee,

16. For example, D. Ure, *General View of the Agriculture of the County of Dumbarton*, London, 1794, 93–4.

17. P.T. Hoffman, D.S. Jacks, P.A. Levin and P.H. Lindert, 'Real Inequality in Europe since 1500', *Journal of Economic History*, 62:2, 2002, 322–55.

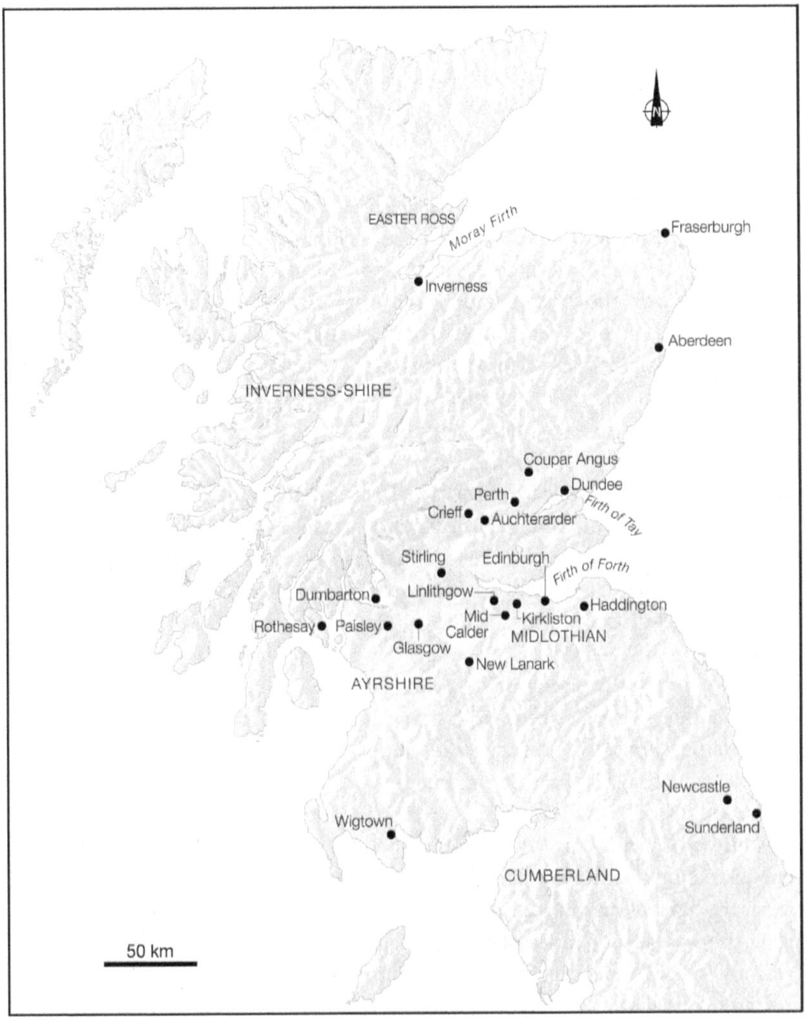

Figure 2. Map of Scotland: Places Mentioned

Edinburgh and Glasgow, and large provincial towns like Dumbarton, Perth or Stirling, the reports claim almost universal use of coal,[18] a situation which reflected

18. See, for example, D.J. Withrington and I.R. Grant, eds. *The Statistical Account of Scotland*, ix, *Dunbartonshire, Stirlingshire, Clackmannanshire*, Wakefield, 1978, 42, (Dumbarton), 619 (Stirling).

the impracticality for their expanding populations of securing their fuel needs from traditional alternative sources and the inadequacy of those alternatives. Cost alone was a deterrent to general adoption of coal by the upper social strata in towns remote from coal producing centres – it was a generally held view that transportation costs of coal by road became unprofitable at distances in excess of twelve miles, except where a dearth of locally available alternatives allowed suppliers to charge inflated prices – but fashion dictated that the new fuel should be burned in circumstances where its use could be observed by the user's peers and inferiors. For example, in the south-western burgh of Wigtown, a small port through which coal was imported from Cumberland though at comparatively high cost, 'the better sort' burned coal in the public rooms of their houses where their sophistication and wealth could be noted by outsiders, but used peat in the kitchen, domestic offices and servants' quarters, where no-one who mattered socially would have ventured.[19] The abolition of duty on coal led to some increase in its use by the urban elites in smaller towns distant from the mines, but general availability and fashion pressure probably accounted more for its prevalence in this social stratum than simple cost considerations.

A concomitant consequence of these prosperous urban classes' switch to coal was a collapse in the market for the traditional fuels supplying groups that had been amongst their largest individual domestic consumers. These same prosperous classes, in both urban and rural contexts in Scotland, also controlled or regulated access to the principal sources of peat or firewood. With their own needs secure in alternative forms, the management or maintenance of the peat mosses and wood reserves ceased to be a priority and, under their direction, many areas of former common fuel supply were turned to alternative use in the period c.1780 to c.1840, usually as private agricultural developments.[20] Alongside this development pressure ran a second issue, that of sole tenancies being granted exclusive access to particular fuel sources. This move triggered conflicts between the new sole tenants and former beneficiaries of what had been common resources.[21] Although this tension is once again seen most clearly in records of the process of consolidation of farms on rural properties, it was also a feature of periurban agricultural areas where the town's former common mosses had been given over to cultivation but former urban users of those common resources were still attempting to secure their fuel needs from them.[22]

While in many areas the transition to coal use had been a necessity driven by the exhaustion of traditional fuel forms, in others the processes of privatisation

19. R.H. Campbell, *Owners and Occupiers. Changes in Rural Society in South-West Scotland before 1914*, Aberdeen, 1991, 14.

20. For examples of such development, see *New Statistical Account of Scotland* [*NSA*], vii, *Renfrew-Argyle*, Edinburgh, 1845. Renfrew: 47, 54–5, 156, 319–20.

21. See, for example, National Archives of Scotland [NAS] GD112/11/3/4/11; NAS GD248/533/3/22–3.

22. Oram, 'Abondance inépuisable?' 39–40.

referred to above made it at once an obligation driven by social and legislative change and a symbol of those same changes. One dimension of the proletarianisation of the rural poor in later eighteenth century England, for example, has been identified as the loss of common rights, including access to windfall and dead wood.[23] While the cutting and transportation of heavy materials like large timbers or peat had traditionally been adult male functions, collection of such lighter fuel types had been primarily undertaken by women and children, who thus performed an important economic function within the family. Removal of rights to windfall and dead wood presented many households with the stark alternatives of fuel poverty through the expenditure of a higher proportion of wage income on the purchase of supplies or a re-deployment of household labour resources to provide additional income to make up the threatened shortfall. For the poorer families, that generally entailed the women and children increasing household earnings through securing work as low-waged industrial labour.

A similar process can be seen in Scotland, but in respect of the urban poor. The exclusive right to the fuel resources of a designated hinterland was one of the privileges held by Scotland's royal burghs (the towns founded by and receiving a charter of privileges from the crown) throughout the Middle Ages. The sustainability of these resources was recognised as a problem in the early sixteenth century, with some burghs facing the exhaustion of accessible peat mosses either as a consequence of the inadequacy of the original grant for meeting the long-term needs of a rising population or of new demands from an emerging industrial sector. The response of most burghs to this problem was prescriptive and usually entailed limitations on the volume of peat individuals were permitted to cut, regulation of the manner in which it was cut and enforcement of restriction of access to the mosses to only those with legal rights to take their fuel there. Continued overuse of the mosses forced burgh councils from the later sixteenth century onwards to impose stricter regulations and ultimately to privatise the common resource by the removal of personal access to peat sources for most town dwellers.[24] Such privatisation was a response to fears concerning the long-term viability of common supplies, further conditioned by a view that unregulated cutting of peat was causing the physical deterioration of the surviving mosses, thereby rendering fuel quality peat useless through water-logging or disintegration. In place of the free-for-all common access, in the late sixteenth century councils introduced leasing arrangements whereby common mosses were divided into blocks and licences to cut in them were awarded to the highest bidders, on the proviso that they would sell the cut peat to their fellow

23. J. Humphries, 'Enclosures, Common Rights, and Women: The Proletarianization of Families in the Late Eighteenth and Early Nineteenth Centuries', *Journal of Economic History*, 50:1, 1990, 17–42, here 32–4.

24. For example at Elgin in 1580, see W. Crammond, ed. *Records of Elgin*, vol. 1, Aberdeen, 1903, 158; Oram, 'Abondance inépuisable?' 33–38.

townsmen at an agreed price.[25] By the late seventeenth century, the legal supply of fuel to most urban consumers in Scotland's larger towns was thus in the hands of licensed commercial operators.

Legal access to the common supply had never been available to all burgh inhabitants in the pre-modern period. While full or 'free' burgesses had enjoyed that privilege, the unenfranchised inhabitants – referred to merely as unfree indwellers (mainly the non-artisans, unskilled labour and servants) – had restricted access to common resources. Most either took their requirements from the common supply illegally or obtained their needs through performing fuel-gathering services for the 'free' burgesses. The unfree indwellers formed the class most disadvantaged by the privatisation of fuel resources. The lease system turned them from direct producers of their own supply into cash-paying customers for a commodity whose price was fixed by an urban elite whose own use of that commodity was in sharp decline, or exposed them to increasingly harsh punishment for illegal peat cutting if they continued to use their own labour to meet their household needs. In large, prosperous towns where waged employment was available, poorer households could offset the cash cost of fuel through new income from paid work, simply converting their time and labour into wages which could be spent as required. That solution, however, was not available everywhere and analysis of wage data from around Scotland reveals that income levels for the poorer households contracted sharply in the later eighteenth and early nineteenth centuries. For them, the proportion of earnings spent on fuel was rising steeply; fuel poverty had become a visible and quantifiable social ill.

In some larger provincial towns, accessibility and affordability of fuel by the poorest members of the community deteriorated in the early 1800s as remaining areas of common fuel supplies within easy access of the urban area were disposed of by burgh councils for alternative uses. The loss of general access to these resources was felt most by the lowest social classes, as it coincided with a period when real wage levels were falling dramatically and living standards were stagnating or declining, especially in urban areas.[26] Exclusion from traditional supply sources, depletion of alternatives and the growing trend for female and child industrial labour added to the pressure on low income households, which were obliged to purchase a higher proportion, if not all, of their fuel requirements from retailers. Wages had not kept pace with fuel prices and the offsetting of household labour expended on fuel-gathering for household needs against paid employment to satisfy those fuel needs through purchase no longer bridged the gulf between income and expenditure.

25. For such leases see Crammond, ed. *Elgin Records*, vol. 1, 158. But, as Crammond, ed. *Elgin Records*, vol. 1, 431, shows, there was general public belief in widespread overcharging and price fixing under this system by the mid-1700s.

26. W.W. Knox, *Industrial Nation: Work, Culture and Society in Scotland 1800–Present*, Edinburgh, 1999, 39.

This financial burden, difficult enough to bear during the warmer months of the year, was greatly exacerbated during winter when household fuel needs increased and rising demand drove up wholesale and retail fuel prices but actual income may have fallen, due to the seasonal nature of much of the employment of the poorer segments of the male labouring community.[27] With the adult males at this social level often being out of employment for around one and a half months over winter, and the only wage income coming from lower-waged female and juvenile earners in families whose normal earning capacity precluded them from saving cash or stockpiling fuel throughout the rest of the year, daily existence for the lowest class of workers must truly have been 'to live on the edge of purgatory'.[28]

Aberdeen provides an extreme illustration of the increasing difficulties faced in the early nineteenth century by low-income families in securing adequate fuel for even basic domestic needs. The burgh had formerly possessed extensive reserves of moss land from which to obtain fuel, but already by the late fourteenth century wealthier households and craftsmen had been replacing peat with coal imported from Northumberland.[29] Even before Aberdeen had begun its rapid physical growth and industrialisation in the late eighteenth century, peat reserves had been badly depleted and cutting rights had been restricted since the early 1600s.[30] Increased employment opportunities in the burgh had enabled poorer households to supplement wage income from which to buy fuel, chiefly English coal, but the seasonality of much of that employment, often related to maritime activities, exposed the lowest segments of the working population to a sharp drop in earnings at times of the year when expenditure on fuel was highest. The main period of slack employment in mid-winter, moreover, was of no use for large-scale fuel-gathering by the males of the household, as the peat mosses were normally waterlogged at that time of year and the saturated peat could not be dried for burning. Consequently by the late eighteenth century the urban poor came to rely on charitable provision of fuel in winter to supplement what they could obtain for themselves.[31]

Scotland's system of poor relief, based on sixteenth century legislation, confirmed in 1672 and only reformed in 1845, exacerbated the problem. It had three basic tenets: responsibility for the poor lay with their parish of origin; the able-bodied received no support; and parishes could impose an assessment. By the early 1800s only around a quarter of parishes had assessments and in most rural and smaller urban parishes the poor fund relied on church collections and

27. Levitt and Smout, *The Scottish Working Class*, 109.

28. Levitt and Smout, *The Scottish Working Class*, 103.

29. D. Ditchburn, *Scotland and Europe: the Medieval Kingdom and its Contacts with Christendom, 1214–1560*, East Linton, 2000, 187.

30. For example, A.M. Munro, ed. *Records of Old Aberdeen, 1157–1891*, vol. 1, Aberdeen, 1899, 38, 77, 91, 93–4, 124–5.

31. *NSA*, xii, *Aberdeen*, Edinburgh 1845, 60–1.

voluntary charitable endowments; levels of support were consequently minimal. As a result, with no equivalent of the English 1662 Law of Settlement or 1697 Settlement Act, which allowed parish authorities to forcibly expel from the parish anyone who might become a burden on the poor fund, and a greater probability of access to charitable relief in larger towns, the rural poor gravitated towards the urban centres. The relatively small scale of the charitable funds, however, meant that fuel doles were given only during times of extreme weather. Several small funds were amalgamated in 1801 to provide a more substantial capital sum, the earned interest on which allowed the trustees to purchase coal cheaply during the summer months for distribution during critical periods in winter. By the 1830s, Aberdeen's United Coal Fund was distributing an average of 179 tons of coal per annum to the poorest families in the burgh at the rate of one quarter boll per person or one sixteenth of an imperial ton per family, a quantity equivalent to a mere 15.5kg and 62kg respectively. Given that over 100,000 tons per annum of coal were being imported by that period into Aberdeen from English and southern Scottish mines, the volume of fuel distributed to the poor as charity was trifling and, conversely, underscores the level of consumption of the economically more prosperous.[32] The inequality in supply and use of fuel in the burgh and its immediate hinterland is further emphasised by the need, on average, for almost 2,900 families to receive fuel in this way each year, a figure representing around twenty to thirty per cent of Aberdeen's population.[33] This figure probably represents a considerable inflation of the number who might normally seek charitable support, for the 1830s and 1840s coincided with one of the sharpest climatic downturns experienced in the northern hemisphere in the post-medieval period. Cool, wet summers and long, cold winters placed particular pressure on poor households, who had to purchase more fuel for a longer period of the year. With only around two per cent of the town's population being officially in receipt of poor relief, set at four shillings per month, the divergence between the numbers of parish supported poor and those receiving private charitable support through this period from the United Coal Fund reveals the scale of socio-economic marginality concealed behind the official statistics.

It is uncertain what percentage of those 2,900 families had an income below six shillings per week or £13 16 shillings to £15 12 shillings per annum, the top-end wage threshold of the 'lowest class of labourers' of the 1844 Poor Law Commission report.[34] This wage level was some forty per cent lower than the national average for labourers, calculated in accordance with the so-called 'Dryfesdale Standard'.[35] The degree of charitable support deemed necessary, however, suggests that there was a

32. *NSA*, xii, 68.

33. During the severe winter of 1837, the amount of coal distributed was effectively doubled to 334 tons, the trustees of the fund authorising a second allocation to recipients.

34. Levitt and Smout, *The Scottish Working Class*, 102–3 and Table 5D.

35. Levitt and Smout, *The Scottish Working Class*, 107–113.

substantial proportion whose earnings were barely adequate, or indeed inadequate, for their needs. With the average wholesale price for a ton of coal in Aberdeen at that date only marginally below 22 shillings, a figure nearly fifty per cent higher than the national mean of 14.6 shillings, and an average labouring family's fuel consumption estimated at equivalent to around 3-3.5 tons annually,[36] to maintain their fuel use at Dryfesdale Standard levels would have entailed an impossibly high expenditure of 20– 25 per cent of income on coal alone, more than twice the accepted modern benchmark for measuring fuel poverty. For the Dryfesdale Standard labourer, expenditure on fuel was estimated to amount to slightly less than ten per cent of earnings, the current accepted threshold above which a family is deemed to be in fuel poverty. Clearly, Aberdeen's urban poor could not begin to contemplate these levels of expenditure and subsisted on a fuel consumption regime fifty per cent lower than modern British minimums.

The position in Aberdeen was affected by circumstances particular both to its location and to the nature of its expansion and industrialisation in the early nineteenth century. Aberdeen's population levels, proportion of industrial workers and low status labourers to skilled workers and professionals and economic attraction to migrant paupers were unique in the region and have skewed the evidence for inequality in access to fuel and social divergence in patterns of fuel use. Elsewhere in northern Scotland the records present less polarised extremes. By the 1790s, especially after the abolition of the duty levied on coastal shipment of coal in 1793, the level of consumption of coal in northern Scottish towns distant from the coal fields began to climb but often remained low relative to the use of other fuels, principally peat, in communities north and west of Aberdeen. By the 1830s, however, while other fuels remained most favoured in landward areas away from the ports, most burghs fringing the Moray Firth reported that coals, generally shipped by sea from north-east England, had become the favoured fuel amongst their inhabitants.[37] Nevertheless, into the 1840s, peat was more commonly used than coal in places where the former remained relatively abundant and cheap in comparison to the latter, like the rapidly expanding Aberdeenshire fishing port of Fraserburgh.[38] The principal issue governing use remained price, which in these northern towns was driven principally by high transportation costs.

Most coal burned in east coast towns north of the Firth of Tay was shipped from Newcastle and Sunderland rather than from Forth Valley mines, while most of the South West was supplied from Cumberland rather than from Ayrshire; but

36. Levitt and Smout, *The Scottish Working Class*, 56.

37. *NSA*, xiii, *Banff-Elgin-Nairn*, Edinburgh, 1845, 27, 62, 177 (reports on the parishes of Banff, 1836, Elgin, 1835, and Forres, 1840); *NSA*, xiv, *Inverness-Ross and Cromarty*, Edinburgh 1845, 35 (report on parish of Inverness, 1835).

38. *NSA*, xii, *Aberdeen*, Edinburgh 1845, 255–56 (report on parish of Fraserburgh by Rev John Cumming).

distance from the English coal fields did not affect price so much as the problems of distribution to landward from the coastal points of import.[39] Even after the abolition of the coastwise duty in 1793, in inland towns remote from mines low- to middle-income households' access to affordable fuel remained problematic. Annual expenditure on coal for day labourers in such communities was some ten to fifteen per cent of earnings. In the southern Perthshire burgh of Auchterarder, for example, the cost of a stone (6.35kg) of coal almost quadrupled in the twelve miles from the nearest mines to the town, while a further ten miles north in Crieff one stone in the summer months cost six times more than at the minehead, rising to nine times the minehead cost in winter.[40] For low-income families in such towns, therefore, coal was a barely affordable luxury.

As Robert Stirling submitted in his report on Crieff for the *Statistical Account of Scotland*, the combined result of high coal prices and a dearth of alternative fuels was the emergence of a sharply segregated hierarchy in fuel use which the generic comments on fuel given for some of the large burghs disguise. Stirling observed that those on the lowest incomes scavenged what they could from locally available resources, principally using broom and waste wood from the town's hinterland; those on middling incomes used a mixture of wood, peat and coal; while the comfortably-off used coal only.[41] It is a pattern of use recognisable in smaller urban communities throughout Scotland in the 1790s, as illustrated in the case of Wigtown mentioned above. Forty-five years later, Stirling's successor reported that the chief economic problem faced by ordinary people in Crieff remained the difficulty of obtaining affordable fuel.[42] Similar problems leading to identical stratification were observed in urban and rural Dunbartonshire in 1811, where it was noted that wood was favoured at opposite ends of the social spectrum: large, high-grade timbers cut into billets at the top end, brushwood and fallen timber, waste off-cuts, decayed stakes and the like by those at the bottom.[43] These accounts suggest that the reports of a headlong rush to coal use and abandonment of traditional fuels described in some of the major population centres may be misleading and that an altogether more subtly nuanced hierarchy of fuel consumption survived in urban communities until well into the 1800s.

It may appear a truism to state that levels of fuel consumption were linked closely with affordability but there are several price related qualifiers other than simple earnings that restricted fuel purchase and use by certain economic classes. A problem for the poor, commonly recognised at the time but overlooked today,

39. Levitt and Smout, *The Scottish Working Class*, 56.

40. *OSA*, xi, *South and East Perthshire, Kinross-shire*, 44–5; *OSA*, xii, *North and West Perthshire*, 283.

41. *OSA*, xii, 297 note.

42. *NSA*, x, *Perth*, Edinburgh 1845, 526 (report on parish of Crieff by Rev William Laing, 1838).

43. A. Whyte and D. Macfarlan, *General View of the Agriculture of the County of Dumbarton*, Glasgow, 1811, 255–6.

was that, whereas wealthy, regular and bulk consumers of coal could either obtain supplies direct from a wholesaler or negotiate discounts from a retailer, low-waged workers could afford to purchase only small quantities, often on an irregular basis and, consequently, only from retailers who charged a *pro rata* higher price. There are striking modern parallels in the higher tariffs charged to customers, usually from socially and economically disadvantaged groups, who, for example, pursue 'pay-as-you-go' metered purchasing strategies and lack the ability to benefit from the discounts available to customers who pay by direct debit. In the small Midlothian town of Kirkliston in 1839, despite its proximity to the Midlothian coal fields, the parish minister observed that:

> Of late, the price of coal has been very high, and to the poor it is always much greater, than to those in better circumstances; for as they are quite unable to pay for a cart-load at a time, they are under the necessity of purchasing what they use from retailers, in very small quantities, and at an enormous additional cost. For this reason very little coal is used by the poor; their fires are chiefly made from the cuttings of the woods.[44]

Competition from industrial bulk consumers also drove up wholesale prices even in areas close to mining districts. In Mid-Calder in the late 1830s, the minister claimed that:

> The vast quantity of coal now annually consumed by the manufacture of iron, and by steam-engines, has raised the price of this necessary article of daily use so high, as to render it scarcely attainable by the poorer classes of the community. The conversion of common peat-moss into a fuel has, therefore, become an object of no inconsiderable importance.[45]

Here, with no trace of irony in the report, access to peat supplies was advocated as a remedy for the social ill of fuel poverty.

The Crieff, Kirkliston and Mid-Calder accounts indicate that while the opportunity to exploit alternative fuel resources satisfactorily was not a viable option for the poorer inhabitants of the larger industrial towns, the semi-rural poor of smaller communities where wage levels and full employment opportunities were much lower were, at the close of the period under review in this essay, still adopting a range of strategies in attempting to satisfy their fuel needs. They exploited licit and illicit sources for combustible materials and it is apparent that there was significant resort to recycling of waste materials or scavenging. Fuel scavenging, where the poorest families simply employed anything that was flammable gathered from any available source, was a particular problem in the immediate rural hinterlands of the towns. The estate manager of the Murrays of Lintrose, a property close to the Perthshire town of Coupar Angus, reported to his employer in 1798 that a plantation on the

44. *NSA*, i, 149.

45. *NSA*, i, 380.

estate 'was verry [*sic*] much hurt by people tearing and breaking down' the trees for firewood.[46] While some rural labouring families made up for the loss of access to common supplies by supplementing their needs through such illicit inroads into stands of brushwood, cuttings from hedges, or the dismantling of fences, these same resources were equally subject to the depredations of urban scavengers. In Dunbartonshire, for example, it was noted that fence posts and palings, even where obviously new and recently erected by farmers, were regularly stolen for use as fuel by the urban poor.[47] Improving landowners also suffered, with woodland being particularly attractive to fuel scavengers from nearby towns. This problem was particularly severe close to towns that lacked a strongly developed wage-based economy and where under-employment or seasonal unemployment was high. At Inverness in 1843, for example, the poor were said to make regular inroads into plantations close to the town to 'plunder' the green wood.[48] There appears to be some correlation between periods of economic depression, with attendant high unemployment, and increased reporting of such attacks on plantations but there has been no analysis of sheriff court records to enable any quantitative assessment of the problem to be made.

Within the towns, all potentially flammable material was exploited by those who could obtain it. Broken or rotten timber from demolished buildings and yards or fences were all marketable commodities but the small splints and trimmings of such wood, deemed too trivial to recover by those contracted to demolish and salvage from the property, were a precious resource for those on the social and economic margins. Labour for undertaking such clearance work, too, could be obtained in return for the promise of fuel. At Stirling, removal of dead trees from the extensive orchard behind the former mansion of the Earl of Mar by men who received the dead wood as payment for their efforts resulted in a lawsuit over ownership of the rotten timber.[49] For such individuals, even the fast-burning decayed wood of the orchard trees was a worthwhile reward for their physical efforts.

Scavenging, however, could not satisfy the needs of the poorest households, especially during the periods of economic recession that badly affected the industrial towns of west central Scotland. The collapse in some workers' earnings during these episodes placed particular strain on already limited supplies of salvageable fuel as more families were driven to scavenge for what they had previously purchased. Increases in crimes of property have long been recognised as a symptom of these periods of economic distress but no research has been undertaken into the level of incidence of court cases involving the theft of fuel or materials that could be used

46. NAS, GD68/2/132.
47. Whyte and Macfarlan, *Agriculture of Dumbarton*, 256.
48. Levitt and Smout, *The Scottish Working Class*, 57.
49. NAS, GD124/6/205.

for fuel. Inroads made on enclosed plantations, for example, may be recorded as episodes of criminal damage to property rather than theft of firewood. An initial survey of Crown Office precognitions for the period 1820–40 revealed a handful of trials for thefts in towns involving fuel but the large bulk and dead weight of most forms of fuel made undetected removal difficult and the low number of prosecutions may indicate that, unlike today, significant fuel theft was a rare occurrence. It is an aspect of this subject, however, that begs further research.

The 1840s was a watershed in the issue of access to fuel for the bulk of the Scottish population. The distress of the poorest households revealed in the accounts of the Aberdeen United Coal Fund in the late 1830s worsened in the short term, as the sharp deterioration in the climate reached its nadir in the mid-1840s. This already adverse situation was worsened by the economic downturn of the early 1840s, which brought significant rises in unemployment in industrial centres. Additional pressures resulted from a further inflow of economic migrants from Highland Scotland and Ireland during the depths of the Famine of the 1840s. This same period, however, witnessed the rapid development of the Scottish railway network, the first national mass and rapid transport system for the conveyance of people and goods. Bulk carriage of coal led to falls in price at depots and the penetration of inland districts by main and spur lines enabled cheap coal to reach new markets. Except in the remotest Highland districts of the mainland, where a non-wage-based agricultural and fishing economy in which household labour was still expended on fuel-gathering for personal use prevailed, coal finally supplanted the alternatives as the fuel of choice. Greater accessibility and falling commodity price, however, did not necessarily mean increased affordability for the poorest households until wage levels began to climb after the recession of the 1840s; fuel poverty remained one of the many ills that plagued the Scottish urban working class. The transition to a wage-based economy that characterised the century of urbanisation and industrialisation after 1750 perhaps freed more Scots from the tyranny of finding their own fuel in an environment of diminishing traditional resources but the growth of the fuel economy presented them with fewer rather than more choices. There is possibly no greater marker of social inequality than the reduced options for meeting their energy needs faced by the bulk of the Scottish people in the early 1800s.

In conclusion, it appears from the evidence reviewed that significant social inequalities in the supply and use of fuel for Scotland's urban population began to emerge in the late sixteenth century. The scale of divergence between those at opposite ends of the urban social hierarchy, however, increased dramatically from the mid-eighteenth century as a range of social, economic and cultural changes caused the rapid transformation of traditional structures of social organisation and mechanisms of fuel provision. The radical reconfiguration of economic systems with the adoption of Improvement agriculture and emerging industrialisation, coupled with trends towards the privatisation of formerly common assets, had their most significant impact on those at opposite extremes of the socio-economic spectrum. At the top end, falling wage levels for servants offset price rises for basic commodities like fuel, while increasing availability of bulk fuel, especially coal, on the wholesale market enabled servant labour time formerly spent on cutting and carting peat to be redirected towards other economic activities. At the lower end, increased employment opportunities in the new industries enabled male heads of household to substitute fuel gathered directly through their own physical efforts with fuel purchased on the market. The risk of adult male seasonal unemployment, falling wages and increased prices for domestic staples, however, coupled with loss of access to traditional common fuel sources and the proletarianisation of female and child labour, pushed up the proportion of income expended on fuel needs. These trends drove many households that lay on the economic margins into fuel poverty and increasing dependence on charitable provision. Significant regional variation is evident in the chronology of these changes and in the level of impact but it is apparent that fuel poverty should be added generally to the list of ills that afflicted the lower social classes in the towns of early industrial Scotland.

Select Bibliography

Cornelisse, C. 'The Economy of Peat and its Environmental Consequences in Holland during the Late Middle Ages', *Jaarboek voor Ecologische Geschiedenis 2005/2006: Water Management, Communities, and Environment. The Low Countries in Comparative Perspective, c.1000–c.1800*, Gent, 2006, 95–121.

Dennison, P. 'The Occupational Structure of Towns c.1100–1700', in M. Mulhern, J. Beech and E. Thompson, eds. *A Compendium of Scottish Ethnology*, vol. 7, *The Working Life of the Scots*, Edinburgh, 2008, 11–26.

Devine, T. *The Scottish Nation 1700–2000*, London, 1999.

Duckham, B.F. *History of the Scottish Coal Industry: 1700–1815*, vol. 1, Newton Abbot, 1970.

Findlay, R.J. '1770–1850: The Age of Industrialisation' in M. Lynch, ed. *The Oxford Companion to Scottish History*, Oxford, 2001, 198–99.

Hatcher, J. *The History of the British Coal Industry*, vol. 1, *Before 1700: Towards the Age of Coal*, Oxford, 1993.

Hoffman, P.T., D.S Jacks, P.A. Levin and P.S. Lindert, 'Real Inequality in Europe since 1500', *Journal of Economic History*, 62, 2, 2002, 322–55.

Humphries, J. 'Enclosures, Common Rights, and Women: The Proletarianization of Families in the Late Eighteenth and Early Nineteenth Centuries', *Journal of Economic History*, 50:1, 1990, 17–42.

Knox, W.W. *Industrial Nation: Work, Culture and Society in Scotland 1800–Present*, Edinburgh, 1999.

Levitt, I., and T.C. Smout, *The State of the Scottish Working Class in 1843. A Statistical and Spatial Enquiry Based on Data from the Poor Law Commission Report of 1844*, Edinburgh, 1979.

Oram, R.D. 'Abondance inépuisable? Crise de l'approvisionnement en combustible et reactions en Écosse du Nord entre environ 1500 et environ 1800', in J-M. Derex and F. Grégoire, eds. 'Histoire économique et sociale de la tourbe et des tourbières', *Æstuaria: cultures et developpement durable*, Cordemais, 2009, 31–44.

Whatley, C.A. 'The Making of Industrial Scotland 1700–1900: Transformation, Change and Continuity', in M. Mulhern, J. Beech and E. Thompson, eds. *A Compendium of Scottish Ethnology*, vol. 7, *The Working Life of the Scots*, Edinburgh, 2008, 27–47.

Chapter Eleven

Environmental Protest Movements against Industrial Waste in Belgium 1850–1914

Wanda Balcers and Chloé Deligne

Collective protest movements against industrial pollution first surfaced in Belgium between 1850 and 1914. This chapter contextualises two such developments and explores these from both a social and a geographical perspective and from the stance of environmental inequalities.

Whilst the current focus of historiography shows an increasing interest in the regulation of pollution in relation to the activities of public authorities, technical measures in the area of sanitation, and industrial effluents, this emphasis obscures the role played by stakeholders who protest at the environmental conditions they experience. Institutional aspects, therefore, have been privileged over those of the individual simply because the archives of public bodies are generally better preserved and easier to access than those of private stakeholders – the individuals and pressure groups most directly affected by the polluters and pollutants. However, some archives do provide evidence of mobilisation against the degradation of natural resources such as vegetation, rivers and streams. An exploration of these archives permits a richer and more complex interpretation of the industrialisation of western societies to emerge. Continuities and disjunctures can be better understood as a result, and this is so in the case of protests against environmental degradation, the basis of this chapter.

The protest movements examined here are not directly linked to one another. Independently, though, they both provide a glimpse of the interconnections between polluter and polluted. Between 1850 and 1870, most protests that took place were by those workers, property owners and farmers most closely located to and directly affected by the actions of a particular factory. These individuals combined to protest against the harmful effects of the air emissions that often adversely affected the surrounding vegetation, mainly crops. From 1880, the social contexts of protests were different; they were articulated by groups of anglers whose indignation was due to a spatial issue that went far beyond their immediate surroundings. The two movements had some points in common. For example, protests originated in urban or suburban working class environments, prompted by an encroachment upon their means of subsistence – crops in the first case and, in the second, the fish resources that constituted major supplements to the diets of the working classes. These movements, however, did not really succeed in making their objections as regards the degradation of animal or plant resources heard at a scientific or political level, or even within workers' movements.

Fuming Suburbs: 1850–60

In Belgium, as elsewhere, the emergence of complaints about the harmful effects of pollution from industrial waste on the air, rivers, streams and ground dates from before the nineteenth century. In the eighteenth century, for example, farmers as users of rivers and streams denounced the effect of contaminated rivers on the health and even the very survival of their livestock. The release of waste in and around major cities was denounced regularly by private individuals concerned about the possibility of resultant infection.[1] Nevertheless, from the 1830s and with the establishment of new chemical and metallurgical industries, the nature of protest changed gradually, taking on an increasingly collective character. No longer was complaint simply the action of private individuals or small groups, but one in which coalitions of interests formed to present a national presence, fostered in no small measure by coverage in the press. Furthermore, rather than express their dissatisfaction in relation to their own personal circumstances, the language used by members of the coalitions assumed a broader significance whereby the impact of pollutants on vegetation, species and river-courses embraced a wider environmental significance.

For the most part, the first movements of this type emerged in the outskirts of cities, close to the prime targets, the chemical and metallurgical factories.[2] During the nineteenth century, for legal reasons[3] as well as for opportunistic ones associ-

1. Among many other examples, see the conflict between Van Ypen, the owner of a paper mill, and the nearby Abbey of Vorst in the south of Brussels, regarding 'the infection by colours and other waste from the Van Ypen mill'. See L. Verniers, *Histoire de Forest-lez-Bruxelles*, Brussels, 1949, 103, and General Archives of the Kingdom of Belgium, Ecclesiastical Archives, file 7082.

2. On these protest movements see the pioneering study by G. Devos, 'Milieuverontreiniging door de industrie omstreeks het midden van de 19de eeuw', in *Bijdragen tot de Geschiedenis, Liber alumnorum Karel van Isacker*, Antwerp, 1980, 347–94; C. Verbruggen, 'Nineteenth Century Reactions to Industrial Pollution in Ghent, the Manchester of the Continent. The case of the chemical industry', in C. Bernhardt and G. Massard-Guilbaud, eds. *Le Démon moderne, la pollution dans les sociétés urbaines et industrielles d'Europe/The Modern Demon, Pollution in Urban and Industrial European Societies*, Clermont-Ferrand, 2002, 377–92. The chemical industries along the River Sambre, which were the object of a major protest movement, were an exception in this case because they were located in a largely rural setting between the urban centres of Namur and Charleroi.

3. Based on the Napoleonic decree of 1810, the different royal decrees renewed in 1824, 1849 and 1863 regulated 'dangerous, insalubrious or bothersome establishments' ['établissements dangereux, insalubres et incommodes'], imposed a certain distance according to criteria based on degree of danger and obliged industrialists to take the necessary measures to reduce the amount of waste that was harmful to the surroundings.

ated with the low cost of land and strategic advantages associated with transport and proximity to dependent industries, these factories were built increasingly on the outskirts of cities.[4] Consequently, they were very often constructed on or near land that had historically been used for market-gardening to supply inhabitants in nearby towns.[5] Therefore, in the suburbs of major Belgian cities such as Ghent, Brussels and Liege, the impact of industrial waste on market-gardening crops, mainly potatoes, fruit trees and livestock, was condemned most frequently.[6]

Although instances of industrial pollution were scattered throughout the country, and included cases in the mid-nineteenth century – of soda works in the Sambre, zinc works in the Saint-Léonard quarter on the outskirts of Liege and pollution by the textile industries along the River Leie in the region of Ghent – where industrialists were accused and sentenced, they generally escaped with only light punishments, which consisted mainly of paying some compensation to their neighbours.[7] Most major cases were hushed up by the creation of committees of inquiry, which eventually proposed measures and recommendations concerning industrialists' activities.[8]

Faced with the proximity of these unpleasant factories, property owners in the surrounding areas tried to enhance the value of their land by means other than agriculture. In some cases, they altered the land use to complement the new industrial activities, for example by developing brickyards or building workers' housing for the factories.[9] In other cases, they simply sold their land to industrialists or to other

4. R. Etienne, 'L'industrie chimique belge (1830–1930)', *Bulletin de la Société belge des Ingénieurs et des Industriels*, XI, 1931, 5–28; C. Verbruggen, 'De reacties op vroege vormen van vervuiling, de chemische nijverheid te Gent, 1820-1892', *Jaarboek voor Ecologische Geschiedenis*, 2001, 112–44, esp. 127–28).

5. This spatial reorganisation is very similar to that observed in Paris by A. Guillerme, A-C. Lefort and G. Jigaudon, *Dangereux, insalubres et incommodes. Paysages industriels en banlieue parisienne XIX–XXe siècles*, Paris, 2004.

6. See, among others, W. Van Hecke, *Histoire du logement en Belgique de 1830 à 1889*, Brussels, 1985; M. Cassiers and M. De Beule, *Bruxelles, 150 d'habitat social et ouvrier*, Brussels, 1989; J. Miller, M. Cassiers, A. Forti *et al.* 'De l'habitation ouvrière au logement social de 1850 à nos jours', La Louvière, 1990.

7. Sometimes the sentences were more severe for the industrialists: the brothers Van der Elst in Saint-Gilles (Brussels) got a (short) prison sentence in 1854 for not respecting the measures imposed to protect the surroundings of their chemical works. L. Peeters, *Les Fabriques de produits chimiques et autres établissements insalubres*, Brussels, 1856, 104–5.

8. It was the same with the report submitted in January 1856 by the committee of inquiry designated to shed light on the dangers of soda works to the surroundings, and the report finished at about the same time by another committee responsible for the same task regarding the zinc works in Liege (*Le Moniteur Belge*, 26th year, Wednesday 27 Feb 1856, no. 58, 655–57 and Parliamentary Documents, House, 1855–56, 1343–44).

9. Land ownership dynamics in the outskirts of growing cities is still a relatively unexplored subject. The case of a certain Taziaux who, around 1860, purchased land near the above-mentioned factory belonging to the brothers Van der Elst (chemical works producing soda, sulphuric acid and

stakeholders, who then themselves built on it to support their industrial activities. In the absence of a genuine public housing policy before the 1890s, other than the provision of rail passes for workers to travel from the countryside to the cities, Belgian workers lived very close to the factories in which they worked, in housing built by private enterprise in accordance with the laws of the market.[10] As a result, authorities were in denial about their responsibilities to intervene and, by largely ignoring complaints about industrial pollution, they reinforced the socio-spatial disparities that existed in the expanding cities. In effect, the official non-interventionist policy endorsed the coexistence of workers' housing and polluting factories in the same areas of the urban fringe. Increased exposure to industrial pollution was therefore added to social inequalities and difficult working conditions.

The proximity of noxious, polluting industries was sometimes denounced by the inhabitants themselves, in particular in the case of La Vieille Montagne factory, an industrial complex for zinc processing built in 1805 in a suburb to the north of the city of Liege. Between 1825 and 1860, this industry was the object of many complaints on the part of the surrounding area's inhabitants who formed an action committee in the 1850s. At this time, in support of their complaints and following studies of the rate of mortality recorded from the 1830s[11] residents denounced the high death rate in their neighbourhood compared to that of other neighbourhoods in the city. Ignoring other explanatory factors, these inhabitants considered the low life expectancy to be the price the working class had to pay for the proximity 'amongst them' of employment in insalubrious factories.[12] These Liegeois workers denounced, yet were resigned to, a form of inequality and injustice.

The surroundings of certain chemical factories, and in particular of soda works, were recognised as dangerous. This was particularly the case around Namur

---

then bleaching powder, as well as nitric acid and ammonium chloride) in a suburb of Brussels. Taziaux bought land with the explicit aim of setting up a brickyard in order to build 52 workers' housing units next to the factory (State Archives in the Brussels–Capital Region, Brabant Provincial Government, Dangerous, insalubrious and bothersome establishments, D 278, 20 Jan 1862, letter from the mayor of the municipality of Saint-Gilles to the Governor of Brabant).

10. In 1813, the industrialist J-J. Dony, founder of the zinc works in Liege later purchased by the company 'Vieille Montagne', bought the land around his factory from the complainants (*Protestation des habitants du Quartier Nord contre l'usine de Saint-Léonard, adressée à la députation permanente du Conseil provincial de Liège*, Liege, 1854, 6).

11. Studies on mortality rates in major Belgian cities began in the 1830s, with the first study on mortality rates in Brussels published in 1842 by Adolphe Quételet, followed quickly by a more in-depth study conducted in 1844 by Edouard Ducpétiaux (E. Ducpétiaux, *De la mortalité à Bruxelles, comparée à celle des autres grandes villes*, Brussels, 1844).

12. *Protestation des habitants du Quartier Nord contre l'usine de Saint-Léonard, adressée à la députation permanente du Conseil provincial de Liège*, Liege, 1854, 14: 'And finally there is the North Quarter, the most decimated of all, as the death rate is 3.05%. It is also the quarter with the highest number of insalubrious factories ... Who could deny that life has become intolerable in the surroundings of such an establishment?'

and Charleroi, which particularly outraged Léon Peeters, a chemist from Brussels who lived in the region of Charleroi. Peeters wrote several works critical of the disregard shown by industrialists and the political class for the problems caused by factory waste and did not hesitate to advocate alliances of workers and farmers against the ravages of industry.[13] Although now rarely cited, and despite arguments based on summaries or phenomena which were not proved scientifically, Peeters' works constitute a milestone in the history of political ecology and triggered a very strong reaction by elites in political, scientific and industrial spheres.[14]

However, at a point in time when workers' movements were in their infancy, the condemnation of working and living conditions in and around industrial plants in Belgian cities did not find lasting political expression.[15] The length and cost of administrative or legal procedures, their modest impact on the measures taken by the industries, the fear of job losses within working class environments themselves and the complexity and difficulty of scientific argumentation got the better of these first collective protest movements which objected to the harmful effects of industries on the health of inhabitants and on their means of subsistence.

Thus, the occasionally vigorous protests that took place in suburban areas in the 1860s faded, one after the other. However, denunciations of a new type emerged. This time, they came from scientific figures who were concerned with the living conditions of the working class and, a few decades later, they resulted in the first organised anti-pollution movement, that of the anglers.

13. L. Peeters, *Les Fabriques de produits chimiques*, 21–2, addressing the farmers: 'Shareholders have understood that there is strength in numbers. Farmers should put this powerful and fruitful principle into practice. You should therefore be ready for the call; join together in vast numbers; remember the two beautiful lines from your popular song: "People, form a holy alliance and take one another by the hand!" Do what is being done in Liege, where the north committee sits permanently in order for Vieille Montagne to be driven out, as it poisons one of the most beautiful suburbs of this city.'

14. Léon Peeters was a chemist, and son of a chemist from Brussels. He studied in Mons, received his degree in 1836 and settled in Wanfercée-Baulet, north-east of Charleroi, in 1840. He was the author of several works complaining about the ravages of industrial pollution, especially affecting the most disadvantaged social classes. See L. Peeters, *Guérison radicale de la maladie de la pomme de terre et d'autres végétaux ou moyen d'en faire disparaître la cause*, Namur, 1855; and his *Les Fabriques de produits chimiques et autres établissements insalubres*, Brussels, 1856. His stance earned him the enmity of many policy-makers and industrialists as well as a short prison sentence at the time of the Sambre upheavals in 1855.

15. J. Neuville, *Naissance et croissance du syndicalisme*, vol. 1: 'L'origine des premiers syndicats', Brussels, 1979; L. Delsinne, *Le parti ouvrier belge des origines à 1894*, 1955; D. Deweerdt, *De Belgische socialistische arbeidersbeweging op zoek naar een eigen vorm*, Antwerp, 1972.

A National and Social Issue: 1860–80

In the 1860s and 1870s a few scientists and politicians took up the fight against the harmful effects of industrial waste on the health of inhabitants and workers. This time, their attention was drawn in particular to the problem of water contamination, as they called it. Their interest was aroused by concerns generated by the epidemics, and specifically the cholera epidemic of 1866, that were scything through the working class populations of major Belgian cities. As with the earlier industrial pollution, death rates in the workforce were not unimportant considerations for the ruling elite.

Among the active figures concerned with water quality for workers were a significant number of people with an interest in fishing and fish-farming. As in France and elsewhere in Europe, proponents of better water quality were closely associated with agendas that straddled the preservation of fish resources, the health of and food for the working classes and the development of the productive capacity of the nation state.[16] In other words, in a single virtuous system, by protecting fish through the eradication of pollution, the population could be provided with healthy food and thus strengthen the nation.[17]

In Belgium, the forerunner of these activists was Edmond de Sélys-Longchamps. Descended from nobility, this famous naturalist and entomologist was also a liberal politician with responsibilities at local, provincial and national level.[18] In academic coteries as well as in political assemblies, de Sélys-Longchamps constantly advocated the need to take an interest in the problem of water quality. To his mind, this concern was related to the preservation of fish stocks, endangered

16. D. Kinsey, 'Seeding the Waters as the Earth: The Epicenter and Peripheries of a Western Aquacultural revolution', *Environmental History*, 11:3, 2006, 527–66, in particular, 527–35.

17. This emphasis is also seen in the first fish farming treaties in Belgium. See J. P. J. Koltz, *Traité de pisciculture et multiplication artificielle des poissons*, Brussels 1858, 9-15; L. Meynne, *De la pêche maritime au point de vue de l'alimentation publique. Lettre à M. Charles Rogier, Ministre de l'Intérieur*, Brussels, 1858.

18. Michel-Edmond de Sélys-Longchamps (Paris, 1813–Liege, 1900) belonged to the old aristocracy. He was a self-taught ornithologist and entomologist, and was a member of many scientific societies, and director of the science course at the Royal Academy of Science, Letters and Fine Arts of Belgium, from 1854 to 1879. As a liberal, he was a council member (1842–1900) of the municipality of Waremme (province of Liege), a provincial council member (district of Waremme, 1846–1848), and a member of the House of Representatives (district of Waremme) from which he resigned from in 1848 following the deportation of Karl Marx and several democrats. He returned to the Senate in 1855 until the end of his life, holding the position of vice-president (1879–80) and president (1880–84). He played an active role during the first liberal congress in 1846 and was one of its promoters in 1894. See M. Caulier-Mathy, *Le monde des parlementaires liégeois, 1831–1893. Essai de socio-biographie*, Brussels, 1996, 259–65; L. Bertrand, *Histoire de la démocratie et du socialisme en Belgique depuis 1830*, 1, Brussels, 1906; E. de Sélys-Longchamps, *Sur la pêche fluviale en Belgique. Discours prononcé à la séance publique de la Classe des Sciences, 16 déc 1866*, Brussels, 1867.

by uncontrolled industrial waste, and to the quality of the population's food. In 1882, in order to promote research in this field, he gave the sum of 3,000 Belgian francs to the Royal Academy of Belgium to reward the best thesis on the theme of 'the search for and proposal of practical means whereby industrial waste from factories could be made inoffensive to fish'.[19]

In addition to the patriotic arguments put forth by those who defended the preservation of rivers and streams were moralising arguments concerning the working class, in keeping with the paternalistic ideas of the bourgeoisie of the time. Thus, for example, the author of the report written in 1886 for the Minister of Agriculture, following the release of young fish into the country's rivers, did not hesitate to make fishing the 'healthy' leisure activity *par excellence*:

> In those who practise it, fishing develops intellectual qualities such as patience and observation, which of course represent an advantage in various social situations ... In terms of health, the value of fishing is impossible to ignore. For example, the worker who has spent the whole week in a workshop goes out on a Sunday to breathe the fresh morning air; ... There could not be a better example of a way to stay healthy and maintain the body's strength. Instead of spending the household budget in the often unhealthy atmosphere of a tavern, he comes home in the evening with an extra meal for his family.[20]

In a speech presented at the Société centrale d'Agriculture two years later, engineer Gustave Defosse said something along the same lines: fishing 'would develop patience and reflection in the worker. The surrounding peacefulness would calm his nerves and muscles'.[21] He also argued that fishing was a beneficial replacement for taverns, where workers developed violent ideas that were destructive to society.

Whatever the views of the authors of these excerpts, it was nonetheless in one of the condemned establishments, a pub in the centre of Liege, that one of the first anglers' associations was born in 1888, opening the way to the creation of the first organised movement at national level against water pollution caused by industry.

Workers and Anglers: 1880–1914

In 1888, in the Café des Mille Colonnes, a popular café in the city centre, the Liege Anglers' Society was founded; other associations in the major industrial centres of

19. 'Prix pour l'assainissement et le peuplement des rivières: concours extraordinaire', in *Bulletin de l'Académie royale des Sciences, des Lettres et des Beaux Arts de Belgique*, t. 3, 11, 14, 16, 17, 1884–1889.

20. *Rapport sur les déversements d'alevins en 1886 adressé à M. le Ministre de l'Agriculture, de l'Industrie et des Travaux Publics*, Brussels, 1886.

21. G. Defosse, *La pisciculture au point de vue de la production, de l'alimentation et de la morale, discours prononcé à la séance du 12 mars 1888 de la Société centrale d'Agriculture de Belgique*, Brussels, 1888.

the country, Ghent and Brussels, soon followed.[22] This chronology is similar to that observed in France, also initially motivated by a desire to protest against river pollution and concerns regarding the disappearance of fish from rivers and streams.[23] The Belgian campaign against river pollution, and the establishment of a federal body at national level developed a few years earlier than in France.

From its inception, the Liege Society, composed mainly of workers, gunsmiths and small shopkeepers, organised a general protest against water pollution in the region. Their particular target was the waters of the River Vesdre, which were highly polluted by the woollen industry of Verviers. The Liege Society circulated petitions, gained support from the municipal authorities, brought its grievances and complaints to provincial level and thus forced the government to take an interest in the problem. With origins in an important urban industrial region and in a working class environment, the Liege Society was soon able raise the profile of the river pollution issue, and to obtain coverage in newspaper columns.

The affinities of certain members of the Liege Society with the political positions of the workers' movement are very noticeable in the articles published in its newspaper, which existed from 1900. The articles routinely denounced the lamentable state of rivers and streams, attributing this state to the inactivity of the ruling class, which had little concern for the health of workers. The article published in the 15 August 1903 issue under the title of 'L'éternelle comédie' ['The same old nonsense'] was representative of their stance. In an effort to condemn the inactivity of different ministers, the author, using a pseudonym, expressed his annoyance:

> The Minister said that it would be looked into, and that is all. Everything is for the best in the best of all possible worlds, and the Belgian public, who are becoming more and more moronic fools, declare that they are satisfied and let things take their course; during this time, our angling societies organise competitions and parties, drink beer and are not very concerned as to whether there will be a single fish left in Belgian waters in a few years; the public health commissions close their eyes and refuse to recognise the horrifying danger of terrible epidemics hovering over populations due to contaminated waters which they and their livestock are forced to drink; we shall continue to send dead fish to our markets due to poisoned residues sent to the river by industrialists. What does this gloomy perspective and dismal picture do to these men [in the government]? The rich only drink mineral water from abroad and do not eat the fish from our rivers; it is much too unrefined a

22. The Anglers' Society Fishing Club of Ghent was founded in 1888; it still existed on the eve of the First World War with Jean Vos and Charles Rinskopt as president and secretary respectively.

23. J-F. Malange, 'Pêche à la ligne et gestion des ressources piscicoles. Le sud-ouest de la France de la fin des années 1880 à la fin des années 1930', *Responsabilité et Environnement. Recherches, débats, actions, Série trimestrielle des Annales des Mines*, 48, 2007, 91–9; J-F. Malange, 'Risque perçu, risque vécu. Les pêcheurs à la ligne et la pollution des cours d'eau en France aux XIXe et XXe siècles', Au fil de l'eau. 3e Séminaire (2007). Maison des Sciences de l'Homme de Clermont-Ferrand, online publications, accessible at: http://www.msh-clermont.fr/article1630.html Accessed May 2010.

> dish for their delicate palate. All of that is good for the ordinary people, and if the people die, they are quite indifferent to it; they have got it easy and look down on the public with truly unbelievable flippancy.[24]

In this revealing excerpt, there are several arguments that are worth noting. The author criticises the general inertia and passivity of stakeholders including the government, angling societies and public health commissions, and accuses industry of being the major, if not the only, cause of river pollution, of killing fish and of poisoning livestock and people. The author then attacks the social inequalities that affect the 'people' who drink the surface waters and eat the fish that live there, contrasting this with the situation of members of the government who can afford bottled mineral water. Thus, the author logically presents industry as being responsible for increasing environmental inequality, so reinforcing existing social inequalities based on income and wealth. Implicit, too, is the idea that workers/anglers are victims for two reasons: firstly, because of the failure of the ruling class to introduce the necessary measures to guarantee healthy water for those of more humble means; and secondly, because of the industries that were damaging the water and fish resources.

The fight to restore the food quality of the working class constituted a common cause for the members of the angling societies, who were by no means entirely from the working class. In addition to a large number of workers and other employees, the Liege Society included industry directors, lawyers, doctors, notaries and researchers.[25] The Society was led by an eminent chemist and engineer, Lucien-Louis De Koninck, professor at the Université de Liège.[26] Support for the anglers'

24. *Epuisette et Carnassière*, 15 Aug 1903, 59–60. 'M. le Ministre a répondu que l'on étudierait et c'est tout. Tout est pour le mieux dans le meilleur des mondes et le public belge, gogo et abruti de plus en plus, se déclare satisfait et laisse faire; pendant ce temps, nos sociétés de pêche à la ligne organisent des concours et des fêtes, boivent du gueuse et du lambic, se souciant fort peu de savoir si d'ici quelques années il y aura encore le moindre petit poisson dans les eaux belges ; les commissions d'hygiène publique ferment les yeux et refusent de reconnaître l'effroyable danger des épidémies terribles qui planent sur les populations du chef des eaux contaminées que, elles-mêmes et leurs bestiaux sont obligés de consommer ; on continuera à envoyer sur nos marchés des poissons crevés par suite des résidus empoisonnés envoyés à la rivière par les industriels. Que fait toute cette triste perspective et ce sombre tableau à ces messieurs [du Gouvernement]? Eux qui sont riches ne boivent que des eaux minérales venant de l'étranger, ils ne mangent pas le poisson de nos rivières; c'est un mets bien trop grossier pour leur palais délicat, c'est bon pour le brave populo tout cela et, ma foi, que le peuple crève, ils ne s'en soucient guère; ils ont l'assiette au beurre et se moquent du public avec une désinvolture réellement incroyable.'

25. *Epuisette et Carnassière*, 2nd year, 1, Jan 1891, list of the members of the society.

26. Lucien-Louis De Koninck (Liege 1844–1921) received a civil engineering degree in mining and applied arts and manufacture from the Université de Liège (1867), as well as a doctorate in natural science from the University of Bonn (1870). His work was focused above all on the study of organic chemistry. After working as an assistant lecturer in the Faculty of Science at the Université de Liège (1876), he obtained the title of lecturer in 1885. He was dean of the faculty twice (1905–06 and 1912–13) and president of the Chemical Society of Belgium (1899), for

movement from key figures from the middle class was even more noticeable in the Société Centrale pour la Protection de la Pêche Fluviale (Central Society for the Protection of River Fishing) founded in 1890.[27] It was established in Brussels with the aim of uniting the efforts of all the local societies and had a socially mixed membership, as seen in the following excerpt from its newspaper in 1893:

> Angling, which is accessible to all budgets, represents a form of relaxation enjoyed by thousands of people from all classes of society. During the summer months, the trains are full and carriages often have to be added. A statistic on this subject would surprise many of those who are responsible for social issues. And it is worth mentioning that in the world of anglers, class consciousness does not exist: there are only brothers.[28]

It was managed until just before World War II by eminent members of the bourgeoisie and the aristocracy, who often owned large amounts of land.[29] The social make-up of these associations was closely identified with the original reasons for their foundation. Thus, in June 1903, on the initiative of Karel Rutgeerts, herbalist and shopkeeper, a 'National Anglers' Meeting' was organised in Aarschot, a small city in Brabant, north of Leuven, where the waters of the River Demer were polluted regularly by waste from the large sugar industry in the region. Once again, the aim of the event was to protest against the procrastination of the political authorities as regards the fight against water pollution.[30] A few months later, in November 1903, anglers in the Brussels region united in defence of their interests. The meet-

---

which he founded a branch in Liege. He was a member of many learned societies (Royal Society of Science of Liege, 1871; Deutsche Chemische Gesellschaft, until 1914) and received many honorary distinctions for his service to science. L. Halkin, *L'université de Liège de 1867 à 1935. Notices biographiques*, vol. 2, *Faculté des Sciences, Ecoles Spéciales, Faculté Technique, Rectorat de l'Université*, Liege, 1936, 132–43.

27. The first fish farming society seems to have emerged in 1863 under the direction of A. Schram. It was probably an association with an essentially scientific purpose, which does not seem to have had any anti-pollution claims.

28. *Pêche et Pisciculture*, 4th year, 11, Nov 1893, 171: 'La pêche à ligne qui est la portée de toutes les bourses constitue un délassement goûté par des milliers de personnes appartenant à toutes les classes de la société. Pendant la saison favorable, les trains sont bondés on est souvent obligé d'ajouter des voitures. Une statistique à ce sujet étonnerait beaucoup ceux qui s'occupent des questions sociales. Et chose digne de remarque, dans le monde des pêcheurs, l'esprit de caste n'existe pas: tous sont frères'.

29. The founding members were the Earl of Briey, who was the president, Earl Van der Straeten-Pontoz, the Prince of Rubempré-Mérode, the Earl of Renesse, Baron Edmond de Sélys-Longchamps, the Baron of Voelmont, the Baron of Rasse, the Esquire of Cannart d'Hamale. The Earl of Briey was also a member of the House of Representatives and a member of the Fish Farming Committee, and later, the Governor of the Province of Luxembourg.

30. In December 1900, the government appointed a committee to research all means of preventing the contamination of rivers and streams. The report was written by Hachez, Director of the National Water Distribution Society, and did not lead to anything apart from a few recommendations.

ing was organised at the Cygne, a very popular pub among Belgian socialists,[31] and was successful in that it led to the founding of a new federation, the Anglers of Belgium, whose main task was to challenge political representatives to put water pollution on the legislative agenda of parliament.[32] Several representatives, all of whom were socialists or progressive liberals, rose to the call. A letter from Louis Bertrand, a major figure in the Belgian socialist movement,[33] promised the support of the Labour Party, and was read at the meeting. The political representatives who were present promised to communicate the anglers' grievances and claims to parliament, including 'point 6', which proposed in plain language to take 'legal action against the unscrupulous industrialists who poured waste from their factories into the rivers and canals, not only poisoning thousands of kilogrammes of fish, but also often causing the death of nearby residents and livestock'.[34] In January 1906, they kept their promise and questioned the Minister of Agriculture on the issue of the poisoning of rivers and canals.[35]

On 11 April 1906, at the Cygne, another large meeting was organised. Among those who attended were the brother of the charismatic priest Adolphe Daens, and Léon Lepage, lawyer, progressive liberal and deputy mayor for public education in the city of Brussels.[36] The new federation, which gathered key catholic, socialist and liberal figures concerned about the emancipation of the working class, emerged from the meeting determined to position water pollution on the political centre stage. Bevernage, a lawyer from Ghent who may have been able to see the River Leie and the River Scheldt from his windows, made it his favourite topic. According to Bevernage, the first goal of the federation had to be the fight against

31. Since 1848, the democratic, followed by socialist, societies gathered in this café in the Grand-Place of Brussels. The Belgian Labour Party was also formed there in 1885. L. Bertrand, *Souvenirs d'un meneur socialiste*, Brussels 1927.

32. *Le Pêcheur belge. Organe des Pêcheurs à la Ligne de Belgique*, 6th year, 22, 1 Nov 1905.

33. L. Bertrand (Molenbeek-Saint-Jean, 1856–Schaerbeek, 1943) a former marble worker who became a journalist with the daily socialist newspaper Le Peuple, before becoming its director from 1896 to 1907. He was one of the major political players in the creation of the Belgian Labour Party (POB) and became its representative in parliament (1900–26). He is also the author of many works on political history, and socialism and cooperation in Belgium, in particular. P. Van den Dungen, *La foi du marbrier Louis Bertrand, 1856-1943: acteur et témoin de la naissance du socialisme belge*, Brussels, 2000.

34. *Le Pêcheur belge. Organe des Pêcheurs à la Ligne de Belgique*, 6th year, 22, 1 Nov 1905, 3; 'Poursuites contre les industriels peu scrupuleux qui déversent dans les fleuves, rivières et canaux les résidus de leurs usines et de leurs fabriques, empoisonnent non seulement des milliers de kilos de poissons mais occasionnent souvent la mort des riverains et du bétail.'

35. At this stage, it is difficult to say whether the words used by these militants, such as corruption, contamination and poisoning, distinguish different types of pollution, for example, chemical v. bacteriological. Considering the nature of the writing, it would seem that they are used generically.

36. *Le Pêcheur belge*, 7th year, 8, 1 April 1906.

pollution and he pressed for the name of the new Anglers of Belgium Federation to be the 'League Against Water Pollution in Rivers'. A few years later, with the slogan 'The health of the people depends on pure water', the commitment was unequivocally more broadly based than that of the anglers, and focused on a policy of high quality water for all users.[37]

In its wake, leagues were formed in most of the country's industrial cities such as Ghent, where a League Against Water Pollution in the Rivers of East Flanders was established in 1907. In Aarschot, the Union of Anglers from the region of Aarschot, founded in 1905, underwent a significant change in 1908, when it became the League Against Water Pollution [*Bond tegen de Besmetting des waters*], under the dynamic leadership of Dr Karel Rutgeerts, before extending its objective in 1913 to include the eradication of all types of pollution, concerning water, ground and air.[38] This followed the precedent in the Netherlands, where Karel Rutgeerts had been a correspondent for the Dutch Association Against Water, Ground and Air Pollution (*Hollandsche Vereniging tegen Water-, Bodem-, en Luchtverontreiniging*) for several years.

The Anglers of Belgium Federation, like the Liege Anglers' Society before it, did not try to mask its specific concern for the working class, which, according to the Federation's estimates, represented close to nine tenths of Belgian anglers at the time.[39] For a few years, it was the spearhead in the political struggle against the industrial pollution of rivers and streams. However, despite the views expressed on occasion by members of the Labour Party, this struggle for improved water quality never figured in the agendas of workers' emancipation movements, whose two main aims were universal suffrage and improved working conditions.

Efforts to improve water quality were interrupted abruptly when the First World War broke out. The conflict also terminated interest in the protection of natural resources and a more general health policy. When the movement to combat water pollution resumed in 1921, it no longer had the same social tones. The prominence of the anglers in earlier campaigning was diminished. The political climate that had led to social reforms such as 'universal' male suffrage, income tax, limitations on daily working hours and the creation of societies of a public nature,

---

37. L. Mineur, 'Rapport sur la pollution des eaux', *VI^de^ internationaal Congres voor Visscherij, Ostende/ VIe Congrès de la Pêche, Rapports et communications*, 18–21 Aug 1913, 10–15. There was also a resemblance here with the French anglers' movements, see S. Frioux and J-F Malange, '"L'eau pure pour tous!" Mobilisations sociales contre la pollution des eaux douces françaises (1908–années 1960)', *Histoire & Sociétés*, 27, 2008, 100–23.

38. K. Rutgeerts, 'Verslag over de Toestand der Rivieren in België en de Besmetting en Zuivering onzer Wateren', *VIe Internationaal Congres der Vischvangst, Oostende op 18, 19 en 20 augustus 1913*, Aarschot, 1913; K. Rutgeerts, *Griefs des Pêcheurs. La Pêche aux filets*, Aarschot, 1914.

39. In 1905, nine-tenths of the 100,000 fishing licence requests were made by people from the working class, according to *Pêche et Pisciculture*, 17th year, 10, 15 May 1906, 129.

for example for low cost housing, moderated the social claims of anglers, who were nevertheless still active.[40]

Conclusions

The study of the first collective protest movements against pollution in Belgium allows two important moments to be put into perspective. Firstly, the 1850s witnessed an initial turning point in the protests against certain types of industrial pollution. For the first time, actions were organised collectively: riots took place in the Sambre valley, committees were formed in Liege and legal action was pursued by communities. For the first time, protesters denounced an inequality with respect to mortality, which they attributed to the presence of factories. The original impetus was directed especially at chemical and metallurgical industries and faded rather quickly after the formation of committees of inquiry in parliament. Secondly, during the following decades, a new type of movement was born: that of anglers protesting against water pollution. This movement became more widespread after 1890, at a time when leisure activities were developing, and was the result of encounters between intellectual circles, alarmed at the consequences of industrial waste in rivers, and an increasing number of anglers. Throughout these years, protest was organised through a multitude of local associations federated at the national level. For twenty years or so, until the First World War, the idea that water pollution was particularly harmful to the health and survival of 'people of modest means', as represented by the majority of anglers, prevailed to varying degrees within these associations. Their claims, therefore, associated the problem of a declining natural resource due to anthropogenic activity and a health problem experienced by a specific part of the population.

As such, both phases play an original part in an extended story of the emergence of environmentalist movements and in development of the contemporary concept of environmental inequalities. Of course, they were not centrally and directly concerned with a type of environmentalism or a fight for environmental justice. These ideas and concepts appeared much more recently, with the construction of ideologies, movements and political parties, that, since the 1970s, have only been able to consider the future of societies with respect to their profound interactions with the environment and natural resources. Nevertheless, by extending their perspective to include more than their immediate surroundings, by mobilising ideas about the

40. On this subject, see W. Balcers and C. Deligne, 'Les sociétés de pêche à la ligne, "conscience" de la pollution des rivières en Belgique (1880–1940)?' in I. Parmentier, dir. *La recherche en histoire de l'environnement: Belgique – Luxembourg – Congo – Rwanda – Burundi. Actes des premières Rencontres de l'Histoire de l'Environnement en Belgique, Namur, décembre 2008*, Namur, 2010, 175–186. The types of public society referred to here were ones created as a consequence of emerging socialist demands: Société Nationale des Habitations à Bon Marché, Société Nationale des Chemins de Fer; Société Nationale de Crédit à l'Industrie.

degradation of natural resources, by denouncing the effect of food sources on the health of the working classes, by building a network of observers and by developing means of action at political level, these associations and their main stakeholders laid the foundations for political reflection and linked the future of some societies to the management of natural resources. In 1919, by proposing that the issue regarding the pollution of the River Espierre, a tributary of the Scheldt from the industrial region of Roubaix–Tourcoing in France, should be resolved at the Peace Conference of the Nations [Conférence de Paix des Nations] the members of the Central Fish Farming Society demonstrated more than ever their comprehension of the highly political character of the issue of the degradation of natural resources.[41]

Select Bibliography

Balcers, W. and C. Deligne, 'Les sociétés de pêche à la ligne, "conscience" de la pollution des rivières en Belgique (1880–1940)?' in I. Parmentier, dir. *La recherche en histoire de l'environnement : Belgique – Luxembourg – Congo – Rwanda – Burundi. Actes des premières Rencontres de l'Histoire de l'Environnement en Belgique, Namur, décembre 2008*, Namur, 2010, 175–186.

Devos, G. 'Milieuverontreiniging door de industrie omstreeks het midden van de 19de eeuw', in *Bijdragen tot de Geschiedenis, Liber alumnorum Karel van Isacker*, Antwerp, 1980, 347–94.

Etienne, R. 'L'industrie chimique belge (1830-–1930)', *Bulletin de la Société belge des Ingénieurs et des Industriels*, 11, 1931, 5–28.

Guillerme, A., A-C. Lefort and G. Jigaudon, *Dangereux, insalubres et incommodes. Paysages industriels en banlieue parisienne XIX–XXe siècles*, Paris, 2004.

Kinsey, D. 'Seeding the Waters as the Earth: The Epicenter and Peripheries of a Western Aquacultural Revolution', *Environmental History*, 11:3, 2006, 527–66.

Malange, J-F. 'Pêche à la ligne et gestion des ressources piscicoles. Le sud-ouest de la France de la fin des années 1880 à la fin des années 1930', *Responsabilité et Environnement. Recherches, débats, actions, Série trimestrielle des Annales des Mines*, 48, 2007, 91–9.

Malange, J-F. 'Risque perçu, risque vécu. Les pêcheurs à la ligne et la pollution des cours d'eau en France aux XIXe et XXe siècles', Au fil de l'eau. 3e Séminaire (2007). Maison des Sciences de l'Homme de Clermont-Ferrand, online publications, accessible at http://www.msh-clermont.fr/article1630.html

Malange, J-F. and S. Frioux, '"L'eau pure pour tous!" Mobilisations sociales contre la pollution des eaux douces françaises (1908–années 1960)', *Histoire & Sociétés*, 27, 2008, 100–23.

Verbruggen, C. 'De reacties op vroege vormen van vervuiling, de chemische nijverheid te Gent, 1820–1892', *Jaarboek voor Ecologische Geschiedenis*, 2001, 127–28.

41. *Pêche et Pisciculture*, 31st year, 1, 15 Jan, 1920, 27.

Verbruggen, C. 'Nineteenth Century Reactions to Industrial Pollution in Ghent, the Manchester of the Continent. The Case of the Chemical Industry', in C. Bernhardt and G. Massard-Guilbaud, eds. *Le démon moderne, la pollution dans les sociétés urbaines et industrielles d'Europe/The Modern Demon, Pollution in Urban and Industrial European Societies*, Clermont-Ferrand, 2002, 377–92.

Chapter Twelve

Technological Choice and Environmental Inequalities: The New England Textile Industry, 1880–1930[1]

Janet Greenlees

There is a wealth of literature about the American cotton textile industry that highlights the centrality of the market to industrial strategy and the importance of specific machines to industrial structure and strategy. Somewhat surprisingly, other forces that shaped the direction of capitalism, particularly locally specific forces, have received little attention. Similarly, the interface between business and the environment, both within and outwith the factory, has been marginalised.[2] A notable omission is the relationship between individual urban peculiarities, business and the workplace environment that created environmental inequalities between towns and firms, making some places healthier places in which to work than others.[3] This business–environment juncture was central to manufacturing because managerial decisions affected not only workers' health, but also output and product quality. Moreover, manufacturing behaviour did not follow any consistent pathway between or within a town or amongst makers of the same product. Therefore, arriving at a conclusion as to *the* dominant influence on technological change in any industry at any given time is difficult. Instead, it is better to comment on the active agents in the process of technological change and the process itself at different firms in various locations, to highlight both the diverse and changing dominant influences on the workplace environment and the importance of the local within these.

---

1. The author thanks the Wellcome Trust (grant no: 066526), the ESRC (grant ref: R000 22 3483) and the Pasold Research Fund for help in funding this research and with conference expenses, and the editors of this volume for helpful comments.

2. The debate on these issues has begun, but significant omissions remain. See C. Sellers, 'Factory as Environment: Industrial Hygiene, Professional Collaboration and the Modern Sciences of Pollution', *Environmental History Review*, 18:1, 1994, 55–83; A.F. McEvoy, 'Working Environments: An Ecological Approach to Industrial Health and Safety, *Technology and Culture*, 36:2, 1995, 145–73.

3. C.M. Rosen argues the importance of integrating business and environmental history in 'The Business–Environment Connection', Anniversary Forum, *Environmental History*, 10:1, 2005, www.historycooperative.org/journals/eh/10.1/rosen.html Accessed 25 March 2009; C.M. Rosen and C. Sellers, 'The Nature of the Firm: Towards an Eco-cultural History of Business', *Business History Review*, 73, 1999, 577–600.

This chapter considers the technologies that determined the quality of the atmosphere in weaving rooms and on a key piece of equipment – the suction shuttle used in weaving. Around the turn of the twentieth century, considerable debate surrounded the relationship between the technologies involved in both processes, the workplace environment and workers' health. Participants included employers, the town council, the community, workers, physicians and the state, all of whom had different priorities. The relationship between these groups in different towns and the resulting dominant force impacted on the quality of the factory environment. Hence, this chapter furthers Philip Scranton's work demonstrating how firms in different cities adopted different manufacturing strategies, with multiple, successful paths to profit and accumulation.[4] It highlights the dominant influences on, and the impact of manufacturing strategy on, the creation of workplace environmental inequalities between cities. Before turning to the individual, urban peculiarities, the legislative and technological parameters in which the local forces were operating are explored.

The interface between business and the factory environment was particularly prominent in Progressive Era Massachusetts due to the state's early industrial legislation. In the federalist United States, the State of Massachusetts was a pioneer of Progressive legislation to aid labour and public health, with the state's largest industry, cotton, the first and primary target for reform. In the workplace, reformers sought to reduce accidents and improve hygiene, sanitation and ventilation in factories and thus focused on the weaving rooms.[5] Reformers convinced the Massachusetts Legislature to regulate mill air quality through a series of Acts between 1907 and 1910. And, in 1911, they secured the first machine legislation to ban a specific piece of technology – the suction shuttle – to prevent the spread of disease.[6] The link between work and health was established.[7] However, legislation was only one step towards substantive reform. Real improvements in the workplace environment

4. P. Scranton, *Proprietary Capitalism: The Textile Manufacture at Philadelphia, 1800–1885,* Cambridge, 1983; P. Scranton, *Figured Tapestry: Production, Markets, and Power in Philadelphia Textiles, 1885–1941,* Cambridge, 1989; P. Scranton, *Endless Novelty: Specialty Production and American Industrialization, 1865–1925,* Princeton, 1997.

5. Especially industrial poisons. See D. Rosner and G. Markowitz, 'The Early Movement for Occupational Safety and Health, 1900–1917', in J.W. Leavitt and R.L. Numbers, eds. *Sickness and Health in America: Readings in the History of Medicine and Public Health,* 2nd edn, Madison, 1985, 507–21.

6. For more on the historical debates about the actual achievements of the Progressive Era, see J. Greenlees, '"Stop Kissing and Steaming!" Tuberculosis and the Occupational Health Movement, 1870–1918', *Urban History* 32:2, 2005, 225–28, 233–39; D. Rodgers, 'In Search of Progressivism', *Reviews in American History,* 10:4, 1982, 113–32; A. Dawley, *Struggles for Justice: Social Responsibility and the Liberal State,* Cambridge, Ma. 1991.

7. W. Mass, *et al.* '"Kiss of Death": Banning the Suction Shuttle in Massachusetts', in C. Levenstein, G.F. DeLaurier and M. Lee Dunn, eds. *The Cotton Dust Papers: Science, Politics, and Power in the 'Discovery' of Byssinosis in the U.S.,* New York, 2002, 15–28.

required employer investment in alternative technologies. To this end, manufacturers needed convincing of the production benefits of change. Nevertheless, textile manufacturers were the first employers in the state who felt political, as well as social pressures, to make technological investments to improve the factory environment, with varying outcomes.

Three of the leading New England cotton towns during the Progressive Era were Lowell, Fall River and Holyoke – all in Massachusetts. All three towns produced bulk staples, but the manufacturing strategies and associated technologies between and within them were uneven, despite some of the rigidities of mass production. The key technologies affecting the weaving environment were twofold. Firstly, the ventilation and humidifying equipment that controlled the hot, humid environment in the weaving rooms raised concerns about relationships between poor ventilation, sanitation, dust inhalation and disease contagion. Secondly, concerns were raised about dust inhalation and the spread of disease through weavers sharing shuttles that required sucking the thread through the eye of the shuttle – a process repeated hundreds of times a day. These were the technologies that the State Legislature sought to control in an effort to improve both the working environment and the health of its citizens.

Manufacturers found the shuttle legislation particularly obtrusive because outsiders sought directly to influence the production process. Yet, by 1911, many alternative devices to the suction shuttle were available which also addressed productivity concerns. These included automatic looms, hand threaders, self-threaders and suction mechanisms.[8] The lack of strong, sustained employer opposition to the legislation suggests that either technological change was not only feasible, but already underway in some firms, or that employers were not expecting rigorous enforcement of the law. Hence, firms' associated technological choices helped create diverse mill environments, with varied health consequences for workers.

The attractiveness of these various technologies changed over the years. One key factor increasing the preference for certain machines at particular times was

8. For example, J.M. Moyers advertised an artificial suction device which drew the thread through the shuttle eye, *Textile Record* (hereafter, *TR)*, July 1892; Dolber advertised a brush device that was inserted through the shuttle eye, caught the thread, and was then drawn back through the eye, *TR*, Oct. 1895, 627; an unknown manufacturer advertised a shuttle whereby the weaver blew through a duct which sent the thread through the shuttle eye, *TR*, Dec. 1900; The New England Shuttle Threader Co. of New Bedford, Ma. advertised a manual threader that sucked the thread through the eye, *TWR*, March 1912, 131. Hand threaders include: Aubé, Cromton and Knowles, *TR*, April 1899, 199; Halls, Sergeson's self threader, *TR*, Jan. 1887; Morin's self-threading shuttle, *TR*, March 1903, 155; Coldwell & Gildard, *Textile World Record (*hereafter, *TWR)*, Nov. 1903, 101–3; The Belisle Loom Shuttle Threader, of J.N. Levasseur-Belisle, of Manchester, NH, *TWR*, Aug. 1909, 125; American Textile Specialty Machinery Co., of Providence, RI, *TWR*, Oct. 1911, 166; Webster hand threaders of Fall River, Ma. *Fall River Evening Herald*, Jan 16, 1911; There were many unnamed devices, for example, *TR*, Jan., 1895, 19; *TR*, Oct. 1898, 577; *TR*, Oct. 1899, 555.

advertising, a key example being the Northrop automatic loom, manufactured by the Draper Company of Hopedale, Massachusetts. The Northrop was *the* major technical innovation in textile manufacturing between the 1890s and 1930s. The battery carriage enabled continuous operation and its auto-stop device improved cloth quality. The Draper Company advertised widely in the trade press.[9] Yet so too did small firms who supplied specific, local markets. These firms did not exit the market but historiography has obscured them due to their comparatively limited advertising and small market share. Indeed, some machine manufacturers never advertised in the trade press and their existence is only revealed through the equipment purchase lists of individual firms.[10] These firms survived either by creating niche markets for their products or by establishing business links with textile manufacturers in a specific area, thereby contributing to the creation of a local mill environment.

Indeed, the adoption of the Northrop loom in New England was slow and inconsistent because it cost almost three times as much as the most expensive non-automatic looms and it had difficulty adapting the automatic mechanism for fancy weaves.[11] Consequently, take-up of the loom remained inconsistent. For example, few Fall River firms invested in Northrops, with only twelve per cent of looms in the city automatic in 1914.[12] Yet, one fifth of the New England cotton industry was located in Fall River and neighbouring New Bedford. Firms in these two cities secured cost advantages with shorter runs of custom order cloth. Indeed, manufacturers throughout New England found the potential labour-cost savings from the Draper looms insufficient to warrant the increased capital costs. Hence, many managers purchased a few Northrops but did not replace their entire compliment of looms.[13] Instead, numerous weaving technologies operated alongside each other, suggesting a complex regional industrial strategy. In terms of bodily damage to workers, the ensuing arrangement of machines and the type of machines allocated to individual

---

9. For example, *TR, TWR, Textile World (TW), Fibre and Fabric* and *The American Cotton and Wool Reporter*.

10. For example, Joseph Lacroix's shuttle firm, based in Fall River during the early twentieth century. The Durfee Mills used his shuttles, but he did not advertise widely.

11. Around the turn of the twentieth century, competitors charged just \$50–\$55 per loom, depending on the company and the loom-width. In contrast, in 1902, a 40-inch Northrop loom cost \$137 and a 30-inch Northrop, \$133. I. Feller, 'The Draper Loom in New England Textiles, 1894–1914: A Study of Diffusion of an Innovation', *Journal of Economic History*, 26:3, 1966, 334. Draper gave a small credit to firms for each old loom turned in to Draper. In 1908, the Draper Co paid \$20 for narrow looms and \$50 for broad looms. HBS: Laconia Mills, Records, V.FB-7, 22 and 23 as cited in Feller, 'Draper Loom', 335.

12. W. Mass, 'Technological Change and Industrial Relations: The Diffusion of Automatic Weaving in the United States and Britain', PhD thesis, Boston College, 1984.

13. Mass, 'Technological Change', chapters 3 and 4; and Feller, 'Draper Loom'.

workers within a department directly impacted on their health, contributing to varied, unequal and constantly evolving work environments.

In contrast, technological variations were limited in ventilation debates. Moreover, from the start, workers' wellbeing was incorporated into debates about the relationship between technology and production. The meetings of the New England Cotton Manufacturers' Association (NECMA) were a key forum for these debates. Here, in 1900, Homer Woodbridge, a professor at Massachusetts Institute of Technology and ventilation engineer, argued that not only did poor factory air quality impair machine operation and hence the quality of the finished product, it also caused drowsiness, lethargy and dullness amongst employees. Furthermore, worker absenteeism fell in a well air-conditioned factory with hygienic conditions.[14] Hence, in 1908, the engineer, GB Wilson, argued that:

> It is unquestionably to the advantage of the mill proprietors to see that provision is made to enable satisfactory conditions to be obtained which will enable them to retain their help during the hot summer months, and at the same time preventing the sources of lessened energy and ill health … A good master should see that his workpeople have good hygienic rooms in which to work, which will not only redound to their health, but also to his own personal profit.[15]

Thus, engineers firmly linked a good working environment with stable labour-relations and workers' wellbeing. Yet, they knew, and mill managers readily admitted, that workers' wellbeing was merely a bonus to the production efficiencies gained by technological investment. 'The history of the world's industries affords no encouragement that working conditions will be spontaneously and vitally improved by the mass of employers, unless the latter can see profit in the change.'[16]

While profit maximisation dominated the New England cotton employers' technological decision-making process, external pressures were not necessarily contradictory to this goal and could even be complimentary. Yet, the dominant thread remains that employers sought to preserve the right to run their business without external interference. Moreover, the multiple paths to profit make it difficult to discuss the Massachusetts industry as a whole. Each manufacturer in each town determined his business strategies, as well as his product, technological

14. S.H. Woodbridge, 'Warming and Ventilation of Mills', *Transactions of the NECMA* 68, April 1900, 225–26. Other engineers supported Woodbridge's argument. For further discussion, see B.B. Peterson, 'Industrial Architecture from the Inside: Textile Mill Design and the Factory Workplace, 1860–1920', in R. Weible, ed. *The Continuing Revolution: A History of Lowell, Massachusetts,* Lowell, 1991, 204–06.

15. G.B. Wilson, *Air-Conditioning: Being a Short Treatise on the Humidification, Ventilation, Cooling and the Hygiene of Textile Factories – Especially with Relation to those in the USA*, New York, 1908, 27, 37.

16. L. Allen, 'Economy of Heating and Ventilating the Machine Shop', *Engineering Magazine,* 21, 1901, 78–9, also cited in D. Nelson, *Managers and Workers: Origins of the Twentieth-Century Factory System in the United States, 1880–1920*, Madison, 1996, 11.

investment, labour choice and factory organisation based on situational particularities. The ensuing decisions created many workplace environments within and between New England cotton manufacturing towns, making certain towns and firms healthier places in which to live and work. This creates a more complex, but richer, picture of the interface between business systems and the environment, and between workers, health and machines.

Local Forces on Technological Change

By 1900, the three mill towns of Lowell, Fall River and Holyoke were the state's leading cotton manufacturing towns. Each town had developed a distinctive character that influenced local employers. Lowell was a town built around textile manufacturing, with the first mill opening in 1822. The same mercantile group owned and controlled the corporations and the employers' early paternalism and their employment of young women are well documented.[17] Employer paternalism faded during the nineteenth century as increasing numbers of immigrants flocked to the town. While the textile employers still influenced the town council, their continued absenteeism and the emergence of a middle class of small businessmen and white-collar workers reshaped the town politics and economy. By the early twentieth century, there was a strong community who sought to improve Lowell as a place to live and work, including inside the factories.[18] The town supported the striking workers in 1912, helping them to secure a wage increase, the right to weigh their own cloth, overtime pay and a management pledge to re-hire the strikers.[19] The textile corporations had become an 'outside force'. They were no longer part of the Lowell community.

In contrast, a handful of families controlled both the Fall River textile firms and the town council. They were more interested in preserving their profit margins than in civic improvements. Their lack of interest was compounded by the large and diverse immigrant population, which prevented the development of a strong sense of community. Hence, the push for social improvements came from outside official structures and had little direct or collective impact on employers. Yet in Holyoke, the paternalistic employers sought to minimise labour disruption and were proactive in improving certain working conditions. This approach, combined

17. See for example, T. Dublin, *Women at Work: The Transformation of Work and Community in Lowell, Massachusetts, 1826–1860,* New York, 1993 (1979); J. Greenlees, *Female Labour Power: Women Workers' Influence on Business Practices in the British and American Cotton Industries, 1780–1860,* Aldershot, 2007.

18. For further details about ethnicity, the Lowell middle classes and the town's varied business structure, see Gray Fitzsimmons' study at http://ecommunity.uml.edu/eth_ent/ Accessed 10 November 2008.

19. P. Richards, 'A Study in Community Power: Lowell, 1912', in Weible, ed. *Continuing Revolution,* 265–79.

with a more diverse local economy than that found in Lowell or Fall River, meant that cotton manufacturers never completely dominated the town. Hence, employer interest in community wellbeing and their own initiatives towards this goal, secured certain improvements to the workplace environment. These characteristics of place that had gradually developed during the nineteenth century influenced employers' decision-making and contributed to the environmental and social inequalities between the three towns.

As the city of Lowell grew during the nineteenth century, residents and the town council increasingly worked together on civic improvements. The local Board of Health (BOH) worked to improve sanitation and public health throughout the city.[20] Water supplies were switched from the polluted Merrimack River to local wells. In the mills, reformers sought to guarantee workers a supply of fresh drinking water to replace the common cup and to ensure a clean water supply for the humidifying and ventilating equipment to help prevent the spread of tuberculosis. These efforts, combined with similar measures elsewhere in the State, helped convince the Legislature to mandate that only clean water be used for humidifiers and to regulate humidity levels and atmospheric temperature.[21] Nevertheless, Lowell mill managers' investment in new humidifying equipment did not always correlate with legislation. Instead, in some firms, humidity and ventilation reforms were already underway, with legislation seeking to standardise the mill environment throughout the state, with varying success.

The early Lowell mills had quickly gained a reputation for their poor air quality and having little or no ventilation.[22] It was only after the Civil War that ventilation improved. New mill buildings with new designs replaced older buildings and some established firms, such as the Tremont and Suffolk, recognised the multiple benefits of updating their ventilation equipment.[23] These included increased production and lower dust levels. Mill managers sought workroom temperature and humidity levels that aided both production efficiencies and workers' wellbeing, recognising the economic benefits of employee comfort. Yet, individual cloth requirements, the number and types of machines and workers per room, room size, the product manufactured, the local weather and the mill building materials all ensured that

20. B.G. Rosenkratnz, 'Cart Before the Horse: Theory, Practice and Professional Image in American Public Health, 1870–1920', *Journal of the History of Medicine and Allied Sciences*, 29, 1974, 55–73; M.V. Melosi, *Garbage in the Cities: Refuse, Reform, and the Environment, 1880–1980*, College Station, 1981. See also Greenlees, '"Stop Kissing and Steaming!"' 234–35.

21. *Annual Report of the State Board of Health for Massachusetts* (*ARSBHM*), 1907, 2; 1909, 766; 1910, 458, 462 and 512.

22. *Lowell Daily News*, May 28, 1902; Kenngott, *Record of a City*, 94–5; Peterson, 'Industrial Architecture', 201; Greenlees, *Female Labour Power*, chapter 7.

23. L. Gross, *The Course of Industrial Decline: The Boott Mills of Lowell, Massachusetts, 1835–1955*, Baltimore, 1993, 136, 138; and, for example, the Tremont and Suffolk: see Peterson, 'Industrial Architecture', 197–212.

machine settings were mill, and even room, specific.[24] Despite continuous debate between employers and employees over the correct settings, some Lowell employers improved atmospheric conditions, creating a better working environment prior to the legislation of the early twentieth century. Others did not.

It was the older firms, such as the Boott, that were most likely to ignore the factory environment, legislation and workers' basic wellbeing.[25] After an external audit in 1916, the Boott was declared to have the worst working conditions in the industry.[26] The antiquated steam heating system and poor humidifying equipment did little to improve either production or working conditions, with some rooms too cold in winter and others too hot. Consequently, workers had more respiratory problems and lost more days to illness than operatives at other Lowell firms.[27] In response to this finding, the auditors recommended technological investment. They argued that 'improved ventilation, less dust and well controlled heat and humidity' would improve workers' wellbeing and they would 'turn off much more and better work'. The expense of the new equipment would quickly be recouped.[28] However, the firm refused to implement any of the recommendations and ignored state legislation. In response, many employees left the firm for superior conditions elsewhere, indicating that the workplace environment was important to them. The lack of technological investment made the Boott a dangerous and unhealthy place in which to work. Moreover, the Boott example reveals that external pressures on employers could be successfully ignored and profitability sustained with outdated machinery.

Community pressures on, and indeed interest in, production machinery were not as forthcoming in Lowell as with air quality. Lowell manufacturers' investment in production machinery was firm-specific and strategies were characterised by gradual investment outwith legislation. Manufacturers' varied choice and pace of technological investment created inequalities in weavers' health within the town. Investment in the Northrop Loom provides a good example. The Tremont and Suffolk Manufacturing Company was an early investor in the new automatic Northrop looms, having installed 1,135 in 1897.[29] In 1902, the Merrimack, Boott, Appleton and Massachusetts firms also purchased Northrops,[30] either to replace old machines or to expand the production of specific types of cloth. However, these

24. G. Cooper, *Air-Conditioning America: Engineers and the Controlled Environment, 1900–1960,* Baltimore, 1998, 40.

25. The Boott was established in 1836.

26. Valentine Report, 1916, 43–4 as cited in Gross, *Course of Industrial Decline,* 136.

27. Valentine Report, 1916, 43–4.

28. Valentine Report, 1916, 43–4.

29. *TR,* Draper Co. advert, Nov. 1897. The Tremont and Suffolk is the only Lowell firm of the 10 listed. See also, *TR,* Draper Co. advert, Feb. 1899. The Tremont & Suffolk is the only Lowell firm of the 15 mentioned by name.

30. Draper Co., *Cotton Chats,* 7, Jan 1902, 3.

investments were not complete weaving room refits. For example, the Merrimack produced approximately ninety per cent of its narrow print cloths on non-automatic looms. By 1903–04, this proportion had only dropped to between 78 and 89 per cent.[31] While these investments pre-empted the 1911 legislation, by 1914 only 36 per cent of the city's looms were Northrops.[32] Clearly, automatic looms were not the city's solution to the legislative requirements. Instead, investment in Northrops was a gradual process that followed changing production plans and included bobbin batteries for self-threading shuttle carriages that attached to most automatic looms. Therefore, despite the gradual improvement to the work environment through technological investment, environmental inequalities remained between and within firms.

The slight resurgence in textile production during and immediately after the First World War saw increased investment in Northrops in Lowell. This secured marginal improvements to the working environment as the number of looms per worker increased. By 1924, three Lowell firms were among the top 35 largest American investors in the Northrop, with the majority of the other leading investors located in the southern states.[33] In contrast, the Boott's early interest in technological improvements quickly waned and by 1931 most of their machinery was obsolete and dated from the pre-First World War era, with much remaining from the 1870s.[34] Shareholder profits were prioritised over productivity, structural improvements and workers' wellbeing. This choice was evident in the key decisions of the late 1920s to run the firm into the ground.[35] Nevertheless, the firm continued operations until 1954, demonstrating that divergent and outdated practices remained profitable and welfare non-essential, and that community pressures and legislation could be successfully ignored. Hence, the workplace environment and occupational health were accorded a low priority; they were negotiable elements in the matrix of business decisions.

Thus, in Lowell, the textile manufacturers acknowledged the link between work and health, but their associated technological investment strategies emphasised the atmospheric conditions, not production. This priority corresponded with those of the community and town council. Hence, certain environmental inequalities were accorded higher priority than others by both business and town residents,

31. Mass, 'Technological Change', 122, 126.

32. Mass, 'Technological Change', 290.

33. The Merrimack, Massachusetts and Tremont and Suffolk companies. Draper Co., *Cotton Chats,* Sept 1924, 255.

34. No new machinery was purchased between 1922 and 1930, despite the fact that new technology could cut costs and the savings quickly recouped outlay. Gross, *Course of Industrial Decline,* 123.

35. Gross, *Course of Industrial Decline,* ch. 7. In addition to antiquated machinery, workers also had to put up with unheated buildings, leaking roofs, broken windows and buckled floors. These conditions created high labour costs and ignored state labour laws and union protests.

albeit for different reasons. Lowell residents' and the BOH's interest in mill workers' wellbeing centred on wages and the collective welfare of city residents through securing good ventilation in all public buildings. The town's collective, economic prosperity and wellbeing took priority over the health of individual citizens. Hence, improvements to weavers' health at some firms were secured despite public and medical concern and were categorised as a negotiable production value in the town.

Multiple Influences on Technological Change and Investment

The Lowell case exhibits a stark contrast with that of Fall River, where the town's economic and political structures were closely knit. The majority of city textile firms were owned by just seven families, with many interrelationships – most notably the Bordens, Durfees and Braytons. During the nineteenth century, these families used their kinship networks to shape the social, economic and political order of the town.[36] The families' patrimony continued into the twentieth century, but the relative harmony that had existed in the mid-nineteenth century vanished as the city's population grew almost 650 per cent between 1860 and 1900, as immigrants flocked to the expanding mills.[37] The changing ethnic composition of the town created ethnic rivalries, which were enhanced by language barriers and class tensions.[38] These internal divisions strengthened as factory conditions deteriorated. Therefore, in contrast to the growing sense of collective community identity found in Lowell, Fall River remained a city divided by ethnicity and class. This lack of unity of purpose limited the potential community influence on manufacturing decisions, while at the same time enabling business to continue its domination of local politics and individual manufacturing preference to prevail, creating considerable inequalities between workplace environments.

The Fall River manufacturers' political and economic might meant that challenges to poor work environments came from people with no mill connections. In Fall River, critics included the local physicians and State Representatives, led by the local BOH physician, Dr Adam MacKnight, the local tuberculosis agent for the State's Anti-Tuberculosis Association. He was a strong advocate of hand-threading shuttles and believed that the old style shuttles increased the risk of tuberculosis con-

36. For more on the inter-relationships between the Fall River mill families, see M. Blewett, *Constant Turmoil: The Politics of Industrial Life in Nineteenth-Century New England*, Amherst, 2000, esp. ch.1.

37. Source: *United States Decennial Census*, 1860–1900, for each decade; *Massachusetts State Census*, 1865–1885, for odd years. Adapted from J. Cumbler, *Working Class Community in Industrial America: Work, Leisure and Struggle in Two Industrial Cities, 1880–1930*, Westport, 1979, Table 8, 113.

38. For more on the rising class tensions, see Blewett, *Constant Turmoil*, ch.6.

tagion.[39] From the early twentieth century, Fall River physicians, led by MacKnight, consistently and repeatedly argued for the legislative ban of the suction shuttle and the regulation of other seemingly dangerous business practices, including steaming and ventilation, in order to improve the work environment, thereby reducing the spread of tuberculosis.[40] The physicians' efforts helped persuade Fall River State Representative Edward Harrington to propose the suction shuttle bill in the State Legislature. Thus, it was the Fall River doctors and local state representatives that challenged the cotton employers' might and sought specific occupational improvements for operatives' wellbeing, albeit from motivations of the collective wellbeing of the cities' residents rather than from concerns about individual workers' health.

The Fall River cotton employers could not escape the legislative campaigns, despite their collective opposition to state interference in industry. Furthermore, their responses were not unilateral in terms of technological choices. Individual firms purchased new machines to replace worn out technologies, meet legislative guidelines, improve efficiency based on current scientific knowledge – which sometimes incorporated operatives' health – or simply because investment was deemed necessary to meet changing production goals. While these motivations were not necessarily mutually exclusive, they made for a more complex and firm-specific accumulation matrix than was found in the Lowell model.

To begin with, whereas Lowell manufacturers invested in ring frames, most Fall River firms were committed to spinning mules, with ninety per cent of the town's spindlage on mules by 1870. Even in the 1920s, seventy per cent of fine yarn continued to be spun on mules in southern Massachusetts and Rhode Island.[41] While the Northrop Loom could have been widely adopted in the city, it was not. Fall River employers preferred traditional looms, with three inter-related explanations for their reluctance to invest in Northrops: product choice, local business connections and managerial preference for making cheap improvements to existing looms while emphasising sharp cotton purchasing.[42] To lower costs, mill managers used many and varied technological modifications and reorganised workloads. For example, they introduced warp-stop motions at $25 per loom and longer bobbins

39. *Fall River Daily Herald*, 9 Jan. 1911.

40. *Massachusetts Commission to Investigate the Inspection of Factories, Workshops, Mercantile Establishments and other Buildings, Hearing 1910* (hereafter *Hearing*), 246–47.

41. Including the Fall River Iron Works, one of the city's largest firms. I. Cohen, *American Management and British Labor: A Comparative Study of the Cotton Spinning Industry*, New York, 1990, 38–54; *Cotton Yarn Association*, 1929, 11, as cited in M.B. Rose, *Firms, Networks and Business Values: The British and American Cotton Industries since 1750*, Cambridge, 2000, 182–83.

42. W. Mass, 'Mechanical and Organizational Innovation: The Drapers and the Automatic Loom', *Business History Review*, 63:3, 1989, 914–17; Scranton, *Figured Tapestry*, 175; T.R. Smith, *The Cotton Textile Industry of Fall River: A Study of Industrial Location*, New York, 1944, 84, 102, 106, 113–16; Cumbler, *Working Class Community*, 201–08; Mass, 'Technological Change', 141–90; Mass, *et al.* 'Kiss of Death', 23–5.

and shuttles that needed replacing less frequently. They also increased the number of looms per weaver.[43] Furthermore, the cotton manufacturers had established close links with local machine builders. Scranton claims that this was the primary reason behind industrialists' refusal to buy Northrop automatic looms.[44] Yet these broad generalities ignore the many variations in firm production strategies found in the city. There was greater firm specificity in Fall River than historians have previously acknowledged, which created greater inequalities between the workplace environments of the city's firms than in Lowell. While only a handful of firms are discussed here, the multiple and varied business strategies concerning technological investment are sufficient to demonstrate both the importance of firm-specific investment strategies within the town, as well as the corresponding environmental inequalities.

The Bourne Mill most neatly fits existing explanations of Fall River's long-term investment strategy. In 1900, it was the first firm to invest in Northrops, purchasing 2,000 in one order, freeing managers from commitments to local technology suppliers. The new Northrops were distributed throughout the weaving rooms and the remaining plain looms were modified. This investment was followed by an increase in the number of looms per weaver to twenty. In response, the weavers walked out, which allowed the Bourne to hire new hands.[45] Thus, the Bourne managers' technological choices secured numerous benefits, including managerial control over the workforce, increased production and cutting costs, but at the expense of aggressively intensifying workloads. Hence, while investment in Northrops improved one aspect of weavers' occupational health through the elimination of the suction shuttle, other conditions worsened. Furthermore, managerial investment did not accompany broader improvements to the working environment.

The Richard Borden Manufacturing Company was another investor in Northrop looms, but the firm proceeded at a more gradual pace than the Bourne. The Richard Borden manufactured print cloth, odd goods and fancies. In 1905, mill managers purchased a hundred Draper automatic looms as replacements for non-automatic looms.[46] They bought more Northrops in 1907 and 1908 to complete the refit of their No.1 mill.[47] By 1909, 49 per cent of the firm's looms were automatic, by 1914, 59 per cent and by 1920, 75 per cent.[48] The measured

43. Mass, 'Technological Change', 167–68; Smith, *Cotton Textile Industry*, 63; Mass, 'Mechanical and Organizational Innovation', 914; Scranton, *Figured Tapestry*, 175.

44. Scranton, *Figured Tapestry*, 175–76, n.101.

45. Scranton, *Figured Tapestry*, 175.

46. It was 1909–10 before the Northrop was considered consistently reliable. Mass, 'Mechanical and Organizational Innovation', 876–929; Feller, 'Draper Loom'.

47. Fall River Historical Society (FRHS), Richard Borden Manufacturing Company, Director's Meetings Record Book, 6 June 1871-14 November 1916; 7 May 1908.

48. It is probable that many of these Northrops were broad looms. Mass, 'Technological Change', 186–87. It was 1916 before the firm replaced their Kilburn and Lincoln looms with machines

adoption of automatic looms does not necessarily imply that the firm flouted the shuttle legislation. Mill managers may have installed self-threading bobbin carriages or provided hand or self-threaders for use with older looms, all of which would have negated the use of the suction shuttle. Instead, the long-term, progressive investment strategy in new weaving looms merely coincided at times with legislation and physicians' reform efforts. It was not directly influenced by it.

In other areas of technological investment, the Richard Borden Mill managers followed legislative requirements, gradually updating ventilation and humidifying equipment. In 1910, the company treasurer ordered a Cramer System of Air Conditioning and Humidifying. Cramer, based in nearby Providence, Rhode Island, advertised more on the basis of its products improving workers' health than on their production benefits.[49] Managers at the Richard Borden agreed that the machine would be installed in the firm's No.2 mill in accordance with legislative guidance, but only 'if nothing objectionable to it [was] found in the bill'.[50] Objections were few and the benefits considerable because in 1912 the firm installed another Cramer humidifying system.[51] While it is impossible to gauge the influence of Cramer's publicity, the Richard Borden's technological investment was multi-faceted. This included long-term business investment and modernisation using regional, rather than local, technology suppliers, but incorporating current scientific understanding and legal compliance. Furthermore, the cautious pace of investment kept costs manageable, while gradually improving certain aspects of the workplace environment.

In contrast, the technological investment strategy at the nearby Durfee Mills was more diverse and included a preference for local suppliers. The Durfee Mills were one of the largest cloth producers in Fall River. Between 1913 and 1919, all weaving equipment purchases met the 1911 legislative guidelines; however, not all weaving machinery was replaced.[52] Purchases of hand-threading shuttles came from several suppliers, including four Fall River firms and one in Boston.[53] Competition between these suppliers was fierce, enabling the Durfee Mills to dictate the terms

---

that definitely met the 1911 requirements. FRHS, Richard Borden Mfg Co., Director's Meetings Record Book, 5 May 1916.

49. S. Cramer, *Useful Information for Cotton Manufacturers. Compiled and Issued by Stuart W. Cramer, Mill Architect and Engineer. Contractor for Cotton Mill Machinery and Equipment*, 2nd ed., vol. IV, Providence, 1909, chapter II, 'Atmospheric Conditions Affecting Health' and chapter III 'Atmospheric Conditions Affecting Manufacturing.'

50. FRHS, Richard Borden Manufacturing Co., Director's Meetings Record Book, 8 Feb 1910.

51. In their No. 1 Mill. FRHS: Richard Borden Mfg Co., Director's Meetings Record Book, 26 Jun 1912.

52. FRHS, Durfee Mills, Summary of Operations. Company managers consistently calculated their purchases based on an average number of looms per weaver at seven between 1909 and 1919.

53. Fall River firms included The Webster Loom Harness Company, the Fall River Bobbin and Shuttle Co., the Waterman Supply Co., and Joseph Lacroix. The American Shuttle Company was based in Boston.

and conditions of purchase. The company treasurer, Sidney Borden, regularly told shuttle suppliers both the purchase price and the delivery date. Furthermore, Borden only purchased small batches of shuttles, in line with current production needs and to prevent surplus stock in case of strikes.[54] Yet, this strategy also minimised dependence on any one shuttle supplier. In contrast to local suppliers for small devices, large machines were sourced from further afield. These included Draper automatic looms, various shuttles, and Northrop batteries for existing Whitin Looms, new Whitin Looms and Mason Looms, among others.[55] This diversified technological investment strategy enabled the firm to keep costs manageable, retain production flexibility without refitting entire weave rooms, maintain close control over costs and to meet legislative requirements. Moreover, as new technologies replaced the old, the workplace environment gradually improved, despite a good working environment not being integral to company strategy.

Economic practicalities also dominated the Stafford Mills' investment strategy. In 1914, they installed a new humidifier from the American Moistening Company (AMC), based in Providence. The Stafford demanded design flexibility so that the machine could be added to or adapted without changing the original installation. The AMC complied, but in return, they required a consistent, available, supply of pure water – to which the Stafford agreed.[56] The late purchase of new humidifying equipment reveals the firm's disregard for environmental legislation. This is further evident from the timing of purchases of weaving equipment. Through 1914, a core supplier of looms was the Fall River-based Kilburn & Lincoln firm. However, as part of their cost-cutting measures, the firm purchased a handful of second-hand Whitin looms from the Hamilton Woollen Company in Amesbury, Massachusetts, for $20 each.[57] Unfortunately, little is known about the firm's shuttle changing technology and both looms could take shuttle changing mechanisms. Yet the dominance of cost to the investment strategy is clear. New Whitin looms were both high quality and high priced, but Whitins could weave similar types of cloth to the Northrop.[58] Moreover, the purchase of used Whitin looms did not

54. FRHS, Durfee Mills, Letters, vol 2, 2/2/1914 – 7/9/1914.

55. FRHS, Durfee Mills, Letters, vol.1, 9/3/1913 – 2/2/1914; vol 2: 2/2/1914 – 7/9/1914; vol. 4, 6/8/1915 – 11/17/1915; vol. 8, 8/21/1918 – 6/27/1919; some of the shuttles for the Draper looms were from Joseph Lacroix. The Whitin Loom, made in Whitinsville, could be converted to take the Draper filling changer and fitted with a warp stop motion cost $50, approximately a third the price of a new Northrop loom. Whitinsville was approximately 45 miles from Fall River and Hopedale about 55 miles away, while Taunton, where Mason Looms were manufactured, was only 18 miles from Fall River.

56. FRHS, Stafford Mills, Directors Records, By-laws and Stockholders Annual Meetings, 30 Dec 1870–25 Oct 1932.

57. FRHS, Stafford Mills, Directors Records, By-laws and Stockholders Annual Meetings, 30 Dec 1870–25 Oct 1932.

58. Mass, 'Technological Change', 218.

form part of a long-term investment strategy. In 1915 the directors recommended no further technological investment in the weaving rooms.[59] As long as current production methods remained profitable, investment in expensive new machines was not considered worthwhile. Hence, at the Stafford, cost and business loyalties dominated company decision-making, with little regard for legislation. Any improvements to the workplace environment were purely incidental.

Similarly, at the Fall River Iron Works, production needs drove investment choices rather than legislation.[60] While some equipment purchases complied with the 1911 shuttle legislation, this was coincidental. By this time, mill managers had exhausted all possible cost-savings from speeding-up their current complement of non-automatic looms. Hence, 325 Northrop looms were purchased that year. Over the next two years, 4,000 more were installed.[61] State legislation was merely a timely coincidence. Technological change and investment were already underway at the firm and machines were chosen merely to meet current production goals. Any corresponding improvement to the workplace environment or legislative requirements was purely coincidental.

Elsewhere in the city, legislation had a greater influence on company technological investment. At the Granite Mills, improvements to the workplace environment both preceded and followed legislation. As early as 1881, mill managers had begun improving factory ventilation and humidity.[62] Moreover, company directors openly admitted that the reason they purchased 440 Northrop looms in 1912 was to comply with the suction shuttle law.[63] Hence, while the Granite Mill managers initially defined their workplace environment as the atmosphere, they later accepted the broader legislative definition. Thus, here, technology did not determine its own history, as McEvoy has suggested.[64] Rather, social and medical arguments and legislative requirements determined its path at different firms.

Clearly, in Fall River, there was a multi-faceted, firm-specific, approach to technological choice and the timing of such investments. External pressures had only a limited impact on company decisions. Indeed, the varied technological choices demonstrated a specificity of influence (reformers, legislation and gradual technological investment outwith external pressures). This is a much more complex

59. FRHS, Stafford Mills, Directors Records, By-laws and Stockholders Annual Meetings, 30 Dec 1870–25 Oct 1932.

60. The Fall River Iron Works was established in 1821 by Richard Borden and Bradford Durfee. It produced a range of iron goods, as well as textiles. Eventually, the firm specialised in manufacturing textile machinery.

61. *Cotton Chats*, various issues, see also Mass, 'Technological Change', 189–90.

62. Smith, *Cotton Textile Industry*, 58.

63. FRHS, Granite Mills, Record Book of the Directors' Meetings, 23 Oct 1911–31 Dec 1928, 23 Dec 1912.

64. McEvoy, 'Working Environments', S146.

model of technological investment in the city than has been previously suggested. It highlights how the firm-specific investment choices helped create environmental inequalities both within and between the city's firms. Lastly, these investment strategies revise the rather simplistic argument that Fall River unions alone deterred the adoption of Northrop looms.[65]

Liberal Employers and the Working Environment

Thus far, analysis has focused on towns where cotton manufacturing dominated the local economy. To fully understand the multiple influences on the development of workplace environmental inequalities, it is necessary to consider towns with more diverse economic structures. Holyoke, Massachusetts, is such a town. While now more famous as the birthplace of volleyball, in the late nineteenth century Holyoke was dominated by two industries, paper and textiles. Yet, neither interest controlled the town council. Thus, while big business influenced local politics, no single interest controlled the city council. This fact allowed the council the autonomy to put general pressure on all city businesses to improve residents' wellbeing. Moreover, because they did not control either the politicians or the local economy, Holyoke businesses had to pay greater lip service to local government pressures than their counterparts elsewhere.

The Lyman Mills were the largest textile firm in the town. Established in 1854, they were an integrated cotton manufacturing company that produced a wide range of goods from coarse sheetings and heavy yarns to fine lawns and fancy goods. This extensive product range meant that the firm sought flexible and diversified machinery. It also meant that technological change could progress gradually, with a mixed pace of modernisation and investment in the separate cloth sectors. Consequently, the quality of the work environment varied throughout the mill and changed as the firm expanded.

The Lyman Mills expanded rapidly along with the town. By 1870, the town's population had doubled that of 1860, with over 10,000 people. By 1880, this number had more than doubled again, with nearly 22,000 people living in the city.[66] Correspondingly, the Lyman Mills expanded their facilities in 1872, 1881

65. Despite historians' disagreements about the manner in which the union response occurred and its relative importance. Smith, *Cotton Textile Industry*, 115–16; see also Mass, 'Technological Change', 141; Mass, 'Mechanical and Organization Innovation', 914–16; *Fall River Daily Globe,* 21 Nov 1914, and *Fall River Daily Globe,* 14 Jan 1915, both as cited in Mass, *et al.* 'Kiss of Death', 23. The firm-specific, rather than collective, city analysis of Fall River environmental inequalities allows for those manufacturers who flouted the suction shuttle law and were then prosecuted in 1915, as well as those who complied with the legislation. J. Cumbler, *Reasonable Use: The People, the Environment and the State, New England, 1790–1860*, New York, 2001, 46.

66. Cumbler, *Reasonable Use*, 46.

and again in 1891. By 1900, the company's buildings stretched for an eighth of a mile and the firm employed over 1,200 people.

The Holyoke labour force was similar to that of Fall River with large-scale immigration and associated problems, including the lack of a sense of community, with language and ethnic divisions limiting collective action.[67] However, ethnic conflicts in Holyoke were never as bitter as in the south-eastern corner of the State; nor were labour relations as hostile. Instead, the Lyman Mills was reputed for good labour relations. The firm responded to strikes elsewhere in the state by giving its own workers a small wage increase. Moreover, in 1907, the firm appointed a local man, James Burke, as mill agent. Burke had worked his way up through the mill ranks, which probably contributed to the relatively good labour relations within the firm and between the firm and the city.

Early in the city's history, politicians tried to improve the working environment and change related business practices. The city's rapid growth meant that local government had to deal with the associated public health concerns of housing, sanitation, fresh water supply and clean air. By the 1870s, well before State factory legislation, Mayor Judge Pearsons regularly encouraged employers to improve the ventilation of their premises. He argued that despite the fact that 'every *tolerable* system of ventilation is expensive... pure air is a *necessity to health*'.[68] The Board of Health agreed and continuously worked towards achieving this goal throughout the city.[69] While the direct influence of these arguments on mill managers is unclear, as early as 1880, the Lyman Mills claimed that '[T]he mills are not run on selfish principles only, the welfare of those employed is carefully considered.'[70] The workplace environment was incorporated into the firm's accumulation matrix.

The firm also looked elsewhere for ideas about best practice. In 1908, the company treasurer, Theop Parsons, corresponded with F.P. Sheldon of Sheldon's mill, hydraulic, steam and electrical engineers in Providence. Sheldon wrote that nearby New Bedford employers had to build new weaving sheds in order to obtain and retain a full complement of weavers because they had refused to work where the lighting and air were poor.[71] Sheldon also complained that around Providence,

---

67. Initially, the Irish flocked to the city and the mills. By 1880 they were surpassed by the French-Canadians. While these remained the two largest immigrant groups in 1900, significant numbers of Poles, Russians, Germans and Austrians had also moved to the city. By 1900, 41.4 per cent of Holyoke's population was foreign-born. Only four other cities exceeded this proportion of immigrants. C. M. Green, *Holyoke, Massachusetts: A Case Study of the Industrial Revolution in America,* New Haven, 1939, 367–68.

68. Emphasis in original. Municipal Registers for the City of Holyoke, *Mayor's Address, 1875*, 23–24.

69. See the Municipal Registers for the City of Holyoke from 1875 onwards, particularly Mayoral addresses and Board of Health Reports.

70. HBS, Lyman Mills Papers, PA-1: Treasurer to Agent, early 1880 (exact date unknown).

71. HBS, Lyman Mills Papers, PO-8-6, Letter from F. P. Sheldon & Co to Mr Theop Parsons, Treasurer of Lyman Mills, 12 May 1908.

'mill help, and especially weavers, are getting more particular and fussy every year'.[72] While it is impossible to know the direct impetus of these events on the Lyman Mills managers, the firm regularly upgraded its humidifying and ventilating system during the nineteenth century and the first two decades of the twentieth century.[73] Once again, these continuous machinery upgrades indicate that here, as elsewhere, factory atmospheric legislation came after improvements to the work environment were already underway.

While concerned about the collective wellbeing of the town's citizens, the Holyoke city government was less interested in the individual. Nevertheless, the Lyman Mills regularly reviewed their existing technology as part of their goal continuously to improve cost efficiencies. Comparatively speaking, Lyman Mills was one of the most cost efficient textile firms in the state. It had developed sophisticated manufacturing cost accounting procedures that were at least fifty years ahead of their time.[74] This constant monitoring of all elements of the production process allowed the firm to target technological investment to meet the current goals of specific product markets and to 'encourage workers to achieve company productivity goals'.[75] This complex accumulation matrix included the monitoring of workers' productivity rates, which determined that productivity increased when operatives were paid piece rate instead of a flat rate.

Another core strategy included targeting several product markets to be better able to weather economic downturns in any one cloth sector. This involved numerous machine suppliers who were located throughout the north-eastern United States and even included a Holyoke Loom developed in Portsmouth, New Hampshire.[76] Thus, technological investment was directly related to both current and future production goals and not business relationships with local machine builders, local government pressures or legislation. This was also evident in the diverse weaving technologies adopted, which included Draper, Whitin and Crompton and Knowles looms, automatic, self and hand-threading devices. Combined, these devices enabled the

72. HBS, Lyman Mills Collection, PO-8-6, F.P. Sheldon to T. Parsons, 12 May 1908.

73. HBS. Lyman Mills Papers, MSS 442: PA-9, Letter book, Agent to Miscellany, Oct. 1881-Aug. 1884; PU-1-1 Factory Inspectors Reports, 1913–20.

74. H.T. Johnson and R.S. Kaplan, *Relevance Lost: The Rise and Fall of Management Accounting*, Boston, 1998, 31.

75. Johnson and Kaplan, *Relevance Lost,* 28–31, quote 31.

76. Including Southbridge, Boston, Hopedale, and Whitinsville, Massachusetts, and Utica, New York. Other examples include shuttle manufacturers based in Southbridge, MA, Boston and Utica, New York. HBS, Lyman Mills Collection: PA-9 Letter Book, Agent to Miscellany, Oct 1881–Aug 1884, 28 Oct 1882, T. Parsons, agent, to Messrs J.N. and N.A. Williams, Utica NY; 28 Mar 1882, Parsons to Litchfield Shuttle Co., Southbridge, Ma.; 20 Apr 1882, Parsons to Sturtevant, Boston.

production of coarse, plain and fancy goods.[77] These machinery contracts were all agreed before the 1911 legislation, with later orders meeting the new requirements.

The Lyman Mills micro-managed their cost accounting procedures with the purchase of various loom technologies. The aim was to keep input costs low. This was essential due to regular labour shortages, particularly of weavers.[78] Nevertheless, this strategy also improved the work environment. Furthermore, this happened outside legislation or external pressures and was a by-product of the goal of increasing profits. This business strategy, combined with the firm's liberal policy towards its workforce in terms of labour relations and welfare, enabled the firm to retain its autonomy without external interference, while also improving the working environment.

After the First World War, micro-management and technological investment continued to dominate company strategy as machines replaced workers. While for the remaining operatives certain working conditions may have improved, the new technologies also brought job losses and new health risks associated with the new machines. Consequently, while improvements to mill conditions may indicate employers' genuine concern for workers' welfare, the investment decisions were directly attached to increasing profits. Indeed, these strategies helped sustain the firm's profitability into the 1920s. Yet in 1927, while still profitable, the directors entered liquidation so that investors would realise a profit from the company's substantial assets.[79]

Overall, of the firms in the three towns, the Lyman Mills had the fewest external pressures on their business practices. Yet, of all the firms, the Lyman Mills adopted the most liberal policy towards their workforce, in terms of labour relations and welfare. While production and profits directed company decision-making, managers improved the working environment and their efforts preceded legislative requirements. Not only did this strategy make the firm a healthier place in which to work, it had the additional benefit of deterring local government intervention in the workplace and secured community support. Thus, of the three cities, Holyoke was probably the healthiest place for cotton operatives to live and work, even if employers were not motivated solely by altruistic concerns. Indeed, workers' wellbeing was not a core motivation behind technological investment at any firm studied. Yet, improvements still occurred when manufacturers, local politicians, physicians and legislators attached environmental importance to business. The ensuing varied

77. HBS, Lyman Mills Collection: MFC-1-2, Machinery Contracts, 1898–1907; MFC-1-3, Machinery Contracts, 1898–1907; MFC-1-4, Machinery Contracts, 1898-1900; PD-7, Letters Agent to/from Treasurer, Agents' Copies, Nov 1911–Dec 1912, PO-10-1, Miscellany Correspondence to Treasurer, Draper Co. to Lyman Mills, Oct 1916; MFC-1-3, Machinery Contracts, 1898–1907.

78. HBS, Lyman Mills Collection: PD-7, Agent to/from Treasurer, Nov. 1911–Dec. 1912, see for example, Jul-Sep 1912; PB-11, Treasurer to Agent, Oct. 1881–Dec. 1882, Oct.

79. Green, *Holyoke, Massachusetts*, 238.

and complex associations created environmental and social inequalities within and between the three towns.

Conclusions

This chapter has emphasised the complexity of the impetuses behind technological investment choices and the resulting occupational and environmental health inequalities between factories within and between cities. It has also demonstrated the varied importance of local government and community pressures, business connections with local machine-makers and individual company priorities. It has shown how multiple external pressures influenced technological investments but that investment choices were rarely a direct response to them. Moreover, the gradual pace of technological change at many firms created complex and varied work environments. Thus, the interaction between workers' bodies, machines and the work environment was inconsistent and continually changing in the three cities. The community influence on technological choice was strongest in Lowell. Yet this only extended to improving the quality of the atmosphere and hence the collective wellbeing of workers, not the health of individuals. Local government pressure on industry was greatest in Holyoke through their encouragement to employers to improve ventilation and the air quality in their factories. The greatest workplace environmental inequalities were found in Fall River. The city mill men's dominance of local politics drove reformers to seek state assistance to improve the mill environment. While successful, the resulting legislation was limited in both strength and enforcement, allowing employers' individual preference to prevail. Despite this, state legislation had the most definitive impact at individual firm level in Fall River, at the Granite Mills.

This chapter has also illustrated how contemporaries clearly recognised the relationship between production costs, bodily damage and workplace inequalities. Firms invested in new technologies and closed on particularly hot days and workers refused to work due to poor ventilation or oppressive heat and humidity. They also moved jobs or switched employers in search of better conditions.[80] In this way, workers were able to preserve some autonomy and agency. They acted for personal occupational health and safety concerns, as well as wages. While the latter took precedence, the former was never neglected, even if it was not a core campaign issue.

The complex relationships between technological investment, the working and business environments, government (both local and state) and the community make the importance of place key to understanding how business has interacted with the environment over time. The relationship between these factors defined the working environment. Furthermore, the importance of local forces and industry in

80. See for example, *Lowell Courier*, 7 and 8 July 1873; *Vox Populi*, 9 Jul 1873; *Lowell Courier*, July 1912; HBS, Lyman Mills Papers, PO-8-6, F.P. Sheldon to T. Parsons, 12 May 1908.

shaping environmental inequalities clearly suggests that further studies are needed if we are to understand the origins of environmental inequalities and how business has interacted with different environments over time.

Select Bibiography

Bahr Peterson, B. 'Industrial Architecture from the Inside: Textile Mill Design and the Factory Workplace, 1860–1920', in R. Weible, ed. *The Continuing Revolution: A History of Lowell, Massachusetts,* Lowell, 1991, 204–6.

Blewett, M. *Constant Turmoil: The Politics of Industrial Life in Nineteenth-Century New England,* Amherst, 2000.

Cumbler, J. *Working Class Community in Industrial America: Work, Leisure and Struggle in Two Industrial Cities, 1880–1930,* Westport, 1979.

Green, C.M. *Holyoke, Massachusetts: A Case Study of the Industrial Revolution in America,* New Haven, 1939.

Greenlees, J. '"Stop kissing and steaming!": Tuberculosis and the Occupational Health Movement, 1870–1918', *Urban History* 32:2, 2005, 223–46.

Gross, L. *The Course of Industrial Decline: The Boott Mills of Lowell, Massachusetts,* 1835–1955, Baltimore, 1993.

Mass, W. 'Mechanical and Organizational Innovation: The Drapers and the Automatic Loom', *Business History Review*, 63:3, 1989, 876–929.

McEvoy, A.F. 'Working Environments: An Ecological Approach to Industrial Health and Safety', *Technology and Culture*, 36:2, 1995, S145-S173.

Meisner Rosen, C. and C. Sellers, 'The Nature of the Firm: Towards an Eco-cultural History of Business', *Business History Review*, 73, 1999, 577–600.

Richards, R. 'A Study in Community Power: Lowell, 1912', in R. Weible, ed. *The Continuing Revolution: A History of Lowell, Massachusetts,* Lowell, 1991, 265–79.

Rosner D. and G. Markowitz, 'The Early Movement for Occupational Safety and Health, 1900–1917', in J.W. Leavitt and R.L. Numbers, eds. *Sickness and Health in America: Readings in the History of Medicine and Public Health,* 2nd edition, Madison, 1985, 507–21.

Sellers, C. 'Factory as Environment: Industrial Hygiene, Professional Collaboration and the Modern sciences of Pollution', *Environmental History Review*, 18:1, 1994, 55–83.

Select Bibliography

Agyeman, J. 'Environmental Justice: From the Margins to the Mainstream?' Town and Country Planning Association, 'Tomorrow' Series, 2000.

Agyeman, J. and B. Evans, '"Just Sustainability": The Emerging Discourse of Environmental Justice in Britain', *The Geographical Journal,* 170:2, 2004, 155–164.

Balcers, W. and C. Deligne, 'Les sociétés de pêche à la ligne, "conscience" de la pollution des rivières en Belgique (1880–1940)?' in I. Parmentier, dir. *La recherche en histoire de l'environnement : Belgique – Luxembourg – Congo – Rwanda – Burundi. Actes des premières Rencontres de l'Histoire de l'Environnement en Belgique, Namur, décembre 2008*, Namur, 2010, 175–186.

Barles S. *L'invention des déchets urbains: France 1790–1970*, Champ Vallon, 2005.

Bärthel, H. *Wasser für Berlin*, Berlin, 1997.

Bayerl, G. 'Konsum, Komfort und Netzwerke. Die Versorgung mit Wasser', in R. Reith and T. Meyer, eds. *Luxus und Konsum – eine historische Annäherung*, Münster, 2003, 130–58.

Been, V. 'Analyzing Evidence of Environmental Justice', *Journal of Land Use and Environmental Law,* 11:1, 1995, 1–36.

Been, V. 'What's Fairness Got to Do With It? Environmental Justice and the Siting of Locally Undesirable Land Uses', *Cornell Law Review*, 78, 1992–93, 1001–85.

Bernhardt, C. 'Die Vertreibung des Wassers aus der Stadt und aus der Planung. Zur Hygienisierung der öffentlichen Räume im 19. Jahrhundert am Beispiel Berlins', in C. Bernhardt ed., *Zur Geschichte der Planung des öffentlichen Raums*, Dortmund, 2005, 71–84.

Bernhardt, C. 'Towards the Socialist Sanitary City: Urban Water Problems in East German New Towns 1945–1970', in D. Schott, B. Luckin, and G. Massard-Guilbaud, eds. *Resources of the City. Contributions to an Environmental History of Modern Europe.* Aldershot, 2005, 185–202.

Bernhardt, C. 'Umweltprobleme in der neueren europäischen Stadtgeschichte', in C. Bernhardt ed. *Environmental Problems in European Cities in the 19th and 20th Century/ Umweltprobleme in europäischen Städten des 19. und 20. Jahrhunderts,* 2nd edition, Münster, 2004, 5–23.

Bernhardt, C. and G. Massard-Guilbaud, eds. *Le Démon moderne, la pollution dans les sociétés urbaines et industrielles d'Europe/The Modern Demon, Pollution in Urban and Industrial European Societies*, Clermont-Ferrand, 2002.

Blanchon, D., S. Moreau, and Y. Veyret, 'Comprendre et construire la justice environnementale', *Annales de Géographie*, 665–66, 2009, 35–60.

Blewett, M. *Constant Turmoil: The Politics of Industrial Life in Nineteenth-Century New England,* Amherst, 2000.

Boardman, B., S. Bullock and D. McLaren, *Equity and the Environment: Guidelines for Green and Socially Just Government*, Catalyst pamphlet 5, 1999.

Boone, C.G. 'Environmental Justice as Process and New Avenues for Research', *Environmental Justice*, 1, 2008, 149–54.

Boone, C.G. 'An Assessment and Explanation of Environmental Inequity in Baltimore', *Urban Geography*, 23, 2002, 581–95.

Brechin, S.R., P. R. Wilshusen, C. L. Fortwangler, and P. C. West, eds. *Contested Nature. Promoting International Biodiversity with Social Justice in the Twenty-first Century*, Albany, 2003.

Bret, B. 'Interpréter les inégalités socio-spatiales à la lumière de la Théorie de la justice de John Rawls', in *Annales de Géographie*, 2009, no. 665–66, 16–34.

Bryant, B. *Environmental Justice: Issues, Policies and Solutions*, Covelo, 1995.

Bullard, R.D. *Dumping in Dixie: Race, Class and Environmental Quality*, Boulder, 1990.

Bullard, R.D. ed. *Confronting Environmental Racism: Voices from the Grassroots*, Boston, 1993.

Büschenfeld, J. *Flüsse und Kloaken. Umweltfragen im Zeitalter der Industrialisierung (1870–1918)*, Stuttgart, 1997.

Byrne, J., L. Glover, and C. Martinez, eds. *Environmental Justice. Discourses in International Political Economy, Energy and Environmental Policy*, vol. 8, New Brunswick/London, 2002.

Campanella, R. *Geographies of New Orleans: Urban Fabrics before the Storm*, Lafayette, 2006.

Charles, L. 'Une appréhension interculturelle des inégalités environnementales', in 'Des inégalités écologiques parmi les hommes', *Écologie et Politique. Sciences, Cultures et Sociétés.* 35, 2007, 79–90.

Charles, L. 'Du milieu à l'environnement', *L'Environnement, question sociale, Dix ans de recherches pour le ministère de l'environnement*, Paris, 2001.

Charles, L. 'Entre milieu et environment', in M. Boyer, G. Herzlich, and B. Maresca, *L'environnement, question sociale*, Paris, 2001.

Charles, L. and B. Kalaora, 'Prégnance et limites d'une approche esthétique de l'environnement', in M. Bédard, dir. *Le paysage, un projet politique*, Québec, 2009.

Charles, L. and B. Kalaora, 'Pensée, sensibilité et action dans la société française autour de la question de la nature', *Annales de Géographie*, 5:2008, 663, 3–25.

Claude, V. 'Une coopération politique dans une mosaïque urbaine, le cas du service de l´eau en banlieue parisienne (1880–1923)', *Genèses*, 4, 2006, 92–111.

Clark, J. '"The Incineration of Refuse is Beautiful": Torquay and the Introduction of Municipal Refuse Destructors', *Urban History*, 34:2, 2007, 255–77.

Colten, C.E. *Perilous Place, Powerful Storms: Hurricane Protection in Coastal Louisiana*, Jackson, 2009.

Colten, C.E. *An Unnatural Metropolis: Wresting New Orleans from Nature*, Baton Rouge, 2005.

Colten, C.E. 'Basin Street Blues: Drainage and Environmental Equity in New Orleans, 1890–1930', *Journal of Historical Geography* 28:2, 2002, 237–57.

Commission for Racial Justice–United Church of Christ, *Toxic Wastes and Race in the United States: A National Report on the Racial and Socio-economic Characteristics of Communities with Hazardous Waste Sites*. New York, 1987

Select Bibliography

Cornelisse, C. 'The Economy of Peat and its Environmental Consequences in Holland during the Late Middle Ages', *Jaarboek voor Ecologische Geschiedenis 2005/2006: Water Management, Communities, and Environment. The Low Countries in Comparative Perspective, c.1000–c.1800*, Gent, 2006, 95–121.

Crawford, C. 'Analyzing Evidence of Environmental Justice: A Suggestion for Professor Been', *Journal of Land Use and Environmental Law*, 12:1, 1996, 103–20.

Cumbler, J. *Working Class Community in Industrial America: Work, Leisure and Struggle in Two Industrial Cities, 1880–1930*, Westport, 1979.

Cutter, S. 'Race, Class and Justice', *Progress in Human Geography*, 19:1, 1995, 111–22.

David, P.A. 'Clio and the Economics of QWERTY', *American Economic Review*, 75, 1985, 332–37.

Davis, M. *City of Quartz: Excavating the Future in Los Angeles*, New York, 1992.

Dobson, A. *Justice and the Environment, Conceptions of Environmental Sustainability*, Oxford, 2003 (first published New York, 1998).

Dowie, M. *Conservation Refugees. A Hundred-Year Conflict between Global Conservation and Native People*, Cambridge, Ma., 2009.

Duchêne, F. and C. Morel-Journel, *De la culture du risque. Paroles riveraines à propos de deux cours d'eau péri-urbains*. La Tour d'Aigues, 2004.

Dunion, K. *Troublemakers. The Struggle for Environmental Justice in Scotland*, Edinburgh, 2003.

Duckham, B.F. *History of the Scottish Coal Industry: 1700–1815*, vol. 1, Newton Abbot, 1970.

Eddison, J. ed. *Romney Marsh: the Debatable Ground*, Oxford, 1995.

Eddison, J., M. Gardiner and A. Long, eds. *Romney Marsh: Environmental Change and Human Occupation in a Coastal Lowland*, Oxford, 1998.

Eiden, C. *Versorgungswirtschaft als regionale Organisation. Die Wasserversorgung Berlins und des Ruhrgebietes zwischen 1850 und 1930*, Essen, 2006.

Elias, N. and J. L Scotson, eds. *The Established and the Outsiders: A Sociological Enquiry into Community Problems*, London, 1994.

England, M. 'When "Good Neighbors" Go Bad: Territorial Geographies of Neighborhood Associations', *Environment and Planning A*, 40, 2008.

Escobedo, F. and D. Nowak, 'Spatial Heterogeneity and Air Pollution Removal by an Urban Forest', *Landscape and Urban Planning*, 90, 2009, 109.

Etienne, R. 'L'industrie chimique belge (1830–1930)', *Bulletin de la Société belge des Ingénieurs et des Industriels*, 11, 1931, 5–28.

Fairclough, A. *Race and Democracy: The Civil Rights Struggle in Louisiana, 1915–1972*, Athens, GA, 1995.

Flanagan, M.A. 'Environmental Justice in the City: A Theme for Urban Environmental History', *Environmental History* 5:2, 2000, 159–64.

Frank, S. and M. Gandy, eds. *Hydropolis. Wasser und die Stadt der Moderne*, Frankfurt, 2006.

Fraser, E.D.G. and W.A. Kenney, 'Cultural Background and Landscape History as Factors Affecting Perceptions of the Urban Forest', *Journal of Arboriculture*, 26:2, 2006, 106–13.

Frioux S. *Les réseaux de la modernité. Amélioration de l'environnement et diffusion de l'innovation dans la France urbaine (fin XIXe-années 1950)*, PhD dissertation, Université Lyon 2, 2009.

Gadgil, M. and R. Guha, *Ecology and Equity. The Use and Abuse of Nature in Contemporary India*, London, 1995.

George, P. *L'Environnement*, Paris, 1971.

Garb, M. 'Drawing the "Color Line": Race and Real Estate in Early Twentieth-Century Chicago', *Journal of Urban History*, 32, 2006, 773–87.

Giraut, F., S. Guyot, and M. Houssay-Holzschuch, 'La nature, les territoires et le politique en Afrique du Sud', *Annales, Sciences Sociales,* 2005:4, 697–717.

Glave, D. and Mark Stoll, eds. *'To Love the Wind and Rain': African Americans and Environmental History*, Pittsburgh, 2006.

Green, C.M. *Holyoke, Massachusetts: A Case Study of the Industrial Revolution in America*, New Haven, 1939.

Greenlees, J. '"Stop kissing and steaming!" Tuberculosis and the Occupational Health Movement, 1870–1918', *Urban History* 32:2, 2005, 223–46.

Guha, R. and J. Martinez-Alier, *Varieties of Environmentalism. Essays North and South*, London, 1997.

Guillerme A., G. Jigaudon, A-C. Lefort, *Dangereux, insalubres et incommodes. Paysages industriels en banlieue parisienne, XIXe–XXe siècle*, Champ Vallon, 2004.

Hallström, J. 'Systemteori och teknik. En introduktion till stora tekniska system', in P. Gyberg and J. Hallström, eds. *Världens gång – teknikens utveckling. Om samspelet mellan teknik, människa och samhälle*, Lund, 2009.

Hallström, J. *Constructing a Pipe-Bound City: A History of Water Supply, Sewerage, and Excreta Removal in Norrköping and Linköping, Sweden, 1860–1910*, Linköping, 2002.

Hamlin, C. *Public Health and Social Justice in the Age of Chadwick: Britain, 1800–1854*, Cambridge, 1998.

Hampson F. O. and J. Reppy, eds. *Earthly Good. Environmental Change and Social Justice*, Ithaca, 1996.

Hardy, A.I. *Ärzte, Ingenieure und städtische Gesundheit. Medizinische Theorien in der Hygienebewegung des 19. Jahrhunderts*, Frankfurt, 2005.

Harvey, D. *Justice, Nature, and the Geography of Difference*, Oxford, 1996.

Harvey, D. 'The Geography of Capitalist Accumulation: A Reconstruction of the Marxian Theory', in R. Peet, ed. *Radical Geography*, Chicago, 1977, 263–92.

Hatcher, J. *The History of the British Coal Industry*, vol. 1, *Before 1700: Towards the Age of Coal*, Oxford, 1993.

Hayden, D. *Redesigning the American Dream: The Future of Housing, Work, and Family Life*, New York, 1984.

Heynen, N. 'The Scalar Production of Injustice within the Urban Forest', *Antipode: A Journal of Radical Geography*, 35:5, 2003, 980–98.

Heynen, N., M. Kaika and E. Swyngedouw, 'Urban Political Ecology: Politicising the Production of Urban Natures.' in N. Heynen, M. Kaika and E. Swyngedouw, eds. *In the Nature of Cities: Urban Political Ecology and the Politics of Urban Metabolism*, Abingdon, 2006.

Hoffman, P.T., D.S Jacks, P.A. Levin and P.S. Lindert, 'Real Inequality in Europe since 1500', *Journal of Economic History*, 62, 2, 2002, 322–55.

Hughes, T.P. 'The Evolution of Large Technological Systems', in W.E. Bijker, T.P. Hughes and T.J. Pinch, eds. *The Social Construction of Technological Systems: New Directions in the Sociology and History of Technology*, London, 1987.

Humphrey, M. *Preservation versus People? Nature, Humanity and Political Philospby*, Oxford, 2002.

Hurley, A. *Environmental Inequalities. Class, Race and Industrial Pollution in Gary, Indiana, 1945–1980*, Chapel Hill, 1995.

Jackson, K. *Crabgrass Frontier: The Suburbanization of the United States*, New York, 1985.

Johnston, B. R. *Who Pays the Price? The Sociocultural Context of Environmental Crisis.* Washington DC, 1994.

Jones, P.M. *Industrial Enlightenment: Science, Technology and Culture in Birmingham and the West Midlands 1760–1820*, Manchester, 2010.

Kinsey, D. 'Seeding the Waters as the Earth: The Epicenter and Peripheries of a Western Aquacultural Revolution', *Environmental History*, 11:3, 2006, 527–66.

Knox, W.W. *Industrial Nation: Work, Culture and Society in Scotland 1800–Present*, Edinburgh, 1999.

de Kraker, A. 'Flood events in the Southwestern Netherlands and Coastal Belgium, 1400–1953', *Hydrological Sciences*, 51:5, 2006, 913–29.

Kymlicka, W. *Contemporary Political Philosophy: An Introduction*, Oxford, 1999. (translated into French as *Les théories de la justice. Une introduction*, Paris, 2003).

Laigle, L. and V. Oehler, *Les enjeux sociaux et environnementaux du développement urbain: la question des inégalités écologiques*, Rapport final, Centre Scientifique et Technique du Bâtiment, Paris, 2004.

Larrère, R. 'L'écologie ou le geste de l'exclusion de l'homme', in A. Roger and F. Guéry, eds. *Maîtres et protecteurs de la nature*, Seyssel, Champvallon, 1991, 172–96.

Lavelle, M. 'Environmental Justice', in World Resources Institute, ed. *The 1994 Information Environmental Almanac*, Boston, 1994, 183–92.

Lawrence, H.W. *City Trees: A Historical Geography from the Renaissance through the Nineteenth Century*, Charlottesville, 2006.

Lefebvre, H. *The Production of Space*, Oxford, 1991.

Long, A.J., S. Hipkin and H. Clarke, eds. *Romney Marsh: Coastal and Landscape Change Through the Ages*, Oxford, 2002.

Low, S. *Behind the Gates: Life, Security, and the Pursuit of Happiness in Fortress America*, New York, 2003.

Lucas, K., G. Walker, M. Eames, H. Fay and M. Poustie, *Environment and Social Justice: Rapid Research and Evidence Review*, Final report, SDRN, DEFRA, 2004.

Luckin, B. 'Environmental Justice, History and the City: The United States and Britain, 1970–2000', in D. Schott, B. Luckin and G. Massard-Guilbaud eds. *Resources of the City: Contributions to an Environmental History of Modern Europe*, Aldershot, 2005.

Malange, J-F. and S. Frioux, '"L'eau pure pour tous!" Mobilisations sociales contre la pollution des eaux douces françaises (1908–années 1960)', *Histoire & Sociétés*, 27, 2008, 100–23.

Martinez-Alier, J. *The Environmentalism of the Poor. A Study of Ecological Conflicts and Valuation,* Cheltenham, 2002.

Martinez-Alier, J. 'Conflits écologiques et languages de valorisation', in 'Des inégalités écologiques parmi les hommes', *Écologie et Politique. Sciences, Cultures et Sociétés.* 35, 2007, 91–107.

Massard-Guilbaud, G. and S. Mosley, eds. *Common Ground. Integrating the Social and Environmental in History*, Newcastle, 2011.

Mayne, A. *The Imagined Slum: Newspaper Representation in Three Cities 1870–1914*, Leicester, 1993.

McDonald, D. ed. *Environmental Justice in South Africa,* Cape Town, 2002.

McDougall, H.A. *Black Baltimore: A New Theory of Community*, Philadelphia, 1993.

McEvoy, A.F. 'Working Environments: An Ecological Approach to Industrial Health and Safety', *Technology and Culture*, 36:2, 1995, S145-S173.

McLeod, H., I. H Langford et al. 'The Relationship between Socio-economic Indicators and Air Pollution in England and Wales: Implications for Environmental Justice', *Regional Environmental Change*, 1:2, 2000, 78–85.

McKenzie, E. *Privatopia: Homeowner Associations and the Rise of Residential Private Government*, New Haven, 1994.

Meisner Rosen, C. and C. Sellers, 'The Nature of the Firm: Towards an Eco-cultural History of Business', *Business History Review*, 73, 1999, 577–600.

Melosi, M. 'Environmental Justice, Ecoracism and Environmental History', in D. Glave and Mark Stoll, eds. *'To Love the Wind and Rain': African Americans and Environmental History*, Pittsburgh, 2006, 120–32.

Melosi, M. *Garbage in the Cities, Refuse, Reform and the Environment, 1880–1980*, Pittsburgh, 2004 (1981).

Melosi, M. 'Environmental Justice, Political Agenda Setting, and the Myths of History', in M. Melosi, *Effluent America: Cities, Industry, Energy and the Environment*, Pittsburgh 2001.

Melosi, M. *The Sanitary City: Urban Infrastructure in America from Colonial Times to the Present*, Baltimore, 2000, 119–21.

Melosi, M. 'Environmental Justice, Political Agenda Setting and the Myths of History', *Journal of Policy History*, 12:1, 2000.

Merchant, C. 'Shades of Darkness: Race and Environmental History', *Environmental History*, 8:3, 2003.

Mokyr, J. *The Gifts of Athena: Historical Origins of the Knowledge Economy*, Princeton, 2002.

Mohajeri, S. *100 Jahre Berliner Wasserversorgung und Abwasserentsorgung 1840–1940*, Stuttgart, 2005.

Mohl, R.A. 'Race and Space in Miami', in A. R. Hirsch and R. A. Mohl, eds. *Urban Policy in Twentieth-Century America*, New Brunswick, 1993, 100–58.

Münch, P. *Stadthygiene im 19. und 20. Jahrhundert. Die Wasserversorgung, Abwasser- und Abfallbeseitigung unter besonderer Berücksichtigung Münchens*, Göttingen, 1993.

Morris, R.J. *Men, Women, and Property in England, 1780–1870: A Social and Economic History of Family Strategies amongst the Leeds Middle Classes*, Cambridge, 2005.

Mosley, S. *The Chimney of the World. A History of Smoke Pollution in Victorian and Edwardian Manchester*, New York, 2008 (2001).

Mosley, S. 'Common Ground: Integrating Social and Environmental History', *Journal of Social History*, 39:3, 2006, 915–33.

Ochojna, A.D. 'The Influence of Local and National Politics on the Development of Urban Passenger Transport in Britain 1850–1900', *Journal of Transport History*, 4, 1978, XX.

Oram, R.D. 'Abondance inépuisable? Crise de l'approvisionnement en combustible et reactions en Écosse du Nord entre environ 1500 et environ 1800', in J-M. Derex and F. Grégoire, eds. 'Histoire économique et sociale de la tourbe et des tourbières', *Æstuaria: cultures et developpement durable*, Cordemais, 2009, 31–44.

Page, M. *The Creative Destruction of Manhattan, 1900–1914*, Chicago, 1999.

Perlin, S., R. Setzer, J. Creason, and K. Sexton, 'Distribution of Industrial Air Emissions by Income and Race in the United States: An Approach Using the Toxic Release Inventory', *Environmental Science Technology*, 29:1, 1995.

Pulido, L. 'Rethinking Environmental Racism: White Privilege and Urban Development in Southern California', *Annals of the Association of American Geographers*, 90, 2000, 12–40.

Purcell, M. 'Neighborhood Activism among Homeowners as a Politics of Space', *Professional Geographer*, 53, 2001, 178–94.

Pye, S., I. Skinner, N. Meyer-Ohlendorf *et al.* 'Addressing the Social Dimensions of Environmental Policy. A Study of the Linkages between Environmental and Social Sustainability in Europe', European Commission Directorate-General 'Employment, Social Affairs and Equal Opportunities', 2008.

Rawls, J. *A Theory of Justice*, Oxford, 1971.

Rodger, R. 'Slums and Suburbs: The Persistence of Residential Apartheid', in P.J. Waller, ed. *The Oxford History of the British Landscape*, Oxford, 2000, 233–68.

Sabel, C. F. and J. Zeitlin, eds. *World of Possibilities: Flexibility and Mass Production in Western Industrialization*, Cambridge, 1997.

Sachs, I. *L'écodéveloppement, stratégie pour le 21e siècle*, Paris, 1980.

Sandler, R., and P. C. Pezzulo, eds. *Environmental Justice and Environmentalism. The Social Justice Challenge to the Environmental Movement*, Cambridge, Ma. 2007.

Schirmer, S.L. *A City Divided: The Racial Landscape of Kansas City 1900–1960*, Columbia, 2002.

Schlosberg, D. *Defining Environmental Justice. Theories, Movements and Nature*, Oxford, 2007.

Schott, D. *Die Vernetzung der Stadt. Kommunale Energiepolitik, öffentlicher Nahverkehr und die 'Produktion' der modernen Stadt: Darmstadt-Mannheim-Mainz 1880–1918*, Darmstadt, 1999.

Schott, D., B. Luckin and G. Massard-Guilbaud eds. *Resources of the City: Contributions to an Environmental History of Modern Europe*, Aldershot, 2005.

Scranton, P. *Endless Novelty: Specialty Production and American Industrialization, 1865–1925*, Princeton, 1997.

Sellers, C. 'Factory as Environment: Industrial Hygiene, Professional Collaboration and the Modern sciences of Pollution', *Environmental History Review*, 18:1, 1994, 55–83.

Sen, A. *The Idea of Justice*, Harvard, 2009.

Simson, J. von, *Kanalisation und Städtehygiene im 19. Jahrhundert*, Düsseldorf, 1983.

Soens, T. *De spade in de dijk? Waterbeheer en rurale samenleving in de Vlaamse kustvlakte (1280–1580)*, Gent, 2009.

Sörlin, S. and P. Warde, 'The Problem of the Problem of Environmental History: A Re-reading of the Field', *Environmental History*, 12:1, 2007, 107–30.

Stabel, P. *Dwarfs among Giants. The Flemish Urban Network in the Late Middle Ages*, Leuven–Apeldoorn, 1997.

Stilgoe, J. *Borderland: Origins of the American Suburb*, New Haven, 1988.

Stippak, M. *Beharrliche Provisorien, Städtische Wasserversorgung und Abwasserentsorgung in Darmstadt und Dessau 1869–1989*, Münster et al. 2010 (originally PhD thesis, Darmstadt University of Technology, Germany, 2008).

Stippak, M. 'The Mental and Practical Impact of Pre-bacteriological Quality Criteria for Water in the 1870s', in R. Heil, A. Kaminski and M. Stippak *et al.* eds. *Tensions and Convergences. Technological and Aesthetic Transformations of Society*, Bielefeld, 2007, 295–303.

Swyngedouw, E. and N. Heynen, 'Urban Political Ecology, Justice and the Politics of Scale', *Antipode*, 35, 2003, 902.

Szreter, S. *Fertility, Class and Gender in Britain 1860–1940*, Cambridge, 1996.

Talarchuk, G.M. 'The Urban Forest of New Orleans: An Exploratory Analysis of Relationships', *Urban Geography*, 18:6, 1997, 65–86.

Tarr J., ed. *Devastation and Renewal, an Environmental History of Pittsburgh and its Region*, Pittsburgh, 2003.

Tarr, J. 'The Metabolism of the Industrial City: The Case of Pittsburgh', *Journal of Urban History* 28:5, 2002, 511–45.

Tarr, J. *The Search for the Ultimate Sink. Urban Pollution in Historical Perspective*, Akron, 1996.

Theys, J. 'Pourquoi les préoccupations sociales et environnementales s'ignorent-elles mutuellement. Un essai d'interprétation à partir du thème des inégalités écologiques', in P. Cornut, T. Bauer and E. Zaccaï, eds. *Environnement et inégalités sociales*, Bruxelles, 2007.

Theys, J. *Les inégalités écologiques, dimension oubliée de l'action publique: entre raisons politiques et explications épistémologiques*, Paris, 2005.

Thoen, E. '"Social Agrosystems" as an Economic Concept to Explain Regional Differences. An Essay Taking the Former County of Flanders as an Example (Middle Ages–19th Century)', in B.J.P. van Bavel and P. Hoppenbrouwers, eds. *Landholding and Land Transfer in the North Sea Area (Late Middle Ages–19th Century)*, Turnhout, 2004, 47–66 (CORN Publication Series 5).

Uekoetter, F. *The Age of Smoke. Environmental Policy in Germany and the United States, 1880–1970*, Pittsburgh, 2009.

Ueland, J. and B. Warf, 'Racialized Topographies: Altitude and Race in Southern Cities', *Geographical Review*, 96:1, 2006, 50–78.

Van Dam, P. and M. van Tielhof, *Waterstaat in stedenland. Het hoogheemraadschap van Rijnland voor 1857*, Utrecht, 2006.

Van de Ven, G.P. *Man-made Lowlands. History of Water Management and Land Reclamation in the Netherlands*, Utrecht, 2004.

Verbruggen, C. 'Nineteenth Century Reactions to Industrial Pollution in Ghent, the Manchester of the Continent. The Case of the Chemical Industry', in C. Bernhardt and G. Massard-Guilbaud, eds. *Le démon moderne, la pollution dans les sociétés urbaines et industrielles d'Europe/The Modern Demon, Pollution in Urban and Industrial European Societies*, Clermont-Ferrand, 2002, 377–92.

Verhulst, A. *Landbouw en Landschap in Middeleeuws Vlaanderen*, Brussels, 1995.

Veyret, Y. *Le développement durable*, Paris, 2007.

Vögele, J. and U. Koppitz, 'Sanitäre Reformen und der epidemiologische Übergang in Deutschland (1850–1920)' in S. Frank and M. Gandy, eds. *Hydropolis. Wasser und die Stadt der Moderne*, Frankfurt, 2006, 75–93.

Washington, S. H. *Packing Them In: An Archaeology of Environmental Racism in Chicago, 1865–1954*, Lanham, 2005.

Whitehand, J.W.R. *The Changing Face of Cities: A Study of Development Cycles and Urban Form*, Oxford 1987.

Williams, N. 'Death in Its Season: Class, Environment and the Mortality of Infants in Nineteenth-century Sheffield', *Social History of Medicine*, 5:1, 1992, 71–94.

Wisner, B., P. Blaikie *et al. At Risk: Natural Hazards, People's Vulnerability, and Disasters*, 2nd ed. New York, 2004.

Würth, G. *Umweltschutz und Umweltzerstörung in der DDR*, Frankfurt, 1985.

Young, I. M. *Justice and the Politics of Difference*, Princeton, 1990.

Zonabend, F. *La Presqu'île au nucléaire*, Paris, 1989.

Index

Index

Index

www.ingramcontent.com/pod-product-compliance
Lightning Source LLC
LaVergne TN
LVHW102008210926
848483LV00010B/165

* 9 7 8 1 8 7 4 2 6 7 6 1 4 *